Advanced Digital System Design using SoC FPGAs

Ross K. Snider

Advanced Digital System Design using SoC FPGAs

An Integrated Hardware/Software Approach

Ross K. Snider
Montana State University
Bozeman, MT, USA

ISBN 978-3-031-15418-8   ISBN 978-3-031-15416-4 (eBook)
https://doi.org/10.1007/978-3-031-15416-4

This Springer imprint is published by the registered company Springer Nature Switzerland AG
The registered company address is: Gewerbestrasse 11, 6330 Cham, Switzerland

To my loving family
My wife Kimberley,
My sons Andrew and Christopher,
My daughters Ashleigh and Emily.

Preface

This textbook arose out of my experiences teaching computer engineering and electrical engineering courses at Montana State University (MSU). Field Programmable Gate Arrays (FPGAs) are digital devices that have been around since the 1980s and are accessible to students, allowing them to create their own custom hardware without the prohibitive expense of creating a custom ASIC or digital chip. FPGAs started as tiny devices where they functioned as "glue" logic, but they have grown to be among the largest digital devices today. As FPGAs grew larger, they started absorbing all sorts of logic functions, including a complete ARM based computer system, i.e., a System-on-Chip (SoC). SoC FPGAs are ideal devices for teaching computer engineering since in the same chip you can create custom hardware in the FPGA fabric and develop software that runs on the ARM CPUs. This allows students to gain a system level understanding of how computers work. They gain this system level knowledge by first creating their own custom hardware in the FPGA fabric and then controlling their hardware by writing a Linux device driver and associated application software. It has been quite satisfying seeing the excitement in student's faces when they finally understand how hardware and software interact.

I also teach the Digital Signal Processing course at MSU, and FPGAs and DSP go well together. DSP is an important application area of FPGAs, and the highest performing DSP is done in FPGAs. This is because there can be thousands of multipliers running in parallel in the FPGA fabric. Not only that, the programmable I/O of FPGAs allow data to be piped directly to the FPGA fabric, get processed, and then piped directly back out. This allows these devices to have the lowest processing latency of any digital device. FPGAs have lower processing latency than CPUs and GPUs due to this custom I/O.

The focus on audio signal processing is a result of my background in auditory neuroscience. Before coming to MSU, I completed a postdoctoral fellowship at Johns Hopkins University in the Laboratory of Auditory Neurophysiology under the direction of Xiaoqin Wang. Thus I'm interested in how the brain processes sound and of course the practical aspects of audio processing using FPGAs. The processing and bandwidth requirements of audio processing fit well within the constraints of the low-cost Cyclone V SoC FPGA family, making audio processing an accessible application area for students. A NIH grant for creating a platform for open speech signal processing was instrumental in developing the audio board targeted in the book. Thus a natural outcome was to have students create their own real-time sound effects processor. This has been very motivating for students.

The culmination of my experiences in FPGAs, DSP, auditory neuroscience, and teaching is this textbook. It is an integrated hardware/software approach to audio signal processing using SoC FPGAs. SoC FPGAs allow the merging of what has typically been the siloed areas of computer hardware and software. This textbook is my attempt at merging these two areas while creating an audio system that students have fun creating and playing with.

Montana State University
Bozeman, Montana
Ross K. Snider
July 1, 2022

Acknowledgments

I would like to thank Trevor Vannoy who helped create some of the lab and Linux material. Trevor was very good and helpful on the Linux front and helped many students debug their systems.

I would like to thank those who had a part in developing the Audio Mini board that was funded by NIH (grant 1R44DC01544301). Ray Weber for the analog interfacing and initial Linux device driver development. Connor Dack for the PCB board refinement that was well laid out. Chris Casebeer, Tyler Davis, and Will Tidd for helping to manage the hardware project and fixing some of the earlier hardware issues. Doug Roberts and Graham Conran who provided the Sensor Logic business that is the home for the Audio Mini board.

Finally, I would like to thank the students in my FPGA courses for trying out new material that didn't always go as planned and who provided great feedback.

Contents

Part I Introductions

Listings

List of VHDL Code

List of C Code

List of MATLAB Code

Acronyms

ACL Access Control List
ADC Analog-to-Digital Converter
ALM Adaptive Logic Module
ASIC Application-Specific Integrated Circuit
ASSP Application-Specific Standard Product
BRAM Block RAM
BSP Board Support Package
CIDR Classless Inter-Domain Routing
CTCFFT Continuous Time Continuous Frequency Fourier Transform
CTFT Continuous Time Fourier Transform
DFT Discrete Fourier Transform
DSP Digital Signal Processing
DTCF Discrete Time Continuous Frequency
DTDF Discrete Time Discrete Frequency
DUT Device Under Test
FFT Fast Fourier Transform
FIR Finite Impulse Response
FPGAs Field Programmable Gate Arrays
FPU Floating Point Units
FT Fourier Transform
GHRD Golden Hardware Reference Design
GPIO General Purpose Input/Ouput
HPS Hard Processor System
IIR Infinite Impulse Response
LEs Logic Elements
LKM Loadable Kernel Module
LUTs Look-Up Tables
MAC Multiply and Accumulate
MBR Master Boot Record
MMU Memory Management Unit
MSB Most Significant Bit

MSEL Mode Select
NAT Network Address Translation
NRE Non-Reoccurring Engineering
PLL Phase Locked Loop
PMU Performance Monitoring Unit
Phandle Pointer Handle
SoC System-on-Chip
TLB Translation Lookaside Buffer
VCP Virtual Com Port
VM Virtual Machine
VSG VHDL Style Guide
WSL Windows Subsystem for Linux

Part I
Introductions

Digital system development using SoC FPGAs is complicated since it covers a lot of ground. The topics include hardware, software, and tool chains. Furthermore, all of these areas are constantly changing. The aim of this introductory section is to give brief introductions of the material in each of these areas that students should know in order to build a working system. Students typically have not been exposed to all these areas that are needed to develop SoC FPGA systems, so introductions to all these areas are provided.

Chapter 1
Preliminaries

1.1 About This Book

This book covers a lot of ground that is constantly shifting. Hardware is constantly changing, software is always being updated, toolchains are always being improved, and development methodologies change over time. So yes, this book is likely to be out of date as soon as it is "printed." So why bother with this book? There are three primary reasons:

Reason 1: Students gain a ***system level*** view of computers where they see how hardware and software interact.

Reason 2: System-on-Chip (SoC) Field Programmable Gate Arrays (FPGAs) are an ideal platform for teaching hardware and software interactions and low cost SoC FPGAs are affordable for students.

Reason 3: Students get jobs. Students have reported great feedback in their job interviews. They report that being able to explain how they created their own custom hardware in the FPGA fabric and being able to explain how they wrote their own device drivers in Linux have impressed interviewers from some very large companies.

This book takes the ***you do not understand it until you build it*** approach to student learning. This means that the labs in the book are the central focus of the book. It is the process of building a complete computer system that cements the various elements together. The chapters exist to support the labs and provide background information that is needed for completion of the labs.

The choices of hardware, software, and methods are all the fault of the author. In defense of these choices, the author will argue that in computer science, if there are choices to be made, they will all be made. This means that one needs to be familiar with the various approaches that exist and pragmatic when it comes to a particular choice. Ultimately, a choice has to be made and this book reflects the biases of the author. Are there better ways? Yes, and this will always be the case in a field that is moving quickly and constantly changing. This book reflects a current

R. K. Snider, *Advanced Digital System Design using SoC FPGAs*,
https://doi.org/10.1007/978-3-031-15416-4_1

snapshot in time with no claim that it is the optimal way of doing things. However, it does accomplish the goal of creating a system-level understanding of computers for students in spite of the fact that one can quibble about the particular choice of hardware, software, or target application.

1.1.1 GitHub Repository

The GitHub repositories associated with the book are listed in Table 1.1.

Table 1.1: GitHub book repositories

Book repository link	Description
ADSD-SoC-FPGA	Primary GitHub site for the book
Code	Code repository
Updates	Update repository. It is anticipated that material in the book will constantly be changing due to advances in hardware, software, toolchains, and methodologies, so check here for updates. **Note:** If something does not work, make sure that you are using the same version of software that the book used. There are no guarantees that newer versions of software or toolchains will work the same way. In fact, using the latest versions of software or toolchains is a good way to break things, so you are on your own if you choose to use different versions than what the book uses. Always expect to have toolchain fights when you upgrade to a new version

1.2 Why Learn About SoC FPGAs?

Why do we learn about *System-on-Chip* (**SoC**) *Field Programmable Gate Arrays* (**FPGAs**)? In short, SoC FPGAs are extremely flexible digital devices that allow you to create custom hardware for embedded computing systems with high performance, high bandwidth, and low deterministic latency.

Computer *engineering* is a discipline about creating systems using computers for a particular purpose. This includes creating both custom hardware and software. Creating custom hardware is what distinguishes computer *engineering* from computer *science*. A rough dividing line between computer engineering and computer science, painted with a broad stroke, is the operating system on a computer. Computer engineering is concerned primarily with everything below the operating system, down to the hardware circuits, and how the computer interfaces to the physical world and other systems. Computer science is concerned about what is theoretically possible, about abstracting computers to make them easier to use, and creating languages with the right amount of useful abstractions for a particular purpose. And while we are painting in broad strokes, *engineers are the people who make science useful.*

In a fantasy world where cost is of no concern, one would create a custom chip, known as an *application-specific integrated circuit* (**ASIC**), for each product developed. However, creating a custom SoC chip that is ideal for a single purpose and that is fabricated using a 7 nm process will cost you hundreds of millions of dollars [1, 2]. This means that creating ASICs that use a leading edge fab process is only economical if you have a large market that supports it. This is because you can spread the *non-reoccurring engineering* (**NRE**) development costs over millions of units.

If your target market does not justify rolling an ASIC and you have multiple customers needing the same functionality, you then create what is known as an *application-specific standard product* (**ASSP**). An example of an ASSP is the AD1939 audio codec[1] from Analog Devices [3] that we use to acquire and play audio signals as explained in Chap. 5. Analog Devices markets this chip to customers who need to convert audio signals from analog to digital and then back to analog. This works well for customers like us who need an audio codec but do not have the deep pockets to fund the infrastructure and expertise to create a mixed-signal audio design. It is challenging to put both digital and analog systems on the same chip since one has to keep the noisy digital system from injecting noise into the analog system.

However, what if you want to develop a *custom* computer hardware system but do not want the cost and development effort associated with developing ASICs or ASSPs? A typical choice is to use the familiar CPU to create your system. CPUs range from cheap microcontrollers where 28.1 billion of them were shipped in 2018 with an average selling price of $0.63 [4] to high-end CPUs such as the Intel Xeon 8180M that cost $13K at introduction [5].

[1] Codec stands for coder-decoder, where the coder is an analog-to-digital converter (ADC) and the decoder is a digital-to-analog converter (DAC).

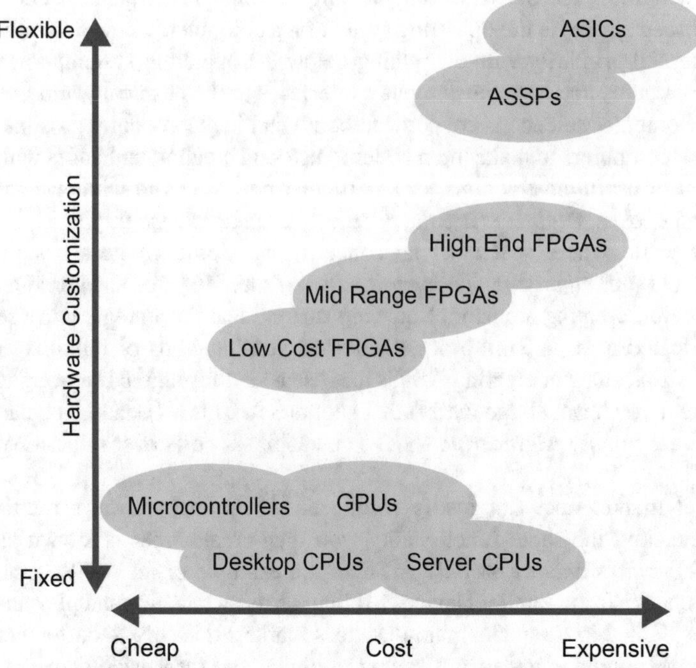

Fig. 1.1: Digital Hardware Devices. ASICs and ASSPs allow great flexibility for creating custom systems but are very expensive to develop. Microcontrollers are very cheap but are limited in their performance and flexibility. FPGAs take the middle ground where the FPGA fabric is programmable, which allows one to create custom hardware without the costs associated with ASICs and ASSPs

When using a CPU, customization is limited to what can be done in software. If you want to develop custom hardware, but without the costs associated with ASICs and ASSPS, this is where FPGAs come into play. FPGAs allow the development of custom hardware, but the NRE for developing the FPGA device has already been spread over thousands of FPGA customers.

What are the *advantages* of FPGAs compared to CPU systems? FPGAs have a programmable hardware fabric and custom I/O. This allows data to enter and exit the device with very low latency. The programmable fabric allows the creation of a custom data plane that gives high performance. An example of this is the creation of a custom interface and data plane for audio processing that we create in this book.

What are the *disadvantages* of FPGAs compared to CPU systems? FPGAs are much harder and take longer to develop. One needs to be familiar with computer architecture and low-level hardware description languages. When designing FPGA logic, one has to have a pretty good idea how the logic will be implemented in the FPGA fabric and what fabric building blocks will be used. If not, the design will be non-optimal and the design will not fit well in the device. Another disadvantage is cost. You will not be using an FPGA for the new toaster oven design that can make use of a cheap micro-controller.

SoC FPGAs are ideal devices for learning computer engineering. In one device you can create custom hardware in the FPGA fabric, create Linux device drivers for your custom hardware, and then control the hardware from a software application in Linux. It allows you to understand how hardware and software interact and how to start thinking about hardware–software co-design where you partition up tasks that are best implemented in hardware and tasks that are best implemented as software. Understanding this system-level design will provide you with computer engineering skills that are in high demand. Knowing how hardware works is a great foundation for being a software developer, which is highlighted in the Alan Kay quote [6] “People who are really serious about software should make their own hardware.”

1.2.1 Further Reading

A good introduction to FPGAs for those that are new to them is the ebook *FPGAs for Dummies* [7].

1.3 Prerequisites

The material covered in this textbook is quite broad, which means that the student should already be familiar with the topics listed below. As an analogy, we are starting out on an expedition to climb a mountain peak and the expedition requires that the expedition members, while not having direct experience climbing mountain peaks themselves, are familiar with camping, starting fires, and cooking outdoors, even when the weather is uncertain and could end up dismal and raining. However, summiting a peak on a clear sunny day where the vista stretches for miles makes the effort to get there all worth it. Being familiar with the topics below will ensure that you are OK camping in the woods by yourself.

1.3.1 Prior Hardware Knowledge

It is assumed that you are familiar with basic digital electronics and computer architecture as sketched below.

- **Basic Logic Gates** that includes CMOS logic and how NAND and NOR gates are used to construct digital logic. How these logic gates are used in both combinational and sequential logic designs. How numbers are represented in digital systems such as 2’s complement numbers?

- **Basic Digital Components** that includes encoders/decoders, multiplexers, flip-flops, registers, adders, multipliers, finite state machines, etc.
- **Basic Computer Architecture** that includes memory (SRAM and DRAM), FIFOs, data path, pipelining, I/O, control, etc.

1.3.2 Prior Software Knowledge

It is assumed that you are familiar with basic programming concepts and have been exposed to the following three languages. Other languages used will be described as we use them (e.g., Python, TCL).

- **VHDL**, which is a hardware description language. This includes the std_logic_vector data type, entity, architecture, concurrent statements, processes, the rising_edge function, if/else and case control statements, etc. A recommended text for this subject matter is *The Designer's Guide to VHDL* by Peter Ashenden [8].
- **C**, which is widely used in embedded systems and in the Linux kernel. There are many online resources that can be uncovered by a quick "C tutorial" Google search.
- **Matlab**, which is the environment that we will use to create our audio processing systems. We will also use **Simulink** for creating our data plane models. A recommended resource for Matlab and Simulink is the Matlab Onramp [9] and Simulink Onramp [10].

1.4 Hardware Needed

The hardware needed for this book is listed in the following sections. The hardware was chosen so that students could purchase their own boards and hardware at minimal cost, which allows them to develop in their own room rather than having to go to a laboratory to use an FPGA board. The majority of students have laptops with Windows 10 installed as the operating system, so this is the PC configuration taken in this book. It is possible to use Linux as the operating system on your laptop or computer, but we will not take this approach. It is assumed that if a user already has Linux running on their laptop, then they are already capable of configuring Linux and installing software on their own if they choose to do so. And, if they have trouble, it is assumed that they are capable of figuring out their own Linux solutions.

1.4.1 Laptop

It is expected that students have their own laptop (or desktop computer) with the following capabilities:

- **Windows 10 Operating System**. Using another operating system is possible, but you are on your own if you choose to do so. This textbook assumes you are using Windows 10 on your laptop or desktop computer.
- **8 GB of RAM** (ideally 16 GB). Currently, 8 GB is the most common RAM size that you will find in laptops, which is adequate for the projects in this textbook. If you have less, it is still possible, but your computer will run slower, especially when using a virtual machine.
- **Wi-Fi**. You will need to connect to the Internet and practically all laptops come with Wi-Fi, so you should be good on this front.
- **Two USB Ports**. You will need two USB ports on your laptop to connect to two different connections on the FPGA board. One USB port will be used to program the FPGA via JTAG, and the other USB port will be used with a terminal window when booting Linux on the SoC FPGA. If you do not have two USB ports, *get a USB adapter for your laptop that has two USB ports and an Ethernet port.*
- **Ethernet Port**. You will need an Ethernet port that is in addition to the Internet Wi-Fi connection on your laptop. This is because you will be connecting to the FPGA board using an Ethernet cable. Some laptops do not have an Ethernet port with an RJ45 jack, so if this is your case, you will need to get an Ethernet adapter for your laptop. If you need to get an Ethernet adapter, get an adapter that has at least two USB ports as well.

1.4.2 DE10-Nano FPGA Board

The FPGA board that is used by this textbook is the DE10-Nano Kit produced by Terasic (www.terasic.com) that contains an Intel Cyclone V SoC FPGA. The reason this board was chosen was because it was the lowest cost SoC FPGA board making it possible for students to get their own board. It provides great value because it contains the largest SoC FPGA (110K LEs) in the low cost FPGA category. A comparison to other low cost SoC FPGA boards can be seen in Table 1.2, where it can be seen that (at the time of this writing) the DE10-Nano had the best value (Fig. 1.2).

Fig. 1.2: The Terasic DE10-Nano SoC FPGA board is the FPGA board used by this textbook. https://www.terasic.com.tw/cgi-bin/page/archive.pl?Language=English&CategoryNo=167&No=1046

1.4.3 Audio Board

The real-time system that students develop in this textbook targets audio signal processing. Since there was no audio codec on the DE10-Nano, nor was there an audio card available, a high fidelity audio board was created for the DE10-Nano. This audio board contains a 24-bit audio codec (Analog Devices AD1939) that can sample up to 192 kHz. Further information on how this audio board was designed and how it can be used is found in Chap. 5 Introduction to the Audio Mini Board. The audio board is shown in Fig. 1.3 and can be purchased from SensorLogic (Audio Mini Link).

1.4.4 Miscellaneous Hardware

Some additional hardware will need to be purchased as well:

- **microSD Card**. The DE10-Nano SoC FPGA board comes with a microSD card that allows it to boot Linux. We will be creating our own version, so you will

Table 1.2: Low cost SoC FPGA boards

Board	Price[1]	SoC FPGA	FPGA Fabric			ARM CPU[5]	
			LEs[2]	DSP Slices[3]	Memory[4]	Speed	DRAM
DE10-Nano	$146	Intel Cyclone V 5CSEBA6U23I7	110 K	112	696 KB	800 MHz	1 GB
DE1-SoC	$175	Intel Cyclone V 5CSEMA5F31C6	85 K	87	496 KB	800 MHz	1 GB
Alchitry Au	$100	Xilinx Artix 7 XC7A35T	33 K	90	225 KB	No ARM CPUs Not a SoC FPGA	256 MB (standalone)
Zybo Z7-10	$159	Xilinx Zynq XC7Z010	17 K	80	270 KB	667 MHz	1 GB
Zybo Z7-20	$239	Xilinx Zynq XC7Z020	53 K	220	630 KB	667 MHz	1 GB

[1]Educational price in 2021 not including shipping. [2]Intel's LE contains an 8-input lookup table. Xilinx's contains a 6-input lookup table. [3]Intel's DSP Slice contains a 27×27 bit multiplier and a 64-bit accumulator. Xilinx's DSP Slice contains a 18×25 bit multiplier and a 48-bit accumulator. [4]Intel's contains 10 Kb M10K blocks. Xilinx's contains 36 kb blocks. [5]Dual Core Cortex-A9

Fig. 1.3: Audio Mini Board that plugs into the DE10-Nano and contains a 24-bit audio codec (Analog Devices AD1939) that can sample up to 192 kHz

need another microSD card so that you can always plug the factory default image back into the DE10-Nano board. Any size greater than 8 GB is fine. Currently, a 32 GB is practically the same cost as an 8 GB, so you might as well get the 32 GB microSD card. You can always allocate the extra space to the Linux root file system on the DE10-Nano.

- **microSD Card Reader**. You will need a USB microSD Card Reader so that you can read/write the microSD card and modify the card images.
- **Type A to Mini-B USB Cable**. The DE10-Nano kit comes with one Type A to Mini-B USB Cable, but we will be using both Mini-B ports on the DE10-Nano, so having an extra cable will be more convenient.
- **Ethernet Cable**. A short Ethernet cable, ~1 foot (or longer), to connect the DE10-Nano board to your laptop.

1.5 Software Needed

The following software will be used in this textbook. Most of the software is free or there are free commercial versions with the exception of Matlab, which has a student version that one must buy if one is not associated with an institution with a Mathworks site license. The list below is given as an overview of the software that will be used. Instructions for setting up the software are found in Sect. 11.1 Software Setup.

- **Windows 10**. It is assumed that the student has a Windows 10 laptop or PC and does not have much experience with Linux.
- **Windows Subsystem for Linux**. *Windows Subsystem for Linux* (**WSL**) comes in two versions, WSL 1 and WSL 2. WSL 1 (and not WSL 2) is required for Intel's Quartus software.
- **VirtualBox**. We will create an Ubuntu *virtual machine* (**VM**) using VirtualBox on Windows 10.
- **Ubuntu 20.04 LTS** . We will install Ubuntu as a virtual machine in VirtualBox.
- **Matlab and Simulink**. The following toolboxes are required in Matlab:
    - HDL Coder
    - Matlab Coder
    - Simulink Coder
    - Fixed-Point Designer
    - DSP System Toolbox (strongly suggested). Required for some example designs.
    - Signal Processing Toolbox (strongly suggested). Required for some example designs.
- **Python**. Version 3.8.x or later
- **Quartus**. The free version Quartus Prime Lite Edition can be used for the Cyclone V FPGA. Note: Quartus requires WSL 1 with Ubuntu 18.04.
- **Putty**. Which is a terminal emulator.

1.6 The Development Landscape

Knowing where you need to be to implement certain development steps, type software commands, or install or run software can be confusing since we will be operating across *two different hardware platforms with two different CPU types:*

Platform 1: **DE10-Nano** FPGA board that contains an ARM CPU inside the Cyclone V SoC FPGA
Platform 2: **Laptop** or PC that contains an x86 CPU

and *three operating systems:*

OS 1: **Windows 10** on a Laptop or PC
OS 2: **Ubuntu VM**, which is Ubuntu Linux running on a virtual machine in VirtualBox that is running on Windows 10, which in turn is using an x86 CPU.
OS 3: **Ubuntu ARM**, which is Ubuntu Linux running on the ARM CPUs inside the Cyclone V SoC FPGA, which is on the DE10-Nano board. Furthermore, the **Root File System** for Ubuntu ARM on the DE10-Nano can be located in two different locations:

a. **On the microSD card** that is inserted into the DE10-Nano board. When Linux uses the root file system on the microSD card, this is known as the **Ship Boot Mode**. This is the setup that comes when you buy the DE10-Nano board, but it is not the setup that you want to develop with.
b. **In an Ubuntu VM folder** served over Ethernet by the Ubuntu VM NFS server. When Linux uses the root file system served by the Ubuntu VM, this is known as the **Developer Boot Mode**. We will be using this boot setup in this book because it is way more convenient to develop with than using the ship boot mode, which is not practical for development.

Each operating system has its own software packages, command lines, and terminal windows. It also means that the DE10-Nano FPGA board and the Laptop/PC can be connected in any one or in all of the following manners:

Connection 1: **USB JTAG** using a USB cable with a Mini-B connector plugged into the USB Blaster port on the left side of the DE10-Nano board. This connection is used to program the FPGA via JTAG.

Connection 2: **USB UART** using a USB cable with a Mini-B connector plugged into the UART port on the right side of the DE10-Nano board. This connection is used to create a terminal window to interact with Linux booting on the DE10-Nano.

Connection 3: **Ethernet** where an Ethernet cable connects the DE10-Nano to the Laptop/PC. This is used so that Linux can boot from the Ubuntu VM when using the Developer Boot Mode.

References

1. Semiconductor Engineering. 10nm-versus-7nm. https://semiengineering.com/10nm-versus-7nm/. Accessed 22 June 2022
2. semiengineering.com. 5nm-vs-3nm. https://semiengineering.com/5nm-vs-3nm/. Accessed 22 June 2022
3. Analog Devices. AD1939 Audio Codec. https://www.analog.com/en/products/ad1939.html. Accessed 22 June 2022
4. IC Insights. Microcontrollers Will Regain Growth After 2019 Slump. https://www.icinsights.com/news/bulletins/Microcontrollers-Will-Regain-Growth-After-2019-Slump/. Accessed 22 June 2022
5. CPU World. Intel Xeon 8180M Specifications. https://www.cpu.world.com/CPUs/Xeon/Intel-Xeon%208180M.html. Accessed 22 June 2022
6. A. Kay, Alan Kay Quote. https://en.wikiquote.org/wiki/Alan_Kay. Accessed 22 June 2022

7. A. Moore, R. Wilson, FPGAs For Dummies (2017). https://www.intel.com/content/dam/support/us/en/programmable/support-resources/bulk-container/pdfs/literature/misc/fpgas-for-dummies-ebook.pdf
8. P.J. Ashenden, *The Designer's Guide to VHDL* (Morgan Kaufmann, Burlington, 2008)
9. MathWorks. MATLAB Onramp. https://www.mathworks.com/learn/tutorials/matlab.onramp.html. Accessed 22 June 2022
10. MathWorks. Simulink Onramp. https://www.mathworks.com/learn/tutorials/simulink-onramp.html. Accessed 22 June 2022

Chapter 2
Introduction to System-on-Chip Field Programmable Gate Arrays

2.1 The Digital Revolution

The digital revolution [1] has changed the course of human history. Its societal impact has been massive as witnessed by the smartphone in everyone's pocket. However, we will not dwell on this topic. Instead we will focus on a digital device called a Field Programmable Gate Array that is known by its acronym *FPGA*. FPGAs have had a long history being invented in the mid-1980s as programmable digital "glue" that could connect other digital parts together into larger systems.

The utility of FPGAs has increased over time due to Moore's law which was a prediction by Gordon Moore in 1965 that the number of transistors that could be put on a silicon chip would double every year [2]. This exponential growth prediction has turned out to be true, although the doubling time has been somewhat variable and has been closer to two years. The historical growth in transistor count can be seen in Fig. 2.1 where the number of transistors that can be put down on a silicon chip is still growing at an exponential rate (top curve of orange triangles). Moore's law is still alive as can be seen in Table 2.1 where in 2019 Intel with their 10 nm process placed 100 million transistors in one square millimeter. Contrast this to Gordon Moore's projection in 1965 that they would be able to put 250,000 components in a square inch. In the near future, Taiwan Semiconductor Manufacturing Company (TSMC) has recently announced that they will be able to put a quarter billion transistors in a square millimeter using their 3 nm process [3].

R. K. Snider, *Advanced Digital System Design using SoC FPGAs*,
https://doi.org/10.1007/978-3-031-15416-4_2

Table 2.1: Transistor density of Fab Process Nodes. Data from [4, 5]

Process node	Intel 14 nm	TSMC 10 nm	TSMC 7 nm	Intel 10 nm	TSMC 5 nm	Intel 7 nm	TSMC 3 nm
Transistor density (millions of transistors per mm^2)	37.2	52.5	91.2	100.7	171.3	237.2	291.2
Year	2017	2017	2018	2019	2020	2021	2022

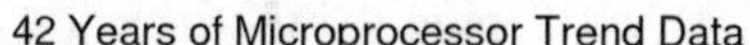

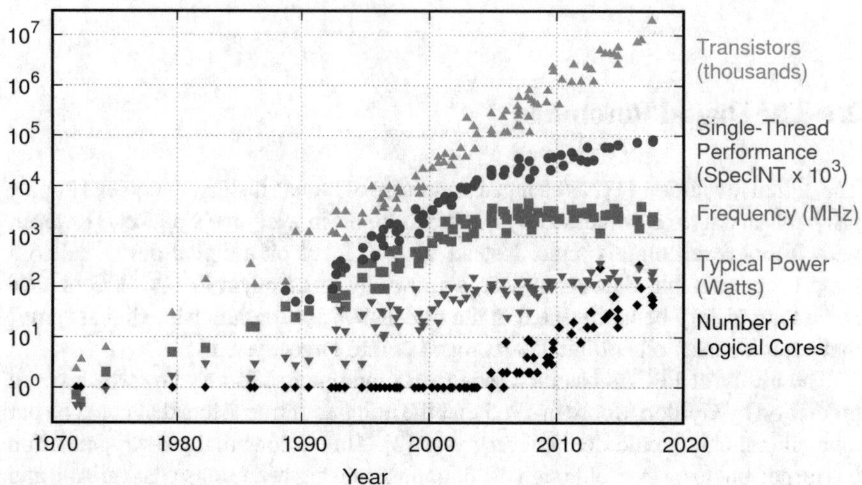

Fig. 2.1: Microprocessor Trend Data. CPU performance is not keeping up with Moore's law. Figure from [6]

What is not keeping up with Moore's law is CPU performance as shown by the blue circles in Fig. 2.1. Single thread performance is plateauing due to CPU clock speeds (green squares) being limited by the amount of power that can be consumed and dissipated in a small area (red triangles). If the transistors get too hot, they will fail. This power dissipation limit is known as the *Power Wall* [7].

CPU power consumption is made up of dynamic and static power

$$P_{CPU} = P_{Dynamic} + P_{Static} \tag{2.1}$$

and dynamic power is comprised of the product of capacitance, voltage, and clock frequency.

$$P_{Dynamic} = CV^2 f \tag{2.2}$$

Capacitance governs how much charge moves each clock cycle and is reduced by having smaller geometries, so each process node advance in Table 2.1 reduces power consumption. The CPU core voltage is also determined by the fab process. Thus the CPU clock frequency is typically the only parameter that can be modified after fabrication. As a result, there is a whole industry devoted to CPU overclocking to get better performance since CPU clock speeds can be set by the motherboard by modifying BIOS parameters. However, power consumption is linearly tied to clock speed in Eq. 2.2, and there is a limit to the amount of heat that can be dissipated, which ultimately places a limit on the clock speed that can be obtained without damaging the CPU. This is the reason that CPU clock speeds are now typically set to around 3-4 GHz and are not increasing (unless you push the envelope and cool your CPU with liquid nitrogen so that you can overclock it to 8.7 GHz, which is the current record [8].)

The divergence between the transistor count curve (orange triangles) and the CPU performance curve (blue circles) in Fig. 2.1 tells us that adding more transistors to a CPU does not help performance much anymore. This means that the only path forward to get more performance is to go parallel. A natural path that CPU vendors are following is to put multiple cores in a CPU. As a result, you can see in the data that once CPU performance started lagging (blue circles), the number of core started picking up (black diamonds). The limit to the number of cores that can be added to a CPU is dictated by how fast data can be moved back and forth from external memory (e.g., DRAM) to a cache associated with a core, which is called memory bandwidth. Unfortunately, DRAM is much slower than what a CPU can run at, so there will become a point with too many cores where they cannot be fed data fast enough. At this point, there will be cores that will just starve from lack of data, so there is no point in adding these cores if the memory bandwidth cannot support them. This memory limitation to core count is known as the *Memory Wall* [7].

What we have been talking about so far are issues related to CPU architectures. There are diminishing returns when additional transistors are added to CPUs. However, are there other architecture that can easily scale with the addition of many transistors? Yes, and I suspect that you have guessed Field Programmable Gate Arrays (FPGAs). What are these devices? How do these scale differently from CPUs?

2.2 Basic FPGA Architecture

FPGAs have basic logic resources that can be connected together via programmable switches that allow arbitrary routing between these logic resources. These logic resources and programmable routing are referred to as the FPGA programmable *fabric*. The basic logic resources are described in the following sections.

2.2.1 External I/O

FPGAs have external pins that can be used as *general purpose input/output* (**GPIO**) pins. These pins support a variety of voltage levels and I/O standards. FPGAs can have as few as 128 GPIO pins (e.g., Cyclone V 5CEA2) to as many as 2,304 GPIO pins (e.g., Stratix 10 GX10M). The GPIO pins allow the FPGA fabric to be directly connected to data sources/sinks, which allows for very low and deterministic latency when connecting external devices to custom hardware in the FPGA fabric. The Cyclone V on the DE10-Nano board contains 288 GPIOs pins connected to the FPGA fabric and 181 GPIO pins connected to the HPS [9]. The DE10-Nano board brings out almost 80 pins in its expansion headers.

2.2.2 Logic Elements

Logic functions are implemented as lookup tables that are programmed when the configuration bitstream is loaded into the FPGA at power-up. A programmable lookup table with eight inputs can implement an arbitrary logic function with eight inputs where each of the 256 addressable bits is programmed as part of the configuration bitstream. Logic elements also include the ability to register the output and they have dedicated circuitry to implement fast adders. FPGAs can have as few as 25,000 LEs (e.g., Cyclone V 5CEA2) to as many as ten million LEs (e.g., Stratix 10 GX10M). The Cyclone V *Adaptive Logic Module* (**ALM**) is shown in Fig. 2.2 and is equivalent to 2.5 *Logic Elements* (**LEs**) of the older 4-input lookup tables [10].

2.2.3 Memory

Memory is commonly used in the FPGA fabric, so dedicated memory blocks are implemented that can be put together to create larger memories. They can also be configured as dual port memories that are useful for creating circular buffers and FIFOs. The amount of memory can be as little as 1760 kilobits (176 M10K blocks) in the Cyclone V 5CEA2 to as much as 253 Mbits (12,950 M20K blocks) in the Stratix 10 GX10M. An M10K block of memory contains 10,000 bits of memory in the FPGA fabric. The Cyclone V on the DE10-Nano board contains 553 M10K blocks of embedded memory in the FPGA fabric.

2.2.4 DSP Blocks

An application area that FPGAs are good at is *Digital Signal Processing* (**DSP**). The general form of the difference equation in DSP is the equation

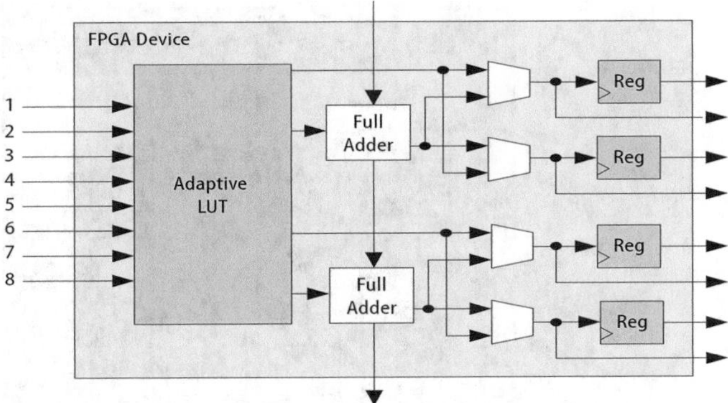

Fig. 2.2: The Cyclone V logic element known as the Adaptive Logic Module (ALM). The eight inputs require that the lookup table stores 256 bits of information. These bits can then be routed and saved in four different registers. Additional logic is contained in the ALM to create fast adders. The Cyclone V on the DE10-Nano board contains 41,509 ALMs or 110,000 LEs. Figure from [11]

$$y[n] = \frac{1}{a_0}\left(\sum_{i=0}^{P} b_i x[n-i] - \sum_{j=1}^{Q} a_j y[n-j]\right)$$

which can make use of multipliers and adders. The Cyclone V has DSP blocks that contain a 27×27 bit multiplier along with a 64-bit accumulator (adder) to implement *Multiply and Accumulate* (**MAC**) operations. The number of DSP blocks can be as few as 25 (e.g., Cyclone V 5CEA2) to as many as 5760 DSP blocks running in parallel (e.g., Stratix 10 GX2800), which gives a peak performance of 23 TMACS (fixed-point) or 9.2 TFLOPS (single-precision floating-point). The Cyclone V on the DE10-Nano board contains 112 DSP blocks that can be used to implement FIR filters such as the one shown in Fig. 2.3. This is why FPGAs can implement high-performance DSP operations since thousands of DSP blocks can be run in parallel and the associated DSP coefficients and delayed signal samples can be stored locally in FPGA memory.

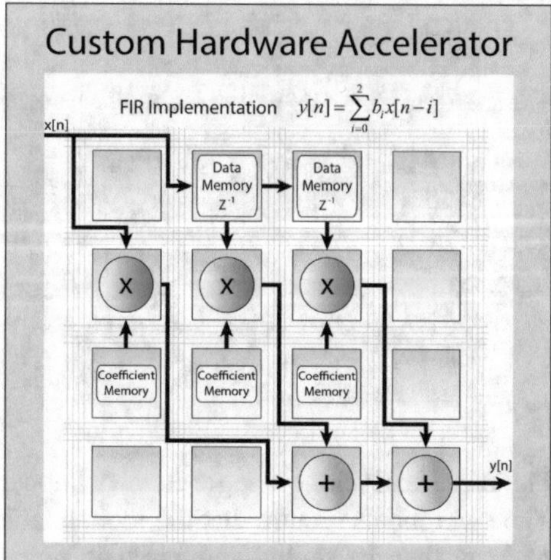

Fig. 2.3 The Cyclone V fabric contains DSP blocks that can be used to implement high performance DSP operations

2.3 SoC FPGA Architecture

System-on-Chip (SoC) Field Programmable Gate Arrays (FPGAs) extend the basic FPGA architecture with the logic, memory, and DSP resources described earlier with a complete ARM based computer system. Thus it has become a complete computer system on a chip as shown in Fig. 2.4. The *FPGA Fabric* allows a hardware designer to create custom hardware that can be controlled by software running on the ARM CPUs. The ARM computer system contains all the peripherals that you would expect to use with a computer. The ARM CPUs along with the peripherals is known as the *Hard Processor System* (**HPS**). It is referred to as a "hard" processor system since the ARM CPUs and peripherals have been implemented in silicon. This is in contrast to earlier systems that could only implement CPUs in the FPGA fabric as custom hardware. Implementing the ARM CPUs in silicon allows the HPS to run much faster than it could be implemented as a "soft" processor system running in the FPGA fabric.

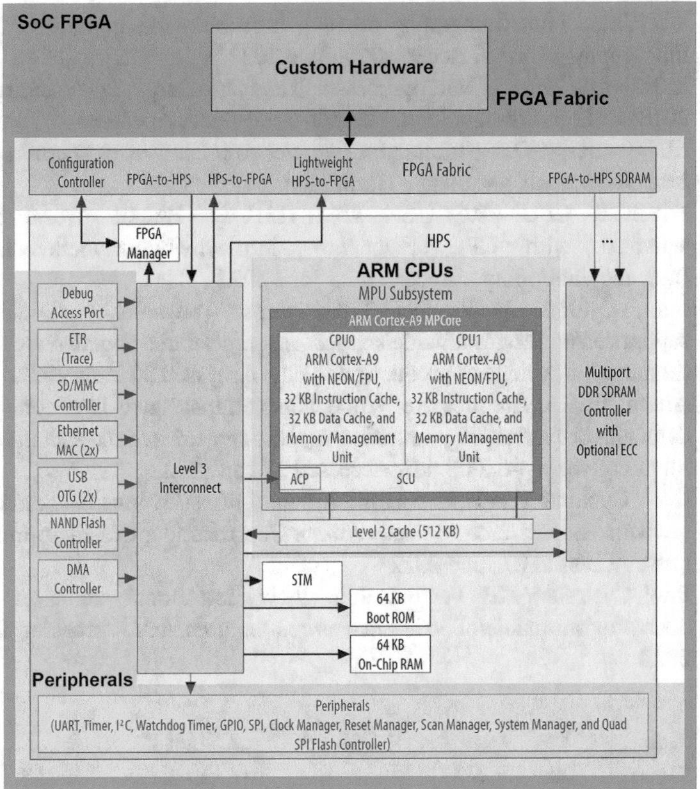

Fig. 2.4: The Cyclone V SoC FPGA contains a complete ARM based computer system with peripherals in addition to the FPGA fabric. The ARM CPUs along with the peripherals are known as the *Hard Processor System*. Figure adapted from [12]

References

1. I. Bojanova, The digital revolution: What's on the Horizon? IT Prof. **16**(1), 8–12 (2014)
2. G.E. Moore, Cramming more components onto integrated circuits. Electronics **38**(8) (1965). https://newsroom.intel.com/wp-content/uploads/sites/11/2018/05/moores-law-electronics.pdf
3. D. Schor, TSMC Ramps 5nm, Discloses 3nm to Pack Over a Quarter-Billion Transistors Per Square Millimeter. https://fuse.wikichip.org/news/3453/tsmc-ramps-5nm-discloses-3nm-to-pack-over-a-quarter-billion-transistors-per-square-millimeter/. Accessed 23 June 2022
4. WikiChip, 5 nm lithography process. https://en.wikichip.org/wiki/5_nm_lithography_process. Accessed 23 June 2022

5. WikiChip, 7 nm lithography process. https://en.wikichip.org/wiki/7_nm_lithography_process. Accessed 23 June 2022
6. K. Rupp, 42 Years of Microprocessor Trend Data. https://www.karlrupp.net/2018/02/42-years-of-microprocessor-trend-data/. Accessed 23 June 2022
7. J.L. Hennessy, D.A. Patterson, *Computer Architecture: A Quantitative Approach* (Morgan Kaufmann, Burlington, 2011)
8. HWBOT, CPU FREQUENCY: HALL OF FAME, World Records achieved with CPU Frequency. https://hwbot.org/benchmark/cpu_frequency/halloffame. Accessed 23 June 2022
9. Intel, Cyclone V Product Table. https://www.intel.com/content/dam/support/us/en/programmable/support-resources/bulk-container/pdfs/literature/pt/cyclone-v-product-table.pdf. Accessed 27 June 2022
10. Altera, FPGA Architecture White Paper. https://www.intel.com/content/dam/support/us/en/programmable/support-resources/bulk-container/pdfs/literature/wp/wp-01003.pdf. Accessed 27 June 2022
11. Intel, Cyclone V Adaptive Logic Module. https://www.intel.com/content/www/us/en/docs/programmable/683694/current/adaptive-logic-module.html. Accessed 27 June 2022
12. Intel, Cyclone V HPS Features. https://www.intel.com/content/www/us/en/docs/programmable/683694/current/hps-features.html. Accessed 27 June 2022

Chapter 3
Introduction to the SoC FPGA Boot Process

3.1 Cyclone V SoC FPGA Boot Process

This chapter covers how the Intel Cyclone V SoC FPGA gets configured when it powers up on the DE10-Nano board. This process includes configuring the FPGA fabric with your custom hardware and booting Linux on the ARM CPUs. The SoC FPGA on the DE10-Nano board contains the ARM CPUs on which we will be running the Linux operating system. This means that the boot process is more involved than the typical FPGA that only has to load the configuration bitstream to configure the FPGA fabric. SoC FPGAs have to boot the CPUs as well. We also make the distinction between the following two booting scenarios:

- **Ship Boot Mode**. This mode is used when the SoC FPGA boots entirely from the microSD card. It is the typical setup that is used when a product ships and the system must be self-contained. However, this boot mode is unusable for development.
- **Developer's Boot Mode**. This mode is used when the SoC FPGA boots over Ethernet and uses a Network File System that is served by a network server, which in our case will be an Ubuntu Virtual Machine (VM). Changing and modifying files is much easier done in an Ubuntu VM directory than having to remove, image, and reinsert a microSD card. Setting up the Developer's Boot Mode is described in Chap. 11 Development Environment Setup (page 193).

The Cyclone V SoC FPGA has three ways to initialize both the FPGA fabric and the ARM CPUs, also known as the Hard Processor System (HPS).

Option 1: Separate FPGA configuration and HPS booting.
Option 2: **HPS boots first and then configures the FPGA fabric. This is the option we will be using.**
Option 3: HPS boots from the FPGA fabric.

The boot process has a number of steps that are listed below. These steps are further described in their own section. The files used during the boot process and the

R. K. Snider, *Advanced Digital System Design using SoC FPGAs*,
https://doi.org/10.1007/978-3-031-15416-4_3

file locations are noted depending on whether the boot mode is the Ship Boot Mode or Developer's Boot Mode.

Boot Step 1: Power-Up or Reset
Boot Step 2: Boot ROM
Boot Step 3: Preloader
Boot Step 4: U-boot
Boot Step 5: Linux
Boot Step 6: Application

3.1.1 Boot Step 1: Power-Up or Reset

The boot process begins when the SoC FPGA powers up or a CPU in the MPU exits from the reset state. (We are assuming a cold reset, i.e., no software registers have been preserved, which happens in a warm reset.) The boot ROM, which is 64 KB in size and hard coded into the HPS silicon and normally located at address range 0xFFFD000-0xFFFDFFFF, is mapped to the reset exception address that is at address 0x0. Thus code starts running from the boot ROM, which has been temporarily mapped to memory address 0x0 upon reset.

3.1.2 Boot Step 2: Boot ROM

Code running from the Boot ROM checks the BSEL FPGA pins (also known as BOOTSEL), which offer multiple methods to obtain the **preloader image**. These options are shown in Fig. 3.1.

Table A-1: BSEL Values for Boot Source Selection

BSEL[2:0] Value	Flash Device
0x0	Reserved
0x1	FPGA (HPS-to-FPGA bridge)
0x2	1.8 V NAND flash memory
0x3	3.3 V NAND flash memory
0x4	1.8 V SD/MMC flash memory with external transceiver
0x5	3.3 V SD/MMC flash memory with internal transceiver

Fig. 3.1: Boot Source Options for the Cyclone V HPS. Table from Intel Cyclone V Hard Processor System Technical Reference Manual (p. A-6)

If we examine the DE10-Nano schematic on sheet five as shown in Fig. 3.2, we can see that these pins have been hard wired to the value of 5. Note that in the schematic there are symbols for all the resistors. However, there are labels DNI next to several of them. This stands for *Do Not Install*. This means that these resistors are not to be installed during assembly of the PCB. If we look at HPS_BOOTSEL0 in the schematic, the resistor to ground has a DNI next to it, so it is not connected to ground. Instead there is a resister connecting HPS_BOOTSEL0 to VCC3P3. This means that HPS_BOOTSEL0 is connected via a pull up resistor (R140) to 3.3 volts, which makes it a logical one. HPS_BOOTSEL1 has the opposite case. It is not connected to 3.3 volts, but instead it is connected to ground via a pull down resistor (R126), which makes it a logical zero. HPS_BOOTSEL2 is connected in a similar fashion as HPS_BOOTSEL0 making it a logical one. Thus the BSEL pins on the DE10-Nano board have been hard wired as 0b101 = 0x5, which means boot from a 3.3V SD/MMC flash memory, i.e., a microSD card [1] by using the pins SDMMC_CMD, SDMMC_D0, and SDMMC_CLK [2]. This means that the Cyclone V SoC FPGA on the DE10-Nano board has been designed to automatically boot from a microSD card.

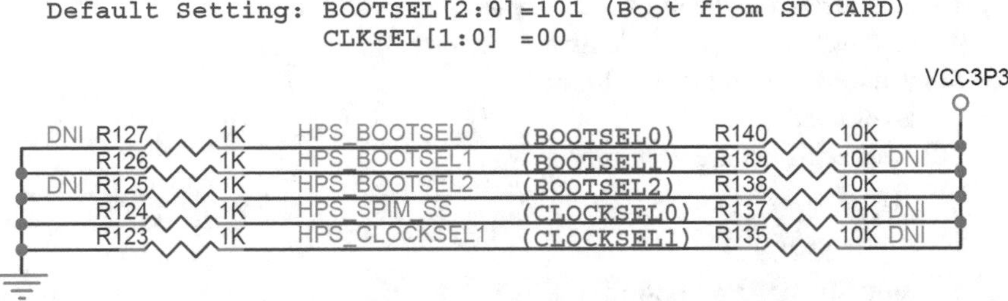

Fig. 3.2: Schematic of boot select signals. From sheet 5 of DE10-Nano schematic

Table 3.1: Summary of boot select signal connections

Schematic name	Voltage connection	FPGA pin connection	Bank
HPS_BOOTSEL0	3.3 V	J17	7A
HPS_BOOTSEL1	0 V	A6	7B
HPS_BOOTSEL2	3.3 V	D15	7B

Code from the Boot ROM then reads the *Master Boot Record* (**MBR**) that is located in the first 512 bytes of the microSD card. The MBR contains information about the partitions (address and size of partitions) on the microSD card. The MBR is scanned for a partition with the partition type field having the value $0 \times A2$, which contains the preloader image. Partition A2 is a custom raw partition with no file system. Most partition tools (such as Windows) will consider this an unknown partition type.

The Boot ROM Code then loads the preloader image from the microSD card into on-chip RAM that is only 64 KB in size. This means that the preloader is restricted to 60 KB (4 KB is reserved). The Boot ROM code then hands control over to the preloader code that is running in the 64 KB on-chip RAM. Thus code has transitioned from running from the hard wired ROM to the on-chip RAM.

Before the preloader runs, the processor (CPU 0) has been set to the following state:

1: Instruction cache is enabled.
2: Branch predictor is enabled.
3: Data cache is disabled.
4: MMU is disabled.
5: Floating point unit is enabled.
6: NEON vector unit is enabled.
7: Processor is in Arm secure supervisor mode.

The boot ROM code sets the Arm Cortex-A9 MPCore registers to the following values:

- **R0** contains the pointer to the shared memory block, which is used to pass information from the boot ROM code to the preloader. The shared memory block is located in the top 4 KB of on-chip RAM.
- **R1** contains the length of the shared memory.
- **R2** is unused and set to 0×0.
- **R3** is reserved.
- All other registers are undefined.

3.1.3 Boot Step 3: Preloader

Once the preloader image has been loaded from the microSD card into on-chip RAM, which is 64 KB in size and located at memory address 0xFFFF_0000-0xFFFF_FFFF, the preloader maps the on-chip RAM to 0×0 so that the exception vectors (interrupts) will use the exception handlers in the preloader image.

The function of the preloader is user-defined. However, typical functions include:

1: **Initializing the SDRAM interface**. The timing parameters specific to the DRAM used on the PCB needs to be set. Setting up the DRAM allows the preloader to load the next stage of the boot software into DRAM since the next stage will not fit into the 60 KB available in the on-chip RAM. In our case the next stage is the open-source boot loader U-boot, which will run from DRAM.
2: **Configuring the HPS I/O pins.**
3: **Initializing the interface that loads the next stage of software** (U-boot).

Once the external DRAM has been set up, the preloader copies the U-boot image from the microSD card into DRAM. The U-boot image also resides in the 0xA2 partition, immediately after the preloader images. The preloader then hands control over to the U-boot code that starts running from DRAM.

Note: The preloader is read from the microSD card in both the Ship Boot Mode and the Developer's Boot Mode, which means the microSD card needs to be present for both booting options. However, the microSD card only needs to be imaged initially once for the Developer's Boot Mode.

Creating the preloader is part of the process when creating the *Board Support Package* (**BSP**) for an FPGA board since it is the hardware designer that knows the timing specifications for the DRAM that was chosen for the board. This BSP is typically created by the board hardware designers. We will not concern ourselves with this since this has already been created for the DE10-Nano board. Interested readers can read more about creating the preloader on Rocketboards.org (Generating and Compiling the Preloader)

3.1.4 Boot Step 4: U-boot

U-boot is an open-source boot loader used to boot Linux on an embedded device. The U-boot image has been copied to DRAM, so it is not constrained in size to 60 KB like the preloader, which has to fit in the on-chip RAM. The particular case of the U-boot instructions mentioned here assumes that the microSD card has been flashed with the image described in Sect. 11.1.3.6 Reimaging the microSD Card with the Developer's Image (page 227).

U-boot allows the SoC FPGA to boot entirely from the microSD card (Ship Boot Mode) or over Ethernet using the Network File System (Developer's Boot Mode). U-boot has environmental variables that can be defined and the choice between Ship Boot Mode and the Developer's Boot Mode is made by setting the U-Boot environmental variable **nfsboot=true** to boot from a Network File System (NFS) when in the Developer's Boot Mode. See Sect. 11.1.3.8 Setting U-boot Variables on the DE10-Nano Board to Boot via NFS/TFTP (page 232) for the instructions on setting up U-boot for the Developer's Boot Mode.

Typically there is a U-boot script (e.g., **u-boot.scr**) that is first read by U-boot. This allows all the file names and locations of these files to be modified by a single script.

The files that are used during the boot process are listed in Table 3.2 and the names and locations of these files can be modified by editing the boot script file. The boot script file name can be changed by modifying the environment variable boot script. The file locations depend on the boot mode. See Fig. 11.13 that illustrates the Developer's Boot Mode setup.

Table 3.2: Order and location of files used during the boot process

Boot order	File/image	Ship boot location	Developer's boot location
1	Preloader Image	microSD Card (Partition 3: Type = A2: Raw Binary)	
2	U-Boot Image	microSD Card (Partition 3: Type = A2: Raw Binary)	
3	u-boot.scr	microSD Card (Partition 1: Type = B: FAT32)	/srv/tftp/de10nano/bootscripts
4	soc_system.rbf	microSD Card (Partition 1: Type = B: FAT32)	/srv/tftp/de10nano/my_project
5	zImage	microSD Card (Partition 1: Type = B: FAT32)	/srv/tftp/de10nano/kernel
6	soc_system.dtb	microSD Card (Partition 1: Type = B: FAT32)	/srv/tftp/de10nano/my_project
7	Linux Root File System	microSD Card (Partition 2: Type = 83: EXT Linux)	/srv/nfs/de10nano/ubuntu-rootfs

3.1.5 Boot Step 5: Linux

U-boot loads **zImage**, the compressed Linux kernel image into DRAM, and the kernel is launched. Linux then reads the device tree blob file **soc_system.dtb** that tells Linux what hardware it is running on. Linux then mounts the root file system that is contained in partition 2 (EXT Linux) on the microSD Card (Ship Boot Mode) or mounts the root file system over the network via NFS (Developer's Boot Mode).

3.1.6 Boot Step 6: Application

The user application can then be started up. This can be done by executing a shell script at boot time. The process for doing this is (very brief outline):

Step 1: Create a shell script that will run your application (e.g., **run_this.sh**).
Step 2: Create a systemd startup script, e.g., **run_this.service** that has the lines:
[Service]
ExecStart=/<path>/run_this.sh
(Note: there are additional lines you will need.)
Step 3: Put **run_this.service** into **`/etc/systemd/system/`**.

References

1. Intel, Cyclone V Hard Processor System Technical Reference Manual. https://www.intel.com/content/dam/www/programmable/us/en/pdfs/literature/hb/cyclone-v/cv_5v4.pdf. Table A-1, BSEL Values for Boot Source Selection, page A-6, Accessed 23 June 2022

2. Intel, Cyclone V Hard Processor System Technical Reference Manual. https://www.intel.com/content/dam/www/programmable/us/en/pdfs/literature/hb/cyclone-v/cv_5v4.pdf. Table A-2, Boot Source I/O, page A-7, Accessed 23 June 2022

Chapter 4
Introduction to the DE10-Nano Board

4.1 DE10-Nano Board

The DE10-Nano board is a low cost FPGA board that contains Intel's Cyclone V SoC FPGA with 110K logic elements. The board is manufactured by Terasic and can be found at here. The block diagram of the board can be seen in Fig. 4.1. The DE10-Nano board contains the SoC FPGA that has the FPGA fabric for creating custom hardware and a complete ARM computer system inside the FPGA.

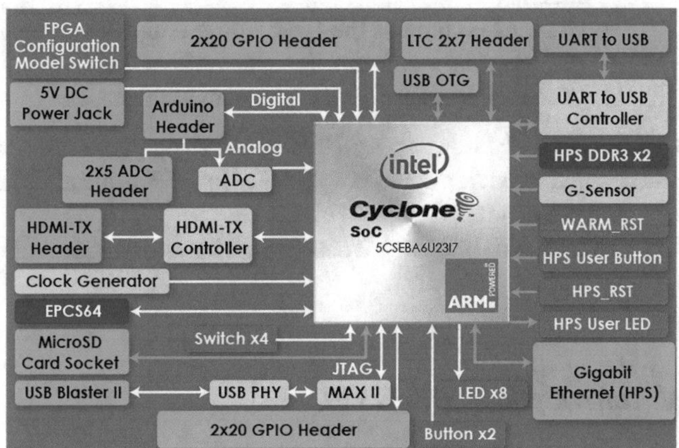

Fig. 4.1: Block Diagram of the DE10-Nano board. Figure from [1]

R. K. Snider, *Advanced Digital System Design using SoC FPGAs*,
https://doi.org/10.1007/978-3-031-15416-4_4

4.1.1 Determining DE10-Nano Board Revision

Download the document Board Revision [2] from Terasic to determine which DE10-Nano board revision you have. You need to know this in order to download the appropriate documentation.

4.1.2 DE10-Nano Information

Information on the DE10-Nano can be found at the following sources:

Source 1: The Terasic Website [3]. Click on the *Resources* tab and then look under the *Documents* section. You can download all the information by scrolling down to the CD-ROM section and selecting the appropriate CD-ROM link (which is why you need to know which board revision you have as determined in Sect. 4.1.1 Determining DE10-Nano Board Revision. When you click on the CD-ROM symbol, it will actually download as a .zip file (registration required). This download contains the following directories:

- `\Datasheet`—contains data sheets for all the parts on the board.
- `\Demonstrations`—contains demonstration projects with source code.
- `\Manual`—contains the **User Manual** and other guides.
- `\Schematic`—contains the board schematics.
- `\Tools`—contains Terasic's system builder.

Source 2: The Rocketboards.org (DE10-Nano) Website [4]. Rocketboards.org is devoted to boards containing Intel's SoC FPGAs and running Linux on these boards.

Source 3: The Intel (DE10-Nano) Website [5]. This site contains a *DE10-Nano Get Started Guide*.

4.1.3 DE10-Nano Cyclone V SoC FPGA

The SoC FPGA device on the DE10-Nano board is labeled **5CSEBA6U23I7**. These letters are interpreted as follows:

5C = Intel Cyclone V
SE = SoC with enhanced logic/memory
B = No hard PCIe or hard memory controller
A6 = 110K LEs
U23 = Ultra Fine Line BGA (UBGA) with 672 pins
I = Industrial temperature range (−40 to 100 °C)
7 = Speed grade. Devices with lower speed grade numbers run faster than devices with higher speed grade numbers (and cost more).

The list of possible options for Cyclone V SE devices can be seen in Figure 1–7 (page 1–22) in the Cyclone V Device Handbook [6], which is shown in Fig. 4.2.

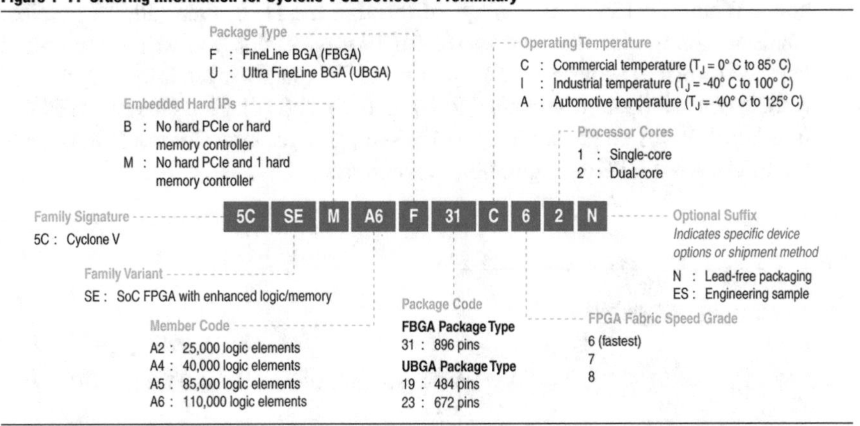

Fig. 4.2: Cyclone V SE Ordering Options

4.1.4 DE10-Nano Configuration Mode Switch Setting

The Cyclone V SoC FPGA must be configured when it is powered up. The information to configure the FPGA is contained in the configuration bitstream (e.g., .sof file) that Quartus creates when a design is compiled. Once the configuration bitstream has been created, it can be loaded into the FPGA in multiple ways depending on where the bitstream is being stored. The ways the bitstream can be loaded into the FPGA are:

1. JTAG via the USB Blaster II that is connected to the USB Mini-B connector on lower left side of board, below the HDMI connector. See Figure 2-1, page 5 of the DE10-Nano user manual. Programming via JTAG is typically done during development from the developer's computer. This is because the bitstream that

has been created by Quartus is located on your computer and you want to try it out on the DE10-Nano board. Loading the bitstream via JTAG typically means that you are iterating on a hardware design in the FPGA fabric. We will occasionally use the JTAG configuration method in this book.

2. **AS** or **Active Serial** configuration from the EPCS128 device. The EPCS128 is a serial flash memory device on the DE10-Nano board that can contain the configuration bitstream. See Table 3-2 on page 12 of the DE10-Nano user manual to set the board in AS mode. This is typically done if there is a single bitstream that will never change, and you will be shipping a product with this bitstream (i.e., Ship Boot Mode).
3. **HPS**, i.e., configured from the Hard Processor System via U-boot when Linux boots. **We will primarily be configuring the FPGA through the HPS** in this book. When the FPGA is configured through the HPS, there are 12 possible configuration modes. Two modes (out of 12 modes) that you will use are listed below. The other modes can be seen in Table 5-1 (page 5-5) in Intel's Cyclone V Hard Processor System Technical Reference Manual [7]. Five pins on the FPGA (see Fig. 4.3), called the *mode select* (**MSEL**) pins, tell the Control Block inside the FPGA what HPS configuration mode to use.

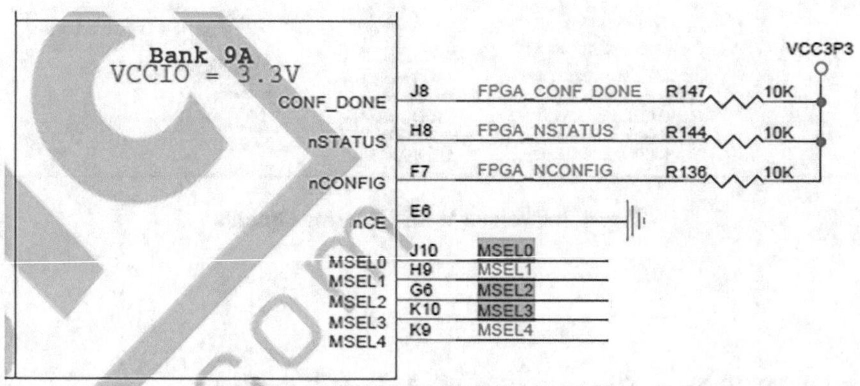

Fig. 4.3: The MSEL pins (MSEL[4:0]) on the FPGA tell the FPGA Control Block what configuration mode to use

The MSEL pins are connected to SW10 (see sheet 7 in the DE10-Nano Board Schematic) and also to pull-up resistors (see Fig. 4.4).

Default Setup MSEL[4:0] = 10010,
AS Fast Mode

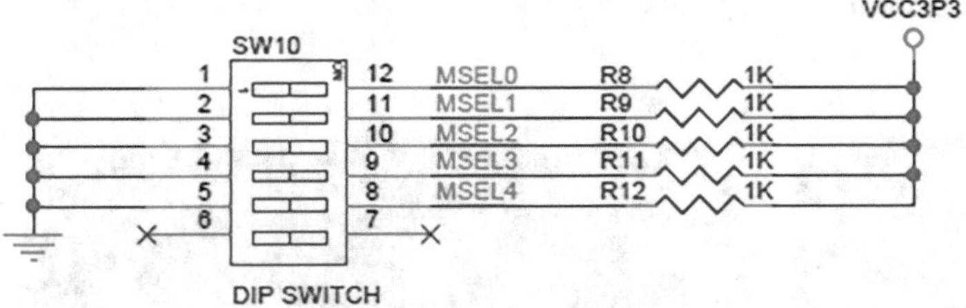

Fig. 4.4: The MSEL pins are connected to SW10 and to pull-up resistors

When a switch in the DIP switch SW10 is turned on, it grounds the MSEL pin. Otherwise the MSEL pin sits at Vcc. Thus a switch in the "ON" position means a "0" on the MSEL pin (negative logic). The location of SW10 on the DE10-Nano board is shown in Fig. 4.5 and marked by the yellow square.

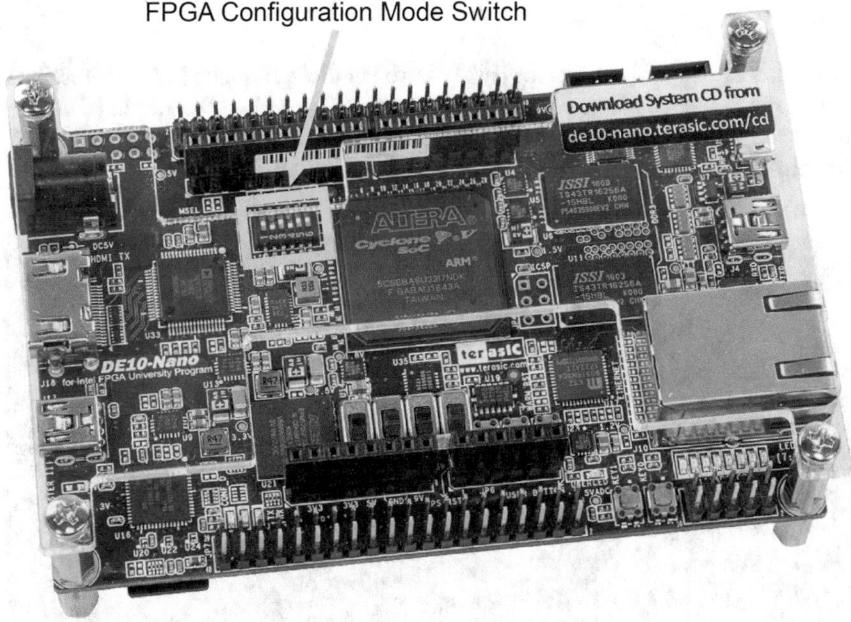

Fig. 4.5: Location of switch SW10 to set the DE10-Nano FPGA Configuration Mode. https://www.terasic.com.tw/cgi-bin/page/archive.pl?Language=English&CategoryNo=167&No=1046

There are two switch settings of the FPGA configuration mode switch SW10 that you need to be aware of. These two switch setting modes are:

Mode 1: FPPx32. This stands for Fast Passive Parallel x32 mode, Compression Enabled, Fast POR. **This is the default setting of the DE10-Nano board when it ships.** See Fig. 4.6 for this SW10 switch setting.

Fig. 4.6: Default switch SW10 setting (FPPx32) when the DE10-Nano board ships

Mode 2: FPPx16. This stands for Fast Passive Parallel x16 mode (no compression). **This is the mode we will use in this book for running embedded Linux (Ubuntu) on the DE10-Nano board.** In this mode, all the switches on SW10 need to be up in the zero (ON) state as shown in Fig. 4.7

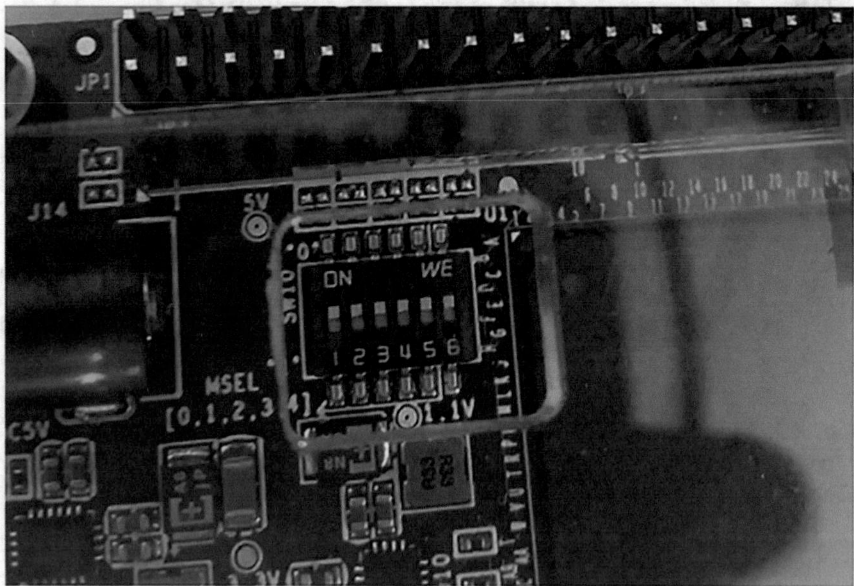

Fig. 4.7: The switch SW10 setting (FPPx16) used in the book when running Ubuntu on the DE10-Nano board. All the switches are in the "ON" position

References

1. Terasic, DE10-Nano Block Diagram. https://www.terasic.com.tw/attachment/archive/1046/image/DE10-Nano_Blockdiagram.jpg. Accessed 27 June 2022
2. Terasic Inc., How to Find the DE10-Nano Board Revision. https://www.terasic.com.tw/cgi-bin/page/archive_download.pl?Language=English&No=1046&FID=5651e4e0b7a3671ffb4494a46a338bf3. Accessed 15 Aug 2021
3. Terasic Inc., DE10-Nano Kit. https://www.terasic.com.tw/cgi-bin/page/archive.pl?Language=English&No=1046&PartNo=4. Accessed 15 Aug 2021
4. RocketBoards.org., DE10-Nano Development Board. https://rocketboards.org/foswiki/Documentation/DE10NanoDevelopmentBoard. Accessed 15 Aug 2021
5. Intel, Terasic DE10-Nano Get Started Guide. https://software.intel.com/content/www/us/en/develop/articles/terasic-de10-nano-get-started-guide.html. Accessed 15 Aug 2021
6. Intel, Cyclone V Device Handbook Volume 1: Device Overview and Datasheet. https://www.intel.com/content/dam/www/programmable/us/en/pdfs/literature/hb/cyclone-v/cv_5v1.pdf. Accessed 15 Aug 2021
7. Intel, Cyclone V Hard Processor System Technical Reference Manual. https://www.intel.com/content/dam/www/programmable/us/en/pdfs/literature/hb/cyclone-v/cv_5v4.pdf. Accessed 15 Aug 2021

Chapter 5
Introduction to the Audio Mini Board

Fig. 5.1: The Audio Mini board plugged into the DE10-Nano FPGA board. The Audio Mini provides high fidelity audio for the DE10-Nano board and contains an Analog Devices' AD1939 192 kHz 24-bit audio codec

The Audio Mini was developed to provide high fidelity audio for the DE10-Nano board. It was developed as part of an NIH grant (1R44DC015443-01) that developed an open speech signal processing platform using FPGAs for high-performance DSP

R. K. Snider, *Advanced Digital System Design using SoC FPGAs*,
https://doi.org/10.1007/978-3-031-15416-4_5

with low and deterministic latency (see (FPGA Open Speech Tools (Frost)). The DE10-Nano board was targeted in order to provide a low cost SoC FPGA platform that could be used to learn how digital signal processing could be implemented in the FPGA fabric and then controlled from Linux running on the ARM CPUs. It allows DSP, FPGAs, and Linux to be combined together in a low cost platform (compared to what some FPGA development boards can cost). The DE10-Nano board did not have audio capability, so we created the Audio Mini add-on board to provide high fidelity audio.

A brief overview of how the audio board was developed is as follows. The first step in the development process was to find all the available audio codecs that were 24-bit, could sample at 192 kHz, and were commercially available (and in stock). It was from this survey that we settled on the Analog Devices AD1939 for the reasons listed in Listing 5.3. We then purchased the evaluation board for the AD1939 (EVAL-AD1939) and connected it to the DE10-Nano FPGA board using handcrafted cables to connect the serial data port and the SPI control port. This allowed us to quickly get a minimal working example together for evaluation purposes. Once we verified that the prototype system worked by loading the appropriate register settings and sending audio data through the FPGA, we undertook the process of creating a custom printed circuit board (PCB) that would plug into the header posts on the DE10-Nano board. The Audio Mini board went through several hardware iterations due to corrections and optimizations and ended up with the final version shown in Fig. 5.1. We used Altium Designer [1] as the design software for the PCB.

5.1 Top Level Block Diagram

The top level schematic of the Audio Mini is shown in Fig. 5.2. The stereo analog signal comes into the board from a 3.5 mm audio jack (3.5 mm LINE IN block, top left) and is sent to the AD1939 audio codec (top center block). The analog stereo signal coming out of the audio codec is sent to the headphone amplifier block (3.5 mm HEADPHONE OUTPUT block, top right). The FPGA connections (FPGA CONNECTORS block, bottom left) connect the header posts on the DE10-Nano board and connect FPGA digital I/O lines to the AD1939 audio codec (serial data, SPI control, clock, and reset), the headphone amplifier (I2C volume control), and four LED and four switches (SUPPORT block lower center). There are also connections to two digital MEMS microphones, but these are not populated.

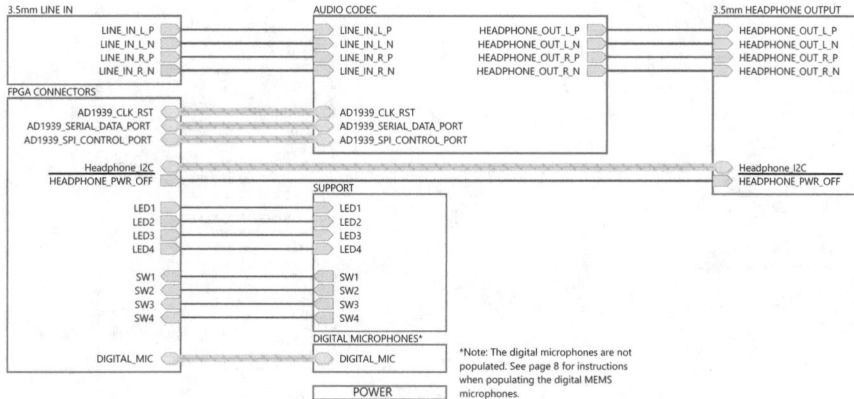

Fig. 5.2: The top level block diagram of Audio Mini board. Schematic figure from [2] (sheet 4)

5.2 Analog Audio Input

The AD1939 audio codec requires analog differential signals as inputs for the audio signals. However, audio connections using a 3.5 mm audio jack have single-ended inputs. Thus the line-in stereo single-ended inputs need to be converted to differential inputs. The conversion is done using Analog Devices ADA4075-2 ultralow noise op-amps [3] and the associated circuit that performs this single-ended to differential conversion is shown in Fig. 5.3. This circuit was adapted from the AD1939 evaluation (eval) board that had multiple audio interfaces. This illustrates one of the reasons for using evaluation boards from the manufacturer. By putting together a prototype system using a reference design provided by the manufacturer, you can see how the manufacturer supports their own device(s). From the manufacturer's reference design, you then add only the circuitry you need from the evaluation board (which has been designed to accommodate many I/O interfaces) into your own PCB. **Note:** The audio input can also handle a microphone input, but doing so will require part changes (instructions are located on sheet 9 of the schematic [2]).

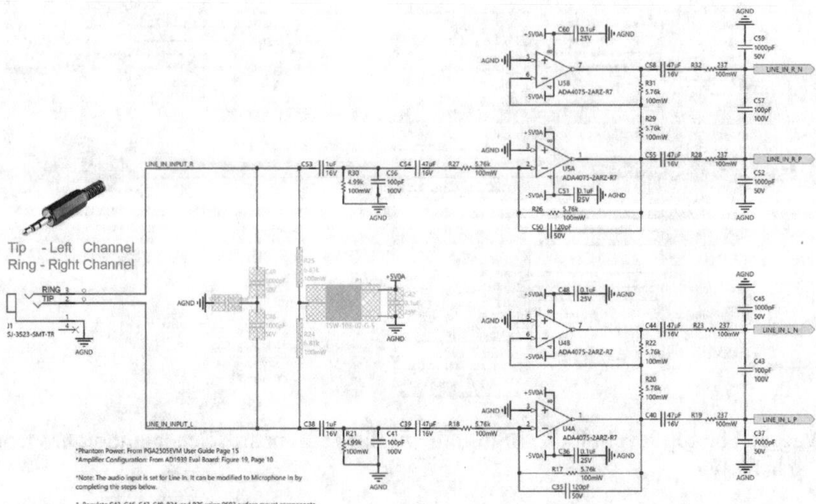

Fig. 5.3: The Audio Mini circuit that converts stereo single-ended inputs to differential inputs required by the AD1939 audio codec. Schematic figure from [2] (sheet 9)

5.3 AD1939 Audio Codec

The heart of the Audio Mini board is the Analog Devices' AD1939 audio codec. The AD1939 was chosen for several reasons:

Reason 1: It is a 24-bit audio codec that can sample up to a sampling rate of 192 kHz, providing high fidelity audio.

Reason 2: It has a clean interfaces for data (serial data) and control (SPI). See Fig. 5.4 that shows that the digital ports connect easily to FPGA I/O pins. This is in contrast to some audio codecs that target embedded audio applications where a DSP processor is also included in the device. We wanted a codec where processing would be done externally in an FPGA and had a straightforward interface.

Reason 3: It has multiple analog audio I/O ports. However, the Audio Mini only implements stereo in and stereo out due to cost considerations.

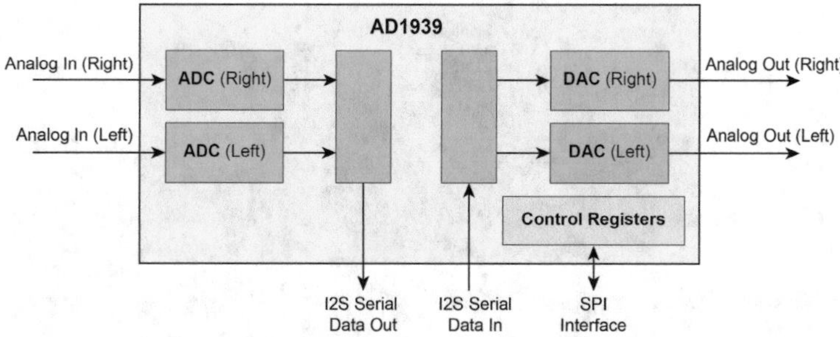

Fig. 5.4: The block diagram of the features used in the Analog Devices AD1939 audio codec

The schematic for the AD1939 audio codec can be seen in Fig. 5.5. Audio signals (stereo left/right) that have been converted to differential inputs (see Sect. 5.2) are piped into the AD1939 and converted to digital signals at one of the three sample rates (48 kHz, 96 kHz, or 192 kHz). The digital samples are then sent to the FPGA using the serial data interface that is comprised of three digital signals (data, bit clock, and left/right framing clock, see Fig. 5.6). Digital samples to be converted to analog are sent by the FPGA in the serial data format. The samples are converted by the AD1939 DAC to analog signals in a differential format and sent to the headphone driver/amplifier (see Sect. 5.4) for amplified stereo output.

The AD1939 is controlled from the SPI or Serial Peripheral Interface [4]. This is used to set internal AD1939 registers that control the sample rate, master clock, and other settings.

The sample rate is controlled by the 12.288 MHz crystal oscillator (see 12.288 MHz Master Clock in Fig. 5.5) that is attached to the AD1939 and drives the internal *Phase Locked Loop* (**PLL**) to create an internal clock that runs at 256 times the sample rate f_s (12.288 MHz/256 = 48 kHz).

The AD1939 is configured to set the ADC bit clock (ABCLK) and ADC left/right framing clock (ALRCLK) as masters (see Table 5.3). We also set the DAC bit clock (DBCLK) and DAC left/right framing clock (DLRCLK) as slaves. This means that the ADC drives all the data clocks in the AD1939. Furthermore, we send the 12.288 MHz master clock (MCLKO) to a clock input pin on the FPGA that can connect to an on-board PLL in the Cyclone V FPGA. The reason we do this is so that we can create a clock in the FPGA fabric using the on-chip PLL where the FPGA fabric clock is a multiple of the 12.288 MHz master clock (e.g., $12.288 \times 8 = 98.304$ MHz). This is so that we can implement ***synchronous*** DSP processing designs for our FPGA fabric data plane processing and avoid issues that arise when crossing clock domains in digital systems.

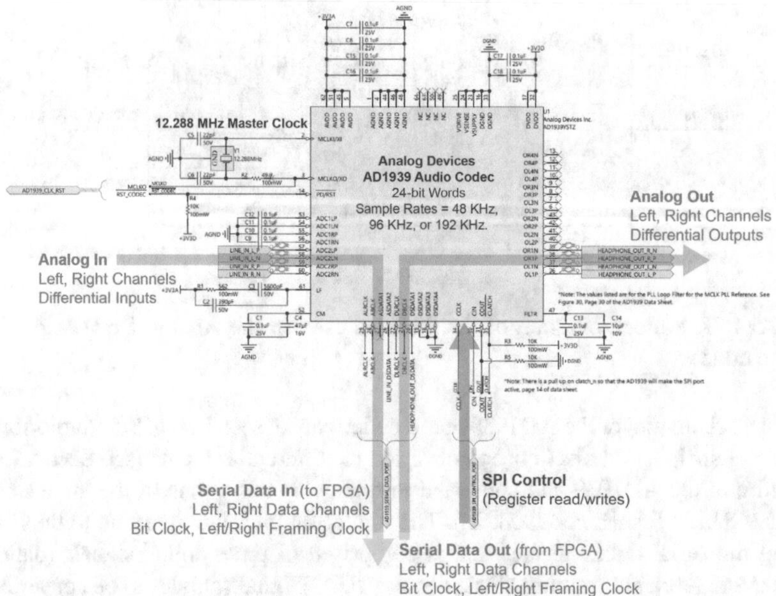

Fig. 5.5: The heart of the Audio Mini, which is the AD1939 audio codec (codec stands for coder–decoder where the coder is the ADC or analog-to-digital converter and the decoder is the DAC or digital-to-analog converter). Analog stereo differential signals are converted to digital serial signals and vice versa. Schematic figure from [2] (sheet 5)

Notice at the top of the schematic in Fig. 5.5 that the AD1939 has two separate 3.3 volt power supplies. This is because the AD1939 is a mixed signal design that contains both analog and digital signals. A mixed signal design is analogous to when you are studying for an exam and want peace and quiet (i.e., need low noise analog signals) and there is a crazy party with very loud music (digital signals) nearby. Hopefully, the building and room you are studying in does not let the party noise through. If the building is cheaply built, you will be distracted by all the noise. In a similar fashion, to keep the digital noisy party signals from injecting noise into the analog signals, the power supplies need to be kept separate. However, it is not practical to have separate power supplies coming into the PCB and we have only a single 5 volt pin coming from the DE10-Nano board with which to power the Audio Mini. The solution is to create separate analog power supplies from the 5 volt input that are well filtered and then kept separate (see power regulation on schematic sheet 7 of [2]).

5.3.1 AD1939 Serial Data Port

5.3.1.1 Data to the FPGA from the AD1939 ADC

The ADC in the AD1939 samples the analog signals and then sends the digital samples out of the serial data port in a serial fashion using three digital signals (SDATA, BCLK, and LRCLK) as shown in Fig. 5.6. The digital signal LRCLK is the left/right framing clock that runs at the sampling rate f_s. When LRCLK is low, the left channel is being sent out and when it is high, the right channel is being transmitted. The digital line BCLK is the bit clock for the serial data line SDATA that allows you to register the sample bits on the rising edge of BCLK. The bit clock BCLK runs 64 times faster than LRCLK providing 32 rising clock edges for the left channel and 32 clock edges for the right channel. The 24-bit sample word fits easily into the 32-bit channel slot and can have different alignment (justification) modes (the different modes can be seen in Figure 23, page 21 of the AD1939 datasheet [5]). We chose the I2S justification mode where the *Most Significant Bit* (**MSB**) of the 24-bit sample word starts on the second rising BCLK edge after an LRCLK edge transition (SDATA delay of 1). The serial data configuration is set in the ADC Control 1 Register (see Table 24 in the datasheet) and is set to the values shown in Table 5.3 (24-bit, Stereo, I2S). These register values for the ADC serial data port are power-up default values, so no power-up configuration needs to be performed.

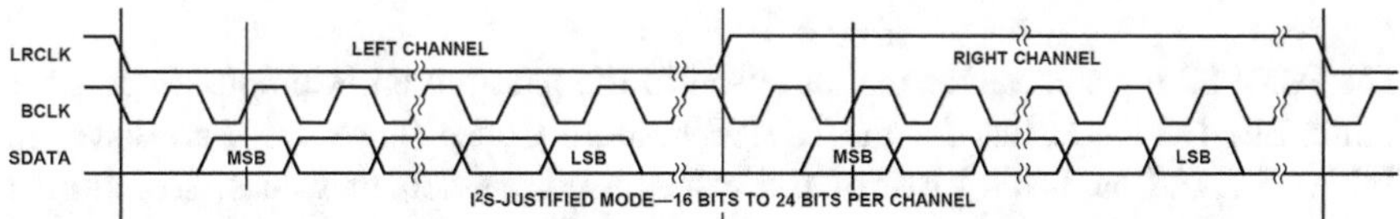

Fig. 5.6: The I2S stereo serial data mode. The serial data interface is comprised of three 1-bit signals. The top signal, LRCLK, is the left/right framing clock that runs at the sample rate f_s. The middle signal, BCLK, is the bit clock for audio bit values that are sent on the bottom signal SDATA. The bit clock BCLK runs 64 times as fast as the left/right framing clock LRCLK. Figure from [5] (Datasheet figure 23, page 21)

5.3.1.2 Data to the AD1939 DAC from the FPGA

The AD1939 DAC is configured (DAC Control 0 register in Table 5.2) to have the same serial data format as the ADC. In the initial board development, this allowed the FPGA to pipe the serial data directly out to the DAC to verify the system was working without converting the serial data to the Platform Designer Avalon streaming format.

5.3.1.3 AD1939 FPGA Data Interfacing

In order for the AD1939 to be used in a Platform Designer system, a VHDL and associated Platform Designer .tcl file was developed so that the AD1939 shows up in the Platform Designer library with streaming interfaces. This was developed as part of the first system developed using the AD1939, which is called unsurprisingly the *passthrough* example. This VHDL and Platform Designer interface is covered in Sect. 1.2 Audio Data Streaming (page 256) and Sect. 1.4 Platform Designer (page 269).

5.3.2 AD1939 SPI Control Port

The AD1939 is controlled from the bit values contained in seventeen registers that are accessed using the SPI control port. The Cyclone V SoC FPGA on DE10-Nano is configured so that the *Hard Process System* (**HPS**) exports a SPI interface, which is connected to the AD1939 SPI port. This Platform Designer HPS configuration is covered in Sect. 1.5.1 Linux SPI Device Driver for the AD1939 Audio Codec (page 289) along with the associated Linux device driver so that the AD1939 registers can be configured after power-up.

5.3.2.1 AD1939 Register Settings

The AD1939 register values that are used in the passthrough example (Sect. 1.5.1 Linux SPI Device Driver for the AD1939 Audio Codec (page 289)) are listed in Tables 5.1, 5.2, and 5.3. Most of the values used are default values, so minimal changes need to be made upon power-up. The values that are not default are listed in blue and have to be set before the system becomes functional.

Table 5.1: AD1939 clock control register settings

PLL and Clock Control 0 (Register Address 0)			
Bit	Value	Function	Description
0	0	Normal Operation	PLL power-down
2:1	00	INPUT 256	MCLKI/XI pin functionality
4:3	00	XTAL oscillator enabled	MCLKO/XO pin
6:5	00	MCLKI/XI	PLL input
7	1	Enable: ADC and DAC active	Internal master clock enable

PLL and Clock Control 1 (Register Address 1)			
Bit	Value	Function	Description
0	0	PLL clock	DAC clock source select
1	0	PLL clock	ADC clock source select
2	0	Enabled	On-chip voltage reference
3	0	0 = Not locked, 1 = Locked	PLL lock indicator (read only)

* Blue values: Non-default values. Must be set after power-up

Table 5.2: AD1939 DAC control register settings (All default values)

DAC Control 0 (Register Address 2)			
Bit	Value	Function	Description
0	0	Normal	Not powered-down
2:1	00	48 kHz	Sample rate
5:3	000	1	SDATA delay (I2S mode)
7:6	00	Stereo	Serial format

DAC Control 1 (Register Address 3)			
Bit	Value	Function	Description
0	0	Latch in mid-cycle (normal)	BCLK active edge (TDM in)
2:1	00	64 (2 channels)	BCLKs per frame
3	0	Left low	LRCLK polarity
4	0	Slave	LRCLK master/slave
5	0	Slave	BCLK master/slave
6	0	DBCLK pin	BCLK source
7	0	Normal	BCLK polarity

DAC Control 2 (Register Address 4)			
Bit	Value	Function	Description
0	0	Unmute	Master Mute
2:1	00	Flat	De-emphasis
4:3	00	24	Word width (bits)
5	0	Noninverted	DAC output polarity

DAC Individual Channel Mutes (Register Address 5)			
Bit	Value	Function	Description
7:0	00000000	0 = Unmute	All Channels Unmuted

DAC L1 Volume (Register Address 6) DAC R1 Volume (Register Address 7) DAC L2 Volume (Register Address 8) DAC R2 Volume (Register Address 9) DAC L3 Volume (Register Address 10) DAC R3 Volume (Register Address 11) DAC L4 Volume (Register Address 12) DAC R4 Volume (Register Address 13)			
Bit	Value	Function	Description
7:0	00000000	No attenuation	DAC volume control ($-3/8$ dB per step)

Table 5.3: AD1939 ADC control register settings

ADC Control 0 (Register Address 14)			
Bit	Value	Function	Description
0	0	Normal	Not Powered-down
1	0	Off	High pass filter
2	0	Unmute	ADC 1L mute
3	0	Unmute	ADC 1R mute
7:6	1	48 kHz	Sample rate

ADC Control 1 (Register Address 15)			
Bit	Value	Function	Description
1:0	00	24	24-bit Word width
4:2	000	1	SDATA delay (I2S mode)
6:5	00	Stereo	Serial format

ADC Control 2 (Register Address 16)			
Bit	Value	Function	Description
0	0	50/50	LRCLK format
1	0	Drive on falling	BCLK polarity
2	0	Left low	LRCLK polarity
3	1*	Master	LRCLK master/slave
5:4	00	64	BCLKs per frame
6	1*	Master	BCLK master/slave
7	1*	Internally generated	BCLK source

* Blue values: Non-default values. Must be set after power-up

5.4 Headphone Analog Audio Output

To convert the differential analog signals coming from the AD1939 DAC to single-ended outputs that a user can use to plug their headphones into, we used the Texas Instrument's TPA6130A2 stereo headphone amplifier with I2C volume control [6]. The associated circuit can be seen in Fig. 5.7.

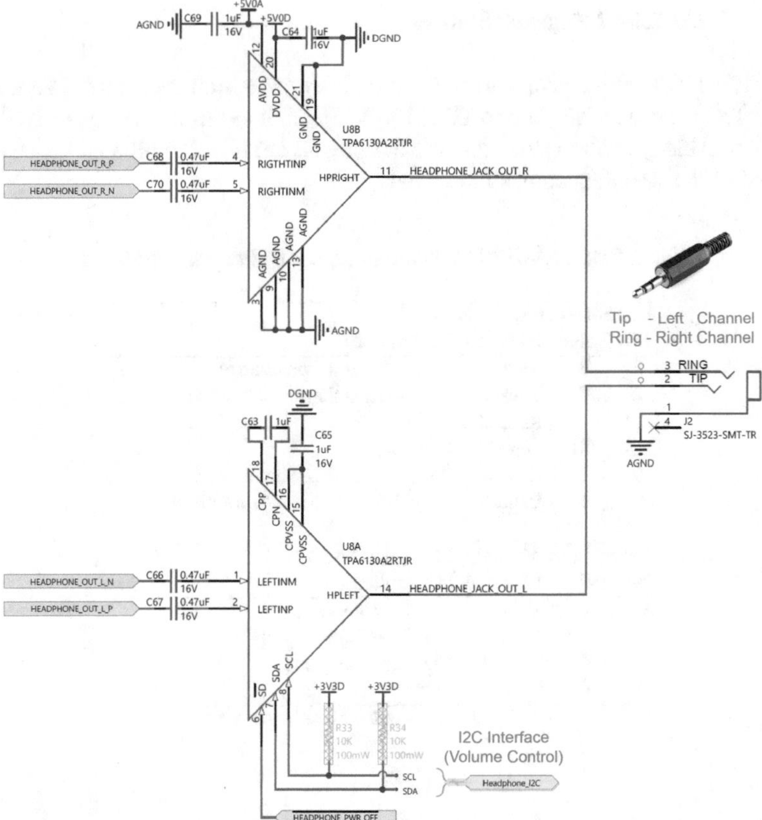

Fig. 5.7: The Texas Instrument's TPA6130A2 headphone amplifier [6] takes the differential output signals from the AD1939, implements volume control (controlled via I2C), and converts them to singled-ended signals suitable for a headphone connection. Schematic figure from [2] (sheet 11)

5.4.1 TPA6130A2 I2C Interface

The TPA6130A2 headphone amplifier is controlled from the bit values contained in two registers that are accessed using the I2C interface. The Cyclone V SoC FPGA on DE10-Nano is configured so that the *Hard Process System* (**HPS**) exports an I2C interface, which is connected to the TPA6130A2 I2C interface. This Platform Designer HPS configuration is covered in Sect. 1.5.2 Linux I2C Device Driver for the TPA6130A2 Headphone Amplifier (page 289) along with the associated Linux device driver so that the TPA6130A2 registers can be configured after power-up.

5.4.1.1 TPA6130A2 Register Settings

The TPA6130A2 register values that are used in the passthrough example (Sect. 1.5.2 Linux I2C Device Driver for the TPA6130A2 Headphone Amplifier (page 289)) are listed in Table 5.4. The values that are not default are listed in blue and have to be set before the system becomes functional.

Table 5.4: TPA6130A2 headphone control register settings

Control Register (Address 1)			
Bit	Value	Function	Description
0	0	SWS	Software shutdown control
1	0	Thermal	1 indicates a thermal shutdown
3:2		Reserved	
5:4	00	Stereo	Mode Select
6	1	Enable	Enable for right channel amplifier
7	1	Enable	Enable for left channel amplifier

Volume and Mute Register (Address 2)			
Bit	Value	Function	Description
5:0	xxxxxx	Volume	000000 = lowest gain 111111 = highest gain
6	0	Unmute	Right channel mute
7	0	Unmute	Left channel mute

* blue values: Non-default values. Must be set after power-up

5.5 DE10-Nano FPGA Connections

The connections from the Audio Mini to the DE10-Nano board can be seen in Fig. 5.8. The signal routing and naming can get confusing, so a "Rosetta Stone" table (Table 5.5) was created to cross reference names and locations depending on a particular reference point (device, DE10-Nano board, Cyclone V FPGA, or VHDL signal name). In the table column 1 gives the schematic signal group name. Column 2 gives the schematic signal name in the signal group. Column 3 gives the manufacturer's datasheet signal name. Column 4 gives the device and the device pin number that the signal is connected to. Column 5 gives the pin number of the DE10-Nano header post (header JP7) that the signal goes through. Column 6 gives the pin number on the Cyclone V FPGA that the signal is connected to. Column 7 gives the GPIO connection name as reference from the DE10-Nano User's manual. Finally, in column 8, the VHDL top level signal name is given that references the signal. This information is typically hidden in the board support file (e.g., FPGA pin assignment file) but is given to show how such board support information is created. The FPGA developer typically only cares about the top level signal names in column 8.

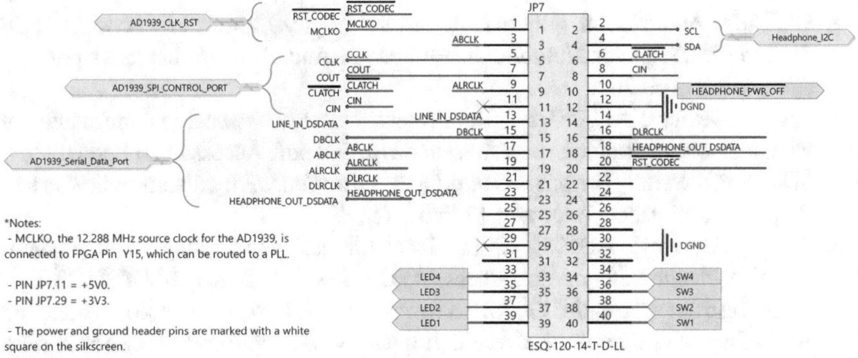
AD1939_CLK_RST
AD1939_SPI_CONTROL_PORT
AD1939_Serial_Data_Port
Headphone_I2C
HEADPHONE_PWR_OFF
JP7
ESQ-120-14-T-D-LL
DGND
LED4
LED3
LED2
LED1
SW4
SW3
SW2
SW1
*Notes:
- MCLKO, the 12.288 MHz source clock for the AD1939, is connected to FPGA Pin Y15, which can be routed to a PLL.
- PIN JP7.11 = +5V0.
- PIN JP7.29 = +3V3.
- The power and ground header pins are marked with a white square on the silkscreen.

Fig. 5.8: The Audio Mini connections to the DE10-Nano board. Schematic figure from [2] (sheet 6)

Table 5.5: Audio Mini FPGA signal connection table

Schematic Signal Group	Schematic Net Name	Device Signal Name	Device/ Pin Number	DE10-Nano JP7 Header Pin Number	FPGA Pin	Terasic GPIO Location	VHDL Signal Name
AD1939_SPI_CONTROL_PORT	CIN	CIN	AD1939/30	8	AF27	GPIO_1[7]	AD1939_spi_CIN
	CLATCH_n	CLATCH_n	AD1939/35	6	AF28	GPIO_1[5]	AD1939_spi_CLATCH_n
	CCLK	CCLK	AD1939/34	5	AG28	GPIO_1[4]	AD1939_spi_CCLK
	COUT	COUT	AD1939/31	7	AE25	GPIO_1[6]	AD1939_spi_COUT
AD1939_SERIAL_DATA_PORT	ABCLK	ABCLK	AD1939/28	3	AA15	GPIO_1[2]	AD1939_ADC_ABCLK
	ALRCLK	ALRCLK	AD1939/29	9	AG26	GPIO_1[8]	AD1939_ADC_ALRCLK
	LINE_IN_DSDATA	ASDATA2	AD1939/26	13	AG25	GPIO_1[10]	AD1939_ADC_ASDATA2
	DBCLK	DBCLK	AD1939/21	15	AH24	GPIO_1[12]	AD1939_DAC_DBCLK
	DLRCLK	DLRCLK	AD1939/22	16	AF25	GPIO_1[13]	AD1939_DAC_DLRCLK
	HEADPHONE_OUT_DSDATA	DSDATA1	AD1939/20	18	AF23	GPIO_1[15]	AD1939_DAC_DSDATA1
AD1939_CLK_RST	MCLK0	MCLK0	AD1939/3	1	Y15	GPIO_1[0]	AD1939_MCLK
	RST_CODEC_n	PD/RST_n	AD1939/14	20	AH22	GPIO_1[17]	AD1939_RST_CODEC_n
HEADPHONE_I2C	SCL	SCL	TPA6130A2/8	2	AC24	GPIO_1[1]	TPA6130_I2C_SCL
	SDA	SDA	TPA6130A2/7	4	AD26	GPIO_1[3]	TPA6130_I2C_SDA
HEADPHONE_PWR_OFF_n	HEADPHONE_PWR_OFF_n	SD_n	TPA6130A2/6	10	AH27	GPIO_1[1]	TPA6130_power_off
LEDS	LED1			39	AE19	GPIO_1[34]	Audio_Mini_LEDs[0]
	LED2			37	AG15	GPIO_1[32]	Audio_Mini_LEDs[1]
	LED3			35	AF18	GPIO_1[30]	Audio_Mini_LEDs[2]
	LED4			33	AG18	GPIO_1[28]	Audio_Mini_LEDs[3]
SWITCHES	SW1			40	AE17	GPIO_1[35]	Audio_Mini_SWITCHES[0]
	SW2			38	AE20	GPIO_1[33]	Audio_Mini_SWITCHES[1]
	SW3			36	AF20	GPIO_1[31]	Audio_Mini_SWITCHES[2]
	SW4			34	AH18	GPIO_1[29]	Audio_Mini_SWITCHES[3]

The top level VHDL file that contains the signal names in the right hand column of Table 5.5 is described in Sect. 1.4.4.2 Hooking Up the soc_system_passthrough System in the Top Level (page 288).

References

1. Altium, Altium Designer. https://www.altium.com/altium-designer/. Accessed 18 Feb 2022

2. C. Dack, Audio Mini Schematics. Audio Logic. https://github.com/ADSD-SoC-FPGA/Documents/blob/main/AudioMini/AudioMini_Schematics.pdf. Accessed 16 Feb 2022
3. Analog Devices, ADA4075-2 Data Sheet. https://www.analog.com/media/en/technical-documentation/data-sheets/ada4075-2.pdf. Accessed 16 Feb 2022
4. Wikipedia, Serial Peripheral Interface. https://en.wikipedia.org/wiki/Serial_Peripheral_Interface. Accessed 17 Feb 2022
5. Analog Devices, AD1939 Data Sheet. https://www.analog.com/media/en/technical-documentation/data-sheets/AD1939.pdf. Accessed 16 Feb 2022
6. Texas Instruments, TPA6130A2 138-mW DIRECTPATH™Stereo Headphone Amplifier with I2C Volume Control. https://www.ti.com/product/TPA6130A2?. Accessed 18 Feb 2022

Chapter 6
Introduction to Intel Quartus Prime

6.1 Intel Quartus Prime Lite Edition

The DE10-Nano FPGA board that we are targeting contains an Intel Cyclone V SoC FPGA. This means that we need to use Intel's Quartus Prime software to create hardware designs for the Cyclone V SoC FPGA. Fortunately, we can use the free version of Quartus Prime with the Cyclone V devices, which is called **Intel Quartus Prime Lite Edition**.

6.1.1 Installing Windows for Subsystem for Linux (WSL)

Quartus, starting with version 19.1, requires installing Windows Subsystem for Linux. There are two versions of WSL: WSL 1 and WSL 2. At the time of this writing, the instructions for Quartus are to install only WSL 1 since WSL 2 is not supported. The procedure for installing WSL 1 is:

Step 1: Go to:
https://docs.microsoft.com/en-us/windows/wsl/install-win10 and follow Microsoft's instructions to install Ubuntu 18.04 LTS for WSL.
Note 1: Windows 10 build version 16215.0 or higher is the recommended operating system version.
Note 2: Install only WSL 1 and skip the instructions for updating WSL 1 to WSL 2. WSL 2 is not supported.

Step 2: After installation has been successfully completed, launch Ubuntu 18.04.

Step 3: Install the distro packages described at: https://www.intel.com/content/www/us/en/docs/programmable/683525/21-3/installing-windows-subsystem-for-linux.html

R. K. Snider, *Advanced Digital System Design using SoC FPGAs*,
https://doi.org/10.1007/978-3-031-15416-4_6

6.1.2 Download and Install Intel's Quartus Prime Lite

Install the free version of Quartus Prime (Lite Edition) from Intel by following these steps:

Step 1: First install Windows Subsystem for Linux by following the instructions in Sect. 6.1.1 Installing Windows for Subsystem for Linux (WSL) (page 55).

Step 2: Get the Quartus Prime Lite Edition from:
https://www.intel.com/content/www/us/en/products/details/fpga/development-tools/quartus-prime/resource.html.

Step 3: Select the Lite Edition.

Step 4: Select the latest version, which is currently version 20.1 .

Step 5: Further down the page, Select the "Individual Files" tab.
You should see a listing of individual files to download. We do not need all these files. The ones we do need are listed below. Click the download arrow at the right to download the files:

- **Quartus Prime (includes Nios II EDS)**
  Note: Also download the Questa-Intel FPGA Edition.
- **Cyclone V device support**
  Note: You do not need any other device support since we will only be targeting the Cyclone V.

Step 6: Select the "Additional Software" tab. Download the files:

- **Quartus Prime Help**.
- **Quartus Prime Programmer and Tools**.
  Note: This includes Signal Tap and System Console that we will use.

Step 7: Install the software by running the downloaded file that starts with "QuartusLiteSetup..." The installation will take some time.
Note 1: The Quartus Lite install executable will see the help and device files and install them as if they are already in the same directory.
Note 2: The "QuartusProgrammerSetup. . . " executable needs to be installed separately.

6.1.3 Quartus File Types

In the Quartus project folder, there are a number of different file types that are identified by their extensions.

A list of Quartus file types that you will run across are:

File Type 1: Quartus Project File (**.qpf**). This file when opened in Quartus automatically loads your project. This file can be created by using the *New Project Wizard* in Quartus.

File Type 2: Quartus Setting File (**.qsf**). This file contains the pin assignments and associates the signal names found in the top level entity with specific I/O pins on the FPGA.

File Type 3: Synopsys Design Constraints File (**.sdc**). **There are two things that must be satisfied for your design to be correct**. *First*, your VHDL logic must be correct, and *second*, the timing of your logic must be correct after the Quartus Fitter places and routes your design in the FPGA fabric. If you do not have a .sdc file in your project, which constrains your timings, your design is wrong (even if it appears to function correctly when you compile it). This means that you must always add a .sdc file to your project. Furthermore, after each Quartus compilation, you need to check the resulting timing because a particular place and route may fail to get the required timing correct. *Do not assume your design is correct just because Quartus compiled your VHDL correctly. It must also meet your timing requirements.*

File Type 4: Top level VHDL file (**.vhd**). This file contains the top level entity that has the signal names that are to be connected to specific I/O pins on the FPGA (i.e., pin assignments) as described in the .qsf file. You will need to set one of your VHDL files in your project as the top level file. This is done in Quartus by selecting *Files* in the drop down list in the *Project Navigator* panel and then right clicking on the desired .vhd file and selecting *Set as Top Level Entity*.

File Type 5: SRAM Object File (**.sof**). This is the configuration file created when a design is compiled and synthesized by Quartus. It is the bitstream that configures the FPGA fabric. This bitstream configuration file is typically loaded into the FPGA by the JTAG programmer.

File Type 6: Raw Binary File (**.rbf**). The raw binary file format contains the same information as the .sof configuration file. This .rbf file format is used by U-boot to configure that fabric when Linux boots up. The .sof file is converted into a .rbf file by using the *Convert Programming File* utility in Quartus. This is found in Quartus by going to the *File* menu and selecting *Convert Programming Files....*

File Type 7: Programmer Object File (**.pof**). The programmer object file format contains the same information as the .sof configuration file. This .pof file format is used to program serial flash devices, which is how FPGAs are typically configured at power-up when they are not SoC devices and do not have embedded Linux.

6.1.4 Converting Programming Files

When Quartus compiles a project, it places the **.sof** bitstream file into the subdirectory `/output_files` under the project directory. This is the file that the Quartus programmer uses to configure the FPGA fabric when it downloads it via JTAG. When this bitstream file needs to be loaded by U-boot, it must first be converted into a **.rbf** file and then placed into the VM directory that TFTP server will use. The directory location and name of the .rbf file are specified in the bootscript file.

The steps for converting a .sof file into a .rbf file, where *soc_system.sof* is the example file name, are:

Step 1: In Quartus, go to *File* → *Convert Programming Files. . . .* and in the *Output programming file* section:

1: Set the *Programming file type:* to **Raw Binary File (.rbf)**.
2: Set the *Mode:* to **Passive Parallel x16**.
Note: This requires the switch positions on the DE10_Nano board to be all in the "On" position (see Fig. 4.7).
3: Set the *File name:* to **soc_system.rbf**.

Step 2: In the section *Input files to convert:*

1: Click on *SOF Data*.
2: Click on the *Add File. . .* button.
3: Navigate to `/output_files`.
4: Select **soc_system.sof**.
5: Click Open.

Step 3: Click Generate. The .rbf file will be written to the Quartus project folder.

6.1.5 Timing

Just because Quartus compiles your VHDL code correctly does not mean that it will run correctly in your FPGA. Not only does the logic in your VHDL code need to be correct, the timing needs to be correct after the design has been placed and routed by the fitter. Getting the timing correct is called **Timing Closure**, and this can take considerable effort when using FPGAs. Fortunately for us, the examples in this book do not push the limits on how fast we are trying to run the clock or how full we are trying to fill the FPGA in regard to the fabric resources being used, which makes placement and routing harder while meeting timing.

Any time you purchase a FPGA, you will need to specify the speed grade of the device. The same device, but with a faster speed grade, typically costs more. In Fig. 4.2, we can see that the Cyclone V has three speed grades (6,7, and 8) where 6 is the fastest speed grade. Every device manufactured is slightly different when manufactured due to process variations, even if it is designed to be an identical part.

This means some parts are faster than others and are tested and labeled with different speed grades. Other factors that affect speed are core voltage (higher Vcc makes the chip faster) and temperature (higher T makes the chip slower). Thus **PVT**, which stands for **Process**, **Voltage**, and **Temperature**, affects how fast a part can run.

After Quartus compiles your design, when you examine the Table of Contents of the Compilation Report, you will see a section called *Timing Analyzer*. If you expand this section, you will see four folders as shown in Fig. 6.1.

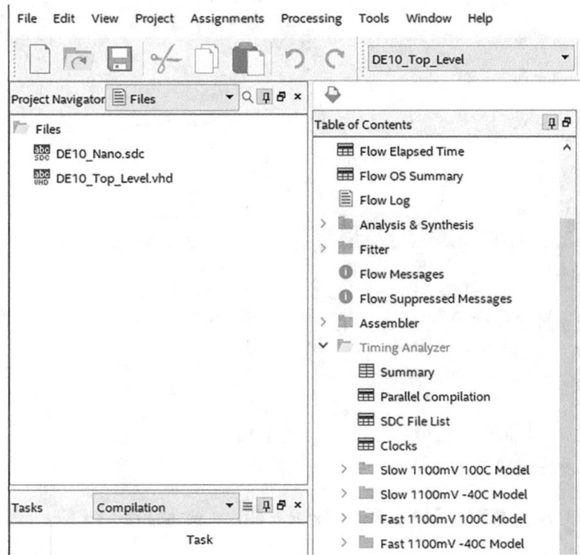

Fig. 6.1: Checking Timing in Quartus

These four folders are associated with the PVT designation where in this case P = {Slow or Fast}, V = 1100 mV = 1.1 V, and T = {−40 C or 100 C}. These cases are known as corner cases since all combinations of PVT for this device will fall in the area enclosed by these corner cases. The slowest device would have the parameters (slowest Process, lowest Voltage, highest Temperature), which in the figure is the **Slow 1100 mV 100 C Model** folder. The fastest device would have the parameters (fastest Process, highest Voltage, lowest Temperature), which in the figure is the **Fast 1100 mV −40 C Model**.

For our initial timing check, we will take the most conservative view of the device that we are targeting where we will assume that we have the slowest of the devices (**Slow 1100 mV 100 C Model**). If we expand this folder, we see the following information (Fig. 6.2).

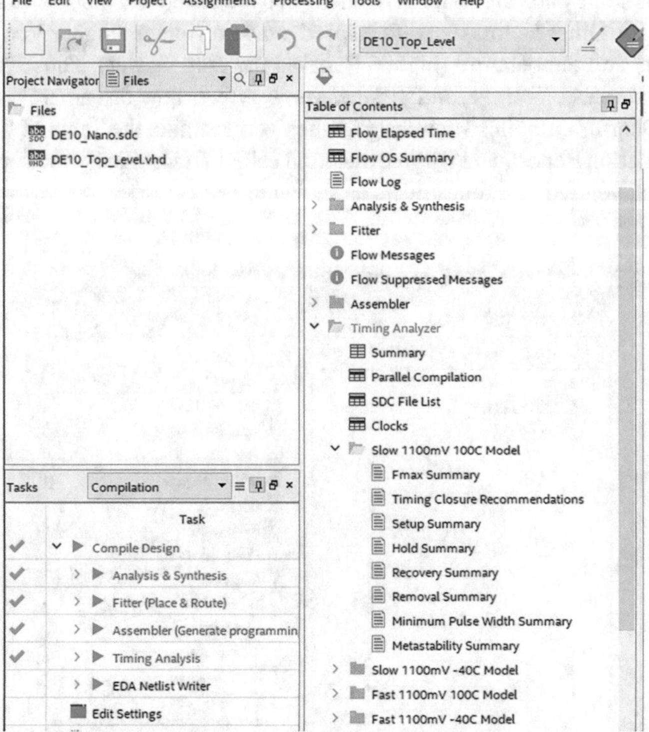

Fig. 6.2: Checking Fmax for the Slowest Model in Quartus

If we click on *Fmax Summary*, this will tell us how fast the FPGA can be clocked if we have the slowest speed grade device. **You should always check Fmax of the slowest model after each compile to see if your design can run at the targeted clock speed.**

6.1.6 Learning Quartus

One of the skills that computer engineers need to develop is the ability to implement lifelong learning also known as continuing education. This is important in the computer industry where technology is moving at a blistering pace. Thus it is necessary to be able to take advantage of training material that exists for learning new skills. We will take advantage of the training material that Intel provides for their Quartus software. If you go to *Intel's FPGA Technical Training Curricula* (click here for the link), you will see a section for FPGA Designers as seen in Fig. 6.3. This training ranges from free online courses to paid instructor-led courses. We only highlight several of the free online courses in Table 6.1 that are related to the Quartus tools

FPGA Designers

Description	Course Catalog		
Traditional FPGA developers code in languages such as Verilog HDL and VHDL. These developers are comfortable with creating FPGAs using the Intel Quartus Prime software, closing timing on complicated hardware circuits and managing complicated I/O interfaces to the FPGA.	Level 100	Level 200	Level 300

Fig. 6.3: Intel's Training Curricula for FPGA Designers

Table 6.1: Intel's free online training courses for Quartus

Course and description	**Level**
Intel **Quartus Prime** Software: Foundation (Standard Edition) (click here) Learn to use the Intel® Quartus® Prime software to develop an FPGA or CPLD design from initial design to device programming.	100
Timing Analyzer: Introduction to Timing Analysis (click here) Closing timing can be one of the most difficult and time-consuming aspects of creating an FPGA design. The Timing Analyzer is used to shorten the process of timing closure. This part of the training introduces you to the basic timing analysis concepts required for understanding how to use the tool. This is part 1. There are 4 parts to watch.	200
Signal Tap Logic Analyzer: Introduction & Getting Started (click here) The Signal Tap embedded logic analyzer (ELA) is a system-level debugging tool that monitors the state of internal FPGA design signals. This is part 1. There are 4 parts to watch.	200
Creating a System Design with **Platform Designer**: Getting Started (click here) The Platform Designer system integration tool, formerly known as Qsys, saves design time and improves productivity by automatically generating interconnect logic to connect intellectual property (IP) functions and subsystems. This is part 1. There are 2 parts to watch.	200
System Console (click here) System Console is an interactive console for system-level debug of Platform Designer systems over JTAG. Based on Tcl, it has a simple set of commands for communicating with various parts of your Platform Designer system.	300

that we will be using so you should browse through all the training that Intel has to offer.

6.2 Platform Designer

6.2.1 Creating an Avalon Memory-Mapped Interface

Any custom hardware that you create for Platform Designer needs to interface to the Avalon interconnect network. The network interface supports both memory-mapped peripherals (covered in this section) and streaming interfaces (covered in Sect. 6.2.3). Memory-mapped peripherals communicate with the Avalon interconnect network and will have specific memory address locations assigned to them. Streaming interfaces are useful for data processing tasks on streaming data such as digital filtering an audio signal.

A memory-mapped interface can function as either a host or an agent. A host can initiate data transfers (i.e., bus transactions), while an agent only responds to data transfers (i.e., responds to a host). The Avalon Interface Specifications can be found at the link Avalon Interface Specifications.

To illustrate how to create and use a memory-mapped interface, control registers for the HPS_LED_Patterns component (see Fig. 12.1) will be created and attached to the HPS lightweight bus. We will only implement a simple memory-mapped agent interface for HPS_LED_Patterns. The Avalon bus agent that we will create will look like Figure 5 and Figure 6 found on pages 13–14 (section 3.1) in the Avalon specification. The agent can handle complicated transfers, so there are additional control signals available. We, however, will only be reading/writing data from Registers 0–3. This means that our interface will be relatively simple. Since the data bus for the HPS lightweight bus is 32 bits, we will create **32-bit registers** where we will place 32 bits of data onto the bus when the ARM CPUs read the register and capture 32 bits of data on CPU writes.

Notice the timing diagram for reads and writes on the Avalon interface that is shown in Figure 7 on page 21 (section 3.5.1) of the Avalon spec. You will see that the data is latched on the rising edge of the clock when the read or write enable is asserted. In this diagram, we will ignore the bus control signal *byteenable* since we will read and write 32 bits and not worry about accessing specific bytes in the 32-bit word. We will also ignore the signals *waitrequest* and *response* since our registers will be able to respond fast enough, and given the simple interface, we do not need to respond with optional read/write status bits.

To interface to the Avalon bus, your HPS_LED_Patterns Platform Designer component should have the following entity (click here):

```
entity HPS_LED_patterns is
    port(
        clk                 : in  std_logic;                     -- system clock
        reset               : in  std_logic;                     -- system reset (assume
```

```
                                    active high, change at top level if needed)
        avs_s1_read       : in  std_logic;                     -- Avalon read control signal
        avs_s1_write      : in  std_logic;                     -- Avalon write control signal
        avs_s1_address    : in  std_logic_vector(1 downto 0);  -- Avalon address; Note: width determines the number of registers created by Platform Designer
        avs_s1_readdata   : out std_logic_vector(31 downto 0); -- Avalon read data bus
        avs_s1_writedata  : in  std_logic_vector(31 downto 0); -- Avalon write data bus
        PB                : in  std_logic;                     -- Pushbutton to change state (assume active high, change at top level if needed) (export in Platform Designer)
        SW                : in  std_logic_vector(3 downto 0);  -- Switches that determine the next state to be selected (export in Platform Designer)
        LED               : out std_logic_vector(7 downto 0)   -- LEDs on the DE10-Nano board (export in Platform Designer)
    );
end entity HPS_LED_patterns;
```

Listing 6.1: Entity of the HPS_LED_patterns component

Notice that the address bus *avs_s1_address* that connects to the Avalon interconnect is only 2 bits and not 32 bits. This is because the signal width is how you define how many 32-bit registers your hardware component has. Platform Designer does the full address decoding for you, and you only need to perform partial address decoding for your registers using the *avs_s1_address* signal in order to determine which register is being selected. Thus, *the number of bits contained in the std_logic_vector signal avs_s1_address will determine how many registers your component has.*

The prefix *avs_s1_* of the Avalon signal names is used because Platform Designer will correctly interpret these signal names as an Avalon agent interface when imported into Platform Designer and saves the trouble of having to assign them to the proper Avalon signal interpretation.

6.2.1.1 Creating Component Registers

To create registers, you first have to name and create signal declarations for them as shown in Listing 6.2 where two registers *left_gain* and *right_gain* are created. These signals are given default values, which will be the values these registers take when the component powers up. These register signals are the internal signals that are used to connect to the ports and logic inside the component's architecture.

```
signal left_gain  : std_logic_vector(31 downto 0) := "00000100110011001100110011001101"; -- 0.3 fixed point value (W=32, F=28)
signal right_gain : std_logic_vector(31 downto 0) := "00000100110011001100110011001101"; -- 0.3 fixed point value (W=32, F=28)
```

Listing 6.2: Example Register Declarations

Next, we need to be able to read these registers from the ARM CPUs, so we need to create a read process as shown in Listing 6.3. This is a synchronous process that checks to see if *avs_s1_read* has been asserted. If it has, it then uses a case statement on *avs_s1_address* to determine which register to read by placing the register's data onto the *avs_s1_readdata* bus. If there are undefined registers in the component's register address space, the read needs to return zeros, which is handled by the when others statement.

```
avalon_register_read : process(clk)
begin
    if rising_edge(clk) and avs_s1_read = '1' then
        case avs_s1_address is
            when "00"   => avs_s1_readdata <= left_gain;
            when "01"   => avs_s1_readdata <= right_gain;
            when others => avs_s1_readdata <= (others => '0'); -- return zeros for unused registers
        end case;
    end if;
end process;
```

Listing 6.3: Example Register Read Process

To write to the registers, we create a write process as shown in Listing 6.4. An asynchronous reset will reset the register values to the given values that are similar to the default power-up values. This synchronous process checks to see if *avs_s1_write* has been asserted. If it has, it then uses a case statement on *avs_s1_address* to determine which register to write to. In this example, the entire 32 bits are being transferred from the *avs_s1_writedata* bus to the register. If the register is smaller, the appropriate signal slice would need to be performed. If there are undefined registers in the component's register address space, the writes are ignored, which is handled by the when others statement.

```
avalon_register_write : process(clk, reset)
begin
    if reset = '1' then
        left_gain   <=  "00000100110011001100110011001101"; -- 0.3  reset default value
        right_gain  <=  "00000100110011001100110011001101"; -- 0.3  reset default value
    elsif rising_edge(clk) and avs_s1_write = '1' then
        case avs_s1_address is
            when "00"   => left_gain   <= avs_s1_writedata(31 downto 0);
            when "01"   => right_gain  <= avs_s1_writedata(31 downto 0);
            when others => null; -- ignore writes to unused registers
        end case;
    end if;
end process;
```

Listing 6.4: Example Register Write Process

Using Listings 6.2, 6.3, and 6.4 as examples, create four registers for the HPS_LED_pattern component. These registers need to be connected to the LED_pattern component signals as follows:

$$Register\ 0 \leftrightarrow HPS\_LED\_control$$
$$Register\ 1 \leftrightarrow SYS\_CLKs\_sec$$
$$Register\ 2 \leftrightarrow LED\_reg$$
$$Register\ 3 \leftrightarrow Base\_rate$$

Note: The default/power-up value for Register 0 ↔ HPS_LED_control should be set to "0" so that the LED_patterns component powers up into the hardware state machine control mode so that the LED patterns will be created without any software intervention. You should also set the default/power-up values for the other signals the same as you created them in Lab 4. In VHDL, you do this in the signal declaration by setting it to the desired initial value when the signal is created.

6.2.2 Creating a Custom Platform Designer Component

Once the VHDL code for HPS_LED_Patterns has been written and is correct, it will be imported in Platform Designer. Do not proceed with Platform Designer until you know that your HPS_LED_Patterns VHDL code is correct.

The steps for creating a Custom Platform Designer Component for HPS_LED_Patterns are:

Step 1: Write the VHDL code for HPS_LED_Patterns.vhd that has the entity shown in 6.1, that implements the Avalon agent interface with the registers you need as described in Sect. 6.2.1.1, and that instantiates the LED_Patterns component from Lab 4. Make sure that this file can be compiled without errors.

Step 2: Copy the VHDL files that you created earlier for Lab 4, i.e., LED_Patterns.vhd, and the others (push-button conditioning files) to the project directory. Quartus will look in the project folder for any files it needs.

Step 3: Open **Platform Designer** in Quartus, and click the **New...** button in the IP Catalog panel (or select File → New Component). The Component Editor window will pop up.

- a: In the **Component Type tab**, enter **HPS_LED_patterns** for both the *Name* and *Display Name*.
- b: Click on the **Files tab**.
  - i: Under the **Synthesis Files** section (not the VHDL or Verilog Simulation Files sections), click **Add File...** . Browse to and open *HPS_LED_patterns.vhd*.
  - ii: Click the **Analyze Synthesis Files** button. You should see the green message Analyzing Synthesis Files: completed successfully. If you do not get this message, it usually means you have a VHDL syntax error in your VHDL code. You will need to correct this before proceeding. Use Quartus to check and correct your VHDL code since the error messages you get from Platform Designer by pressing the "Analyze Synthesis Files" button are typically unhelpful. The assumption is that these files have already been correctly written before being added to Platform Designer.
  - iii: In the messages window, you will see some error messages that we will fix next, so ignore them for now.
- c: Click on the **Signals & Interfaces tab**. The component editor most likely misinterpreted the LEDs, push button, and switches signals as another Avalon interface and made the wrong interpretation regarding these signals.
  - i: Click on the Avalon_slave_0 interface (assuming this was the interpretation).
    - 1: On the right by **Name:**, rename *Avalon_slave_0* to **export**.
    - 2: Change **Type:** by selecting **Conduit** in the drop down menu. Conduit means the signal will be brought out of the soc_system component where the signal will be added to the entity.
    - 3: Change **Associated Reset**: from none to **reset**.

ii: Click on LEDs to highlight and change **Signal Type** to **LEDs** (rename to new custom type).
iii: Click on push button to highlight and change **Signal Type** to **push button** (rename to new custom type).
iv: Click on switch to highlight and change **Signal Type** to **switches** (rename to new custom type).
v: Click on s1 (Avalon Memory-Mapped Slave) and change **Associated Reset**: from none to **reset**.
vi: If there are still errors or warnings, fix them by the process outlined above.

d: Click on the **Block Symbol tab**. You should see all the signals in your entity that have now been interpreted correctly as shown in Fig. 6.4.

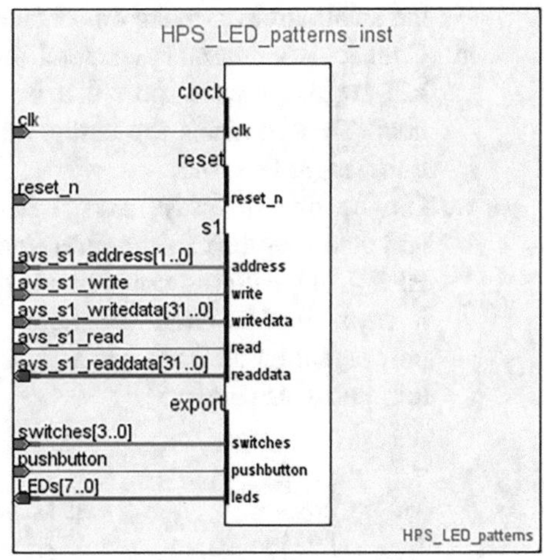

Fig. 6.4: HPS_LED_Patterns component in Platform Designer with signals interpreted correctly

e: Click the **Finish...** button. It will ask you if you want to save the .tcl script HPS_LED_patterns_hw.tcl to your project directory. Click **Yes,Save**. The .tcl file is what allows the new custom component to show up in the IP Catalog panel when Platform Designer opens.

Step 4: In Platform Designer and in the IP Catalog panel, under Project, you should now see the new component name **HPS_LED_patterns**.

a: Add the component to the Platform Designer system.
i: Click on **HPS_LED_patterns** in the IP Catalog panel.
ii: Click on the "+ Add... " button.
iii: Click Finish.

b: Scroll down so you see the component and the bus/signal connection options.

i: Connect **clock** to **clk**. Highlight clk by clicking on clk in the clk_hps component. In gray, you will see that this can be connected to clock. Click on the small circle to make this connection, which will turn the circle black. Since we are adding the memory-mapped interface to the lightweight HPS bus, the clock really needs to be connected to the same clock that is feeding the *h2f_ls_axi_clock* clock input signal in the hps component, which in this case is clk from clk_hps.

ii: Connect **reset** to **clk_reset** Highlight clk_reset by clicking on it in the clk_hps component. In gray, you will see that this can be connected to reset. Click on the small circle to make this connection.

iii: Connect the memory-mapped interface **s1** to the **h2f_lw_axi_master** port that is on the hps component. This connects the components registers to the lightweight HPS bus.

iv: On the line that says **export** in the *Name* column and also **Conduit** in the *Description* column for the HPS_LED_patterns component, double click where it says **Double click to export**. Rename this export signal name from *hps_led_patterns_0_export* to **led_patterns** (Fig. 6.5).

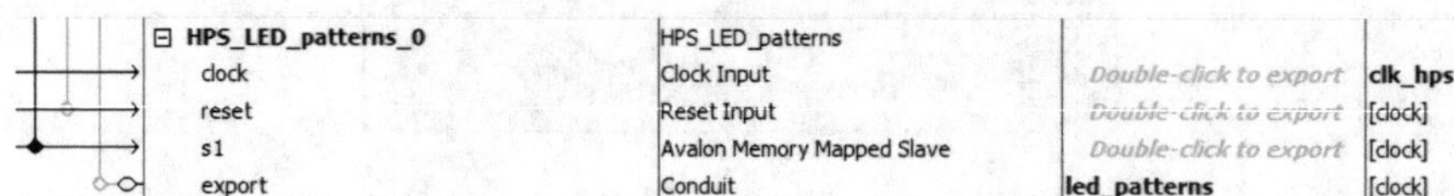

Fig. 6.5: Exporting the led_patterns signal in Platform Designer

c: The base address for the Platform Designer system will likely not be correct, and you will see an error saying that the component overlaps another component. Change the **Base Address** of the HPS_LED_patterns component in Platform Designer by selecting in the menu System →. Assign Base Addresses. Make a note of the base address of the HPS_LED_patterns component since you will be using this information later. You will notice that there are two components with the same base address of 0. This is OK since they are on different buses.

d: Click the Platform Designer button **Finish** that is found at the lower right corner of Platform Designer, which will save the Platform Designer design. Click *close* when it is done saving.

Click **Yes** to regenerate the Platform Designer System, which will take several minutes.

e: A pop-up window will remind you that the generated .qip file needs to be added to the project. However, this has already been done for you. Click "OK."

Step 5: From the Platform Designer menu bar, select Generate →. Show Instantiation Template...

i: Select the HDL language to be **VDHL**.

ii: Click the **Copy** button and paste into a text editor. This gives you the component declaration and instantiation template that needs to be placed in the top level of your Quartus project. **Note: Do not actually paste this into the top level.** We are only using it to see what changed. You do not want to reenter all the instantiation connections for the DRAM, etc.

iii: Notice that there are three new signals in the soc_system component declaration. Add these signals into the soc_system component declaration in DE10Nano_System.vhd.

iv: These signals need to be connected at the top level when the component is instantiated. Connect these signals as you did in Lab 4.

Step 6: Using Quartus, compile the design.

Step 7: Use the programmer to download the bitstream to the DE10 board.

6.2.3 Creating an Avalon Streaming Interface

This is covered in Sect. 1.2 Audio Data Streaming (page 256) in Sects. 1.2.1, 1.2.1.2, 1.2.1.3, and 1.4.1, where an Avalon Streaming Sink and Source is created in Platform Designer for the AD1939 audio codec that is on the Audio Mini-board (Fig. 6.6).

6.3 System Console

System Console is a tool in Quartus that allows you to connect to hardware in the FPGA fabric using the JTAG interface. This allows you to bypass the ARM CPUs but yet interact with your hardware component as if the CPUs were reading and writing registers in your custom component. It allows you to test your hardware before layering software on top of it. If you created your hardware component, hooked it up in Platform Designer, wrote C code to interact with it, and it did not work, where is the problem? Is it your hardware or software? You could end up spending a lot of time debugging and not even be in the right ballpark. However, if you tested your

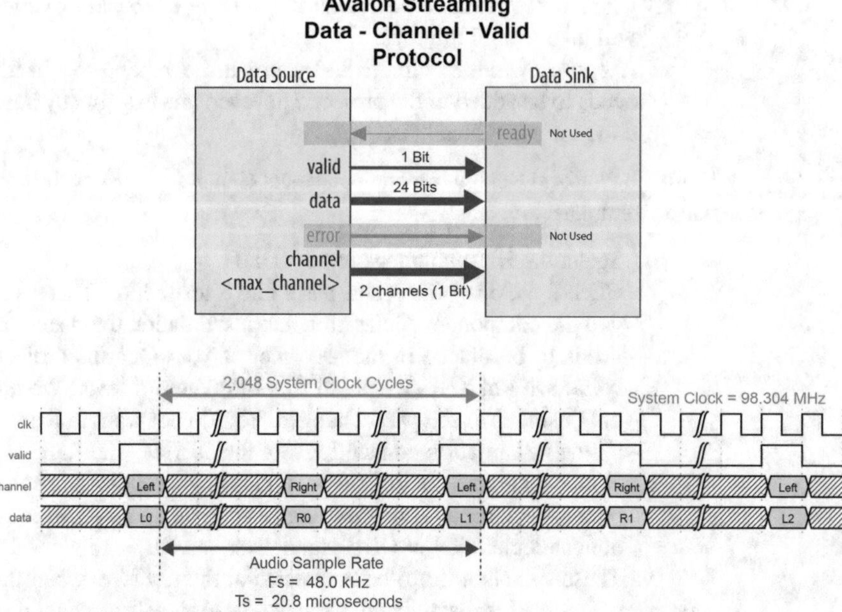

Fig. 6.6: Data–channel–valid protocol for the Avalon Streaming Interface. Left and right audio samples share the same data bus. The valid signal marks when there are valid samples. The channel signal tells which channel (left or right) the sample belongs to. When the FPGA fabric system clock is 98.304 MHz, the valid signal is asserted every 2048 system clock cycles when the sample rate of the AD1939 audio codec is 48 kHz. The ready signal is not used since the downstream components can handle the sample rate by design (no back pressure needed to stall upstream components). The error signal is not used in audio processing designs in the book. Source/sink block diagram from [1]. Signal waveforms were created by WaveDrom [2], and the associated JSON file can be seen here

custom component with System Console and verified that the hardware was working correctly, you would know that it was your software that was the problem if you could not interact with your component. **Note:** The example used in this section is associated with Lab 7.

6.3.1 The General Flow for Using System Console

The general procedure for using System Console is outlined below. More information on using System Console can be found in the User Guide. The overview steps for using System Console are (further details follow):

Step 1: Add the required components to Platform Designer. At a minimum, this includes two components:

1: Your custom component that you are testing, which has a memory-mapped Avalon interface
2: The *JTAG to Avalon Master Bridge* component

Step 2: In Platform Designer, connect the host interface of the *JTAG to Avalon Master Bridge* to the memory-mapped interface of your custom component.
Step 3: Regenerate the Platform Designer system with the new components and compile the design in Quartus.
Step 4: Connect the board and program the FPGA.
Step 5: Start System Console (In Quartus: Tools → System Debugging Tools → System Console).
Step 6: Locate and open a master service path (details in Sect. 6.3.3).
Step 7: Perform the desired operations, which typically involve reading and writing register memory locations in order to test the custom component.
Step 8: Close the master service.

6.3.2 Modifying the Design in Platform Designer

Open Platform Designer and load your system from Lab 6. In the IP Catalog panel (upper left in Platform Designer) and under the Library section, select and add the **JTAG to Avalon Master Bridge** component to your Platform Designer system. This component can be found in the Library at: *Library → Basic Functions → Bridges and Adapters → Memory-Mapped → JTAG to Avalon Master Bridge* as shown in Fig. 6.7.

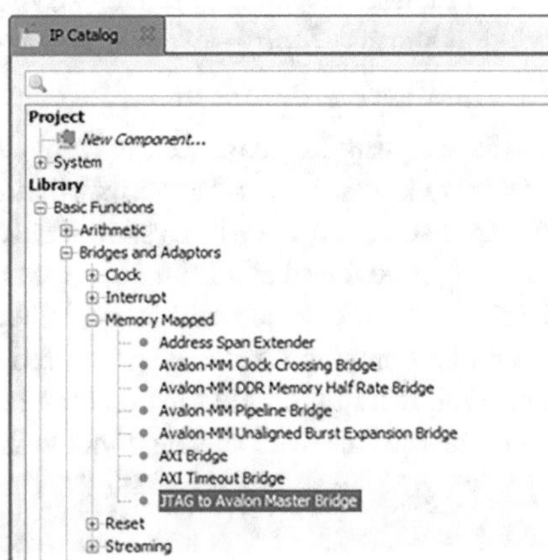

Fig. 6.7: Location of *JTAG to Avalon Master Bridge* component in Platform Designer's library

When you add the component, rename it to **jtag_mm1** since we will need to refer to its name later, and it is possible to have more than one of these components (e.g., if one had more than one clock domain), which is why we have the 1 suffix and mm stands for memory-mapped. In Platform Designer, connect up the jtag_mm1 component in the following manner:

Connection 1: **clk** of *jtag_mm1* ↔ **clk** of *clk_hps* (i.e., the clock being fed to the component under test).
Connection 2: **clk_reset** of *jtag_mm1* ↔ **clk_reset** of *clk_hps* (i.e., the reset being fed to the component under tested).
Connection 3: **master** of *jtag_mm1* ↔ **s1** (memory-mapped interface of *your component* that is most likely named HPS_LED_patterns_0). The memory-mapped interface s1 will also be attached to the HPS bus signal *h2f_lw_axi_master* in the HPS component, so you will be connecting to the lightweight bus as well.
Connection 4: Leave the **master_reset** of *jtag_mm1* **unconnected**.

Regenerate the Platform Designer system and recompile the system in Quartus. This accomplishes steps Step 1: to Step 3: in Sect. 6.3.1.

6.3.3 Using System Console

After the system has been:

1: Compiled by Quartus
2: The DE10-Nano board powered
3: The configuration bitstream downloaded to the FPGA fabric

Start System Console, which can be started in two ways:

Method 1: From within Platform Designer: *Tools → System Console*
Method 2: From within Quartus: *Tools → System Debugging Tools → System Console*

System Console uses the scripting language **Tcl**, or **Tool Command Language**, which is pronounced *tickle*. One can create Tcl scripts that automate testing and data collection. Scripts can also graph data in System Console (using Tk). More information on Tcl (and Tk) can be found at: Tcl main page and Learning Tcl.

We will use several Tcl commands, which will be explained as we use them. Tcl command explanations will be contained in the grey colored text box, and System Console interactions will be shown in the light blue colored text box.

Type the following command in the Tcl Console window in System Console:

```
get_service_paths master
```

Listing 6.5: System Console Tcl Command: get_service_paths master that lists all the master services available which *could* result in the output:

```
/devices/5CSEBA6(.|ES)|5CSEMA6|..@2#USB-1#DE-SoC/(link)/JTAG/
    alt_sld_fab_sldfabric.node_1/phy_0/f2sdram_only_master.
    master
/devices/5CSEBA6(.|ES)|5CSEMA6|..@2#USB-1#DE-SoC/(link)/JTAG/
    alt_sld_fab_sldfabric.node_2/phy_1/fpga_only_master.master
/devices/5CSEBA6(.|ES)|5CSEMA6|..@2#USB-1#DE-SoC/(link)/JTAG/
    alt_sld_fab_sldfabric.node_3/phy_2/hps_only_master.master
```

In this example (unlikely that you will see this), three JTAG to Avalon Master Bridge components show up. This means that you would need to know which master to select and connect to.

Our design is different, and we are only interested in **jtag_mm1** since it is the *JTAG to Avalon Master Bridge* we added and connected to the memory-mapped port of HPS_LED_patterns_0. If you have multiple masters, you need to determine what index to use to select **jtag_mm1**. (**Note:** The masters are listed and indexed starting from index 0). This is why we gave it a specific name jtag_mm1 when we added it to Platform Designer, so we could easily see it in the list if there are multiple masters. In the following example, we will assume that it is the first master listed with an index of zero. If not, use the appropriate index value.

Understanding the Tcl command: get_service_paths <service-type>

System Console uses a virtual file system to organize the available services. Board connection, device type, and IP names are all part of a service path. Instances of services are referred to by their unique service path in the file system. You get the paths for a particular service with the command get_service_paths <service-type>.

System Console automatically discovers most services at startup where it scans for all JTAG and USB-based service instances and collects these service paths.

The **Tcl Command: get_service_paths** ***master*** looks for *master* service types and returns all that are found.

In the Tcl Console window type:

```
% lindex [ get_service_paths master ] 0
```

which will result in

```
/devices/5CSEBA6(.|ES)|5CSEMA6|..@2#USB-1#DE-SoC/(link)/JTAG/
    alt_sld_fab_sldfabric.node_2/phy_1/jtag_mm1.master
```

The Tcl command has selected the **jtag_mm1.master** service, which has an index of 0 in the list.

Understanding the Tcl command: lindex [get_service_paths master] 0

The **Tcl Command: lindex** is a built-in command that retrieves an element from a list.

The **Tcl syntax: [ ]** (square brackets) allows a script to be embedded at any position of any word by enclosing it in brackets. The embedded script is passed verbatim, without any processing, to the interpreter for execution, and the result is inserted in its place in the embedding script. This means that **[get_service_paths master]** will get executed, which results in a list that is used by **lindex**.

Thus the **Tcl Command: lindex [get_service_paths master] 0** first finds all the master service paths and returns them in a list, which is then used by lindex to select element 0, which is the master service associated with the first master listed, which in our case is **jtag_mm1**.

Let us create a variable in Tcl that contains this information since we do not want to type:

```
/devices/5CSEBA6(.|ES)|5CSEMA6|..@2#USB-1#DE-SoC/(link)/JTAG/
    alt_sld_fab_sldfabric.node_2/phy_1/jtag_mm1.master
```

every time we want to use **jtag_mm1**. Let us save this in a variable called **m**, so now type:

```
% set m [lindex [ get_service_paths master ] 0]
```

Listing 6.6: The set command saves the selected service path to the variable m.

Understanding the Tcl command: set

set is a built-in command that reads and writes variables. It returns the value of the variable named varName or, if a value is given, stores that value to the named variable, first creating the variable if it does not already exist.

This means that the **Tcl Command: set m [lindex [get_service_paths master] 0]** extracts a list item as described earlier and creates a variable m that stores this path information. **Note:** We can see what is in the variable (i.e., read the variable) by just calling set with no value associated with the variable name (an unfortunate function naming convention instead of using a keyword like read).

In summary:

Write *value* to variable m: set m *value*

Read *value* from variable m: set m

To see what m now contains type:

```
% set m
```

which returns:

```
/devices/5CSEBA6(.|ES)|5CSEMA6|..@2#USB-1#DE-SoC/(link)/JTAG/
    alt_sld_fab_sldfabric.node_2/phy_1/jtag_mm1.master
```

Since System Console services are contained in a virtual file system, we first need to open the file or service to use it.

Let us now open this master service for **jtag_mm1**, so type:

```
% open_service master $m
```

Listing 6.7: Services in System Console are virtual files that need to be opened to be used.

Understanding the Tcl command: open_service master $m

The **Tcl Command: open_service** *<service_type>* *<service_path>* is a built-in command that opens a service of *<service_type>* pointed to by the path *<service_path>*.

The **Tcl syntax: $name** When $ occurs in a word and is followed by a sequence of standard ASCII characters and that sequence is the name of a variable, the value of the variable replaces the $ and the variable name. This means that $m is replaced by the service path to jtag_mm1.

Thus the **Tcl Command: open_service master $m** opens a master service with the path specified in $m that points to **jtag_mm1**.

We can now read the registers we have created in our custom component. Recall from Lab 6 that Register 0 is the HPS_LED_control register, Register 1 defines how many system clocks occur are one second, Register 2 is what gets displayed on the LEDs in software mode, and Register 3 controls the base rate of the LED patterns when under state machine control. Also recall what base address of HPS_LED_patterns_0 was set to in platform Designer. (If you have forgotten, you need to open the system in Platform Designer to check the base address.) Assuming the base address has been set to 0x0 for this example and that we want to read one 32-bit word (0x1) from Register 0, we type:

```
% master_read_32  $m  0x0  0x1
```

Listing 6.8: Reading a 32-bit word from memory.

which results in:

```
0x00000000
```

In order to figure out what memory address to type in for a particular register in System Console for your custom component, read section 7.2 The View of Memory from System Console (page 89).

Understanding the Tcl command: master_read_32 $m 0x0 0x1

The **Tcl Command: master_read_32** *<service_path>* *<base_address>* *<number_of_32-bit_words>* is a built-in command that uses the service pointed to by *<service_path>* that reads from memory starting at *<base_address>* and reads *<number_of_32-bit_words>*.

Thus the **Tcl command: master_read_32 $m 0x0 0x1** uses **jtag_mm1** to read from memory location zero and reads one 32-bit word.

Register2 should be a read/write register, so let us first see what is in the register and then write a new value to the register. Since our commands start at the base address, we need to read three 32-bit words. Type the command:

```
master_read_32  $m  0x0  0x3
```

which results in:

```
0x00000000 0x00000000 0x00000000
```

We see that register 2 (third value since we are counting from Register 0) has returned a zero. Let us write a value of 1 to register 2 (and zeros to registers 0 and 1) by typing the command:

```
master_write_32  $m  0x0  0x0  0x0  0x1
```

Listing 6.9: Writing three 32-bit values to memory.

Understanding the Tcl command: master_write_32 $m 0x0 0x0 0x0 0x1

The **Tcl Command: master_write_32** *<service_path>* *<base_address>* *<list_of_32-bit_values>* is a built-in command that uses the service pointed to by *<service_path>* that writes to memory starting at *<base_address>* and writes all the values in *<list_of_32-bit_values>*.

Thus the **Tcl command: master_write_32 $m 0x0 0x0 0x0 0x1** uses the **jtag_mm1** connection to write to memory location starting at zero the three values in the list (0x0 0x0 0x1). This means we will set Register0 = 0, Register1 = 0, Register2 = 1.

Now if we read our registers by typing the command:

```
master_read_32  $m  0x0  0x3
```

we can see that Register 2 has been set to 1:

```
0x00000000 0x00000000 0x00000001
```

Once we are done testing the Platform Designer component by reading and writing registers, we close the service by typing:

```
% close_service master $m
```

Listing 6.10: Closing the service.

6.3.4 Summary of System Console Commands

The steps for interacting with the registers in your component are:

Step 1: Add a *JTAG to Avalon Master Bridge* component to your Platform Designer system, name it a recognizable name such as **jtag_mm1**, and connect it to the slave port of the component that contains the registers that you want to interact with. Pay attention to the base_address of the component since you will need this address for System Console commands.

Step 2: Get the list of all the master services in System Console by typing the command:

```
% get_service_paths master
```

and note what the index is for **jtag_mm1**. (Note: The index values start at zero.)

Step 3: Save the service path for **jtag_mm1** in the variable m (here we assume the index value is zero) by typing the command:

```
% set m [lindex [get_service_paths master] 0]
```

Step 4: Open the service by typing the command:

```
% open_service master $m
```

Step 5: Read the 32-bit registers by typing:

```
% master_read_32 $m <address> <number_of_words>
```

Step 6: Write to the registers by typing:

```
% master_write_32 $m <address> <list_of_words>
```

Step 7: Close the service by typing:

```
% close_service master $m
```

6.4 Creating IP in Quartus

6.4.1 Creating a ROM IP Component

When designing a digital system, we put together various logic building blocks. We would like to avoid having to create every building block ourselves, so this is where the Quartus IP Catalog is helpful. We can select and use various building blocks from a library of commonly used blocks. In this example, we will create a ROM that has the memory size and values of our choosing. The ROM that we will create is used in Sect. 8.4 where we create and verify a hardware component that computes the reciprocal square root of a fixed-point value. The ROM is used to compute the initial guess of the solution, which we need for successful convergence when using Newton's method to find the solution.

When Quartus is opened, we can see the *IP Catalog* on the right-hand side. If we open the IP Catalog to *Installed IP* → *Library* → *Basic Functions* → *On Chip Memory*, we can select different memory types as seen in Fig. 6.8. This includes the 1-PORT ROM that we will use.

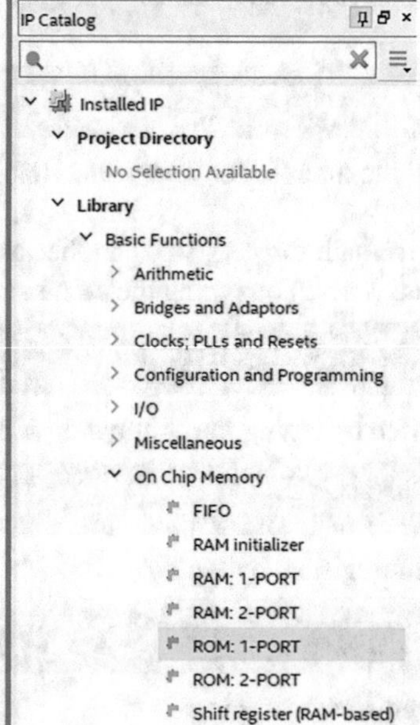

Fig. 6.8: Selecting a ROM from the Quartus IP Catalog

However, before we create this 1-PORT ROM using the IP Catalog, we first need to create a *Memory Initialization File* (**.mif**) that specifies what should be in this memory. Otherwise, when the Quartus MegaWizard that creates this ROM asks us for this file, it will complain if we do not have it ready.

6.4.1.1 Creating the ROM Memory Initialization File

The memory initialization file that we need to create can be seen in Listing 6.11 where only the first 15 lines are shown (click here for the full file listing). The file starts with a header that describes the file contents. The *DEPTH* parameter specifies how many memory words are there in the file, and this needs to be a power of two. The *WIDTH* parameter specifies how many bits are in each memory word. The *ADDRESS_RADIX* parameter specifies in what radix the memory addresses will be written. In this file, it is specified as *BIN* or binary. Other possible radix options are *DEC*, *OCT*, or *HEX*. We use binary in this specific case since the binary address relates directly to a fractional word slice in our design, and it is easier to compare this VHDL signal slice with the associated memory address when it is in a binary radix. The *DATA_RADIX* parameter specifies in what radix the data or memory values will be written. In this file, it is specified as *BIN* or binary. The *CONTENT* and *BEGIN* parameters mark where the (address : value) pairs start. Address specifies the memory address where the paired value will be stored at. Comments can follow the (address : value) pair, and they start with two dashes. After all the (address : value) pairs have been listed, there is the final *END* parameter.

```
DEPTH = 256;
WIDTH = 12;
ADDRESS_RADIX = BIN;
DATA_RADIX = BIN;
CONTENT
BEGIN
00000000 : 100000000000   -- 1 : (1)^(-3/2) = 1
00000001 : 011111110100   -- 2 : (1.0039)^(-3/2) = 0.99414
00000010 : 011111101000   -- 3 : (1.0078)^(-3/2) = 0.98828
00000011 : 011111011101   -- 4 : (1.0117)^(-3/2) = 0.98291
00000100 : 011111010001   -- 5 : (1.0156)^(-3/2) = 0.97705
00000101 : 011111000101   -- 6 : (1.0195)^(-3/2) = 0.97119
00000110 : 011110111010   -- 7 : (1.0234)^(-3/2) = 0.96582
00000111 : 011110101111   -- 8 : (1.0273)^(-3/2) = 0.96045
00001000 : 011110100100   -- 9 : (1.0313)^(-3/2) = 0.95508
```

Listing 6.11: File Contents : ROM.mif

The memory initialization file is created using two Matlab files, which are **mif_gen.m** (click here) and **mif_gen_ROM_rsqrt.m** (click here). The Matlab function **mif_gen.m** takes in four arguments.

```
function mif_gen(filename,array,memory_size,comments)
```

Listing 6.12: Matlab function for generation of the memory initialization (.mif) file

The first argument *filename* specifies the name of the .mif file to be written. The second argument *array* is an array of fixed-point objects that contain the memory values. The third argument *memory_size* is a two-element vector that specifies the number of memory words and the size of a memory word in bits. The fourth argument *comments* is optional and contains a character array of comments that will be placed after each (address : value) pair.

In Listing 6.13, the function first checks to see if the memory size specified is consistent with the data array.

```
if length(array) ~= memory_size(1)
   error('Length of array passed to mif_gen() does not match memory size')
end
if rem(memory_size(1),log2(memory_size(1))) ~= 0
    error('Length of Memory should be a power of 2')
end
a = array(1);
if a.WordLength ~= memory_size(2)
    error('Word length of array does not match memory size')
end
```

Listing 6.13: Error checking the function inputs

After the error checking is passed, it opens the .mif file in write mode

```
fid = fopen([filename '.mif'],'w');
```

Listing 6.14: Opening the file

and then writes the .mif file header as shown in Listing 6.15.

```
line = ['DEPTH = ' num2str(memory_size(1)) ';']; fprintf(fid,'%s\n',line); % The size of memory in words
line = ['WIDTH = ' num2str(memory_size(2)) ';']; fprintf(fid,'%s\n',line); % The size of the word in bits
line = ['ADDRESS_RADIX = BIN;']; fprintf(fid,'%s\n',line); % The radix for address values
line = ['DATA_RADIX = BIN;']; fprintf(fid,'%s\n',line); % The radix for data values
line = ['CONTENT']; fprintf(fid,'%s\n',line);
line = ['BEGIN']; fprintf(fid,'%s\n',line);  % start of (address : data) pairs
```

Listing 6.15: Creating the Memory Initialization File Header

After the file header is written, a for loop goes through the array where it constructs the (address : value) pairs. It creates a fixed-point object for the address and places this in the line string. It then concatenates this address string with the binary string of the associated value. If comments are supplied, it concatenates the comment associated with the pair and writes it to the file. Otherwise, it adds a generic comment.

```
address_bits = log2(memory_size(1));
for index = 1:memory_size(1)
    address = fi(index-1,0,address_bits,0);
    a = array(index);
    line = [address.bin ' : ' a.bin];
```

```
    if nargin <= 3
        line = [line '  -- array(' num2str(index) ') = ' num2str(array(index)) ];
    else
        line = [line comments(index,:)];
    end
    fprintf(fid,'%s\n',line);
end
```

Listing 6.16: The memory initialization file structure

The values for the ROM are created in the Matlab script *mif_gen_ROM_sqrt.m*. As part of the computation to create the initial guess for the reciprocal square root function, which is then refined by Newton's method, the ROM is used as a lookup table in place of the nonlinear function $y = x^{-3/2}$, where $1 \leq x < 2$. The fractional part of x is used as the address into the ROM.

The size of the ROM is specified by the parameters *Nbits_address*, which is the size of the ROM address port, and *Nbits_word_length*, which is the size of the memory words in bits contained in the ROM. These values are design choices for the reciprocal square root hardware component that affect the precision of the computation. However, the memory does not need to be large, in terms of either the number of words or the word size in bits, because we do not need fine precision for the starting point y_0 for Newton's method. We just need a rough guess that will be refined by the Newton iterations.

```
Nbits_address         = 8;  % address size
Nbits_word_length     = 12; % size of word in memory (number of bits)
Nbits_word_fraction = Nbits_word_length-1;   % The number of fractional bits in result.
Nwords                = 2^Nbits_address; % Number of words in memory
```

Listing 6.17: Defining the memory size

Since the address into the ROM comes from the fractional bits of x, where $1 \leq x < 2$, we need to generate all the possible address values for this ROM. We do this as shown in Listing 6.18 by creating an index i that goes from 0 to $2^{Nbits\_address} - 1$. We then create a fixed-point object from the index value with the value interpreted as an unsigned integer (S = 0, W = Nbits_address, F = 0) and get the binary string that represents the memory address (fa_bits). The implicit value that this address represents is x that has a 1 in the one's place. We get this value by creating a fixed-point object container (value is temporarily zero) and where the interpretation is $W = Nbits\_address + 1$ and $F = Nbits\_address$. We then prefix a "1" to the address string and assign it to this fixed-point container, which gives us the value x that we want when Matlab makes the assignment and updates the fixed-point object.

The next step is to get the value $y = x^{-3/2}$. This is simply done by converting the fixed-point object *fb* to a double and raising it to the $-3/2$ power. We then convert y

back to another fixed-point object *a* with the interpretation we want to place in the ROM (W = Nbits_word_length, F = Nbits_word_fraction). This fixed-point object is then stored in the array of all the fixed-point objects. We also create comments that explain what this memory location represents.

```
for i=0:(Nwords-1)  % Need to compute each memory entry (i.e. memory size)
    fa = fi(i,0,Nbits_address,0);  % fixed point object for address
    fa_bits = fa.bin;              % Memory Address as a binary string
    fb = fi(0, 0, Nbits_address+1, Nbits_address);  % Set number of bits for result, i.e. we are creating the value 1.address_bits
    fb.bin = ['1' fa_bits]; % set the value using the binary representation. The address is our input value 1 <= x_beta < 2  where the leading 1 has been added.
    a = fi(double(fb)^(-3/2),0,Nbits_word_length,Nbits_word_fraction);  % compute (x_beta)^(-3/2) and convert to fixed-point with the desired number of fraction bits
    array(i+1) = a;
    comments   = char(comments, ['  -- ' num2str(i+1) ' : (' num2str(double(fb)) ')^(-3/2) = ' num2str(a)]);
end
```

Listing 6.18: Creating the memory array

Having created everything we need for the memory initialization file, we can specify the file name we want ("ROM") and then pass the array we created along with the memory size and comments to *mif_gen* to have it create the memory initialization file.

```
filename    = 'ROM';
memory_size = [Nwords Nbits_word_length];
mif_gen(filename,array,memory_size,comments)
```

Listing 6.19: Generating the ROM values

We then save the array of fixed-point objects into a .mat file so that we can easily access the ROM values in Matlab by loading in the .mat file when we perform verification of any component that uses the ROM. Otherwise, we would have to parse the .mif file.

```
save([filename '.mat'], 'array')
```

Listing 6.20: Saving the ROM values in a .mat file

6.4.1.2 Creating the ROM IP

Now that we have the memory initialization file we need, we can go ahead and create the ROM IP in Quartus. With the IP Catalog opened as in Fig. 6.8, double click on *ROM: 1-PORT*. A window pops up that asks us what we should name the ROM IP that will be created and where the generated files should be placed. **Note 1:** It will be easiest if you first open a project that has already been created that targets the DE10-Nano when you create this ROM since we want to target the Cyclone V. The location would then be this project folder (you do not need to add the IP component to the project). **Note 2:** Copy the *ROM.mif* memory initialization file into this directory. Name the IP variation ROM and select VHDL as the generated file type. When you click OK, the window in Fig. 6.9 pops up (it may take a few moments to appear).

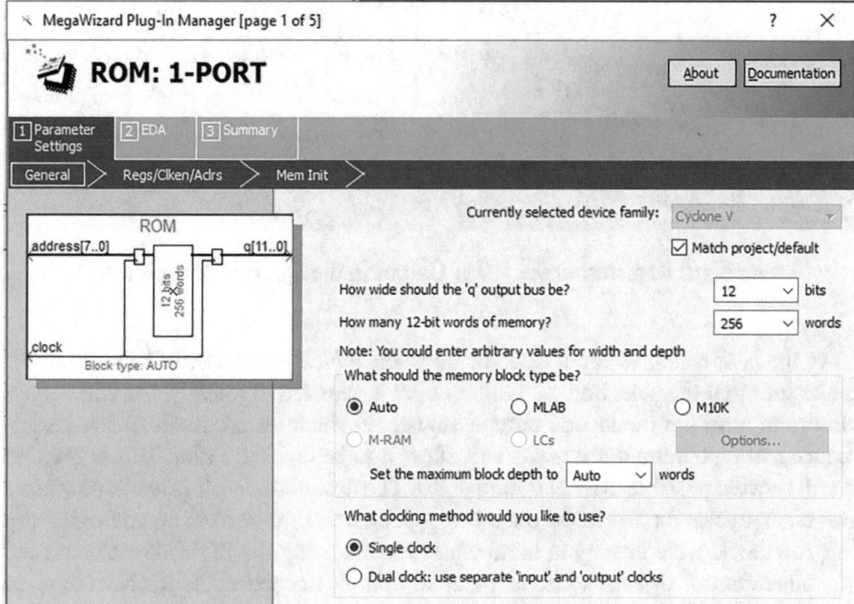

Fig. 6.9: Defining the ROM IP Memory Size

In the first panel, we select the memory size, which must be the same size as the memory initialization file we created. The first question *How wide should the "q" output bus be?* needs to be the same as what we set *Nbits_word_length* to in *mif_gen_ROM_rsqrt.m*, which in this example is 12 bits. The second question *How many 12-bit words of memory?* needs to be consistent with *Nbits_address*, specifically $2^{Nbits\_address}$, which in this example is $2^8 = 256$ memory words.

For the question *What should the memory block type be?*, keep it as *Auto*, so that the fitter can select whatever memory type is available. Keep the clocking method as *single clock* since we are not using the memory in a design with multiple clock domains. Click Next to go to the next panel (Fig. 6.10).

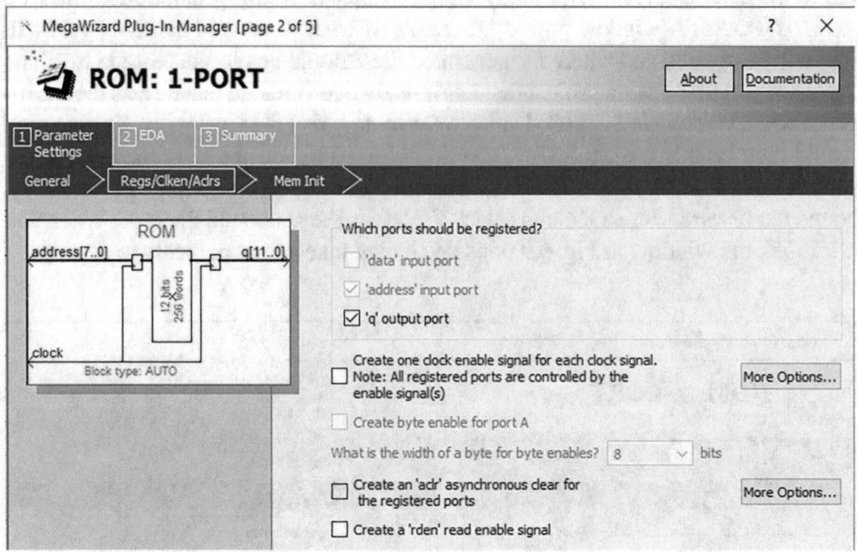

Fig. 6.10: Registering the ROM Output in the Quartus MegaWizard

In the next panel, where it asks the question *Which ports should be registered?*, make sure that the selection *"q" output port* is selected. Typically, we will always choose to have the inputs and outputs registered since we are interested in performance and pipelining our designs will allow it to be clocked faster. This is because it will shorten possible critical timing paths. The design trade-off is that it now takes two clock cycles for the ROM output to appear when we send in an address. Keep this two clock cycle latency in mind when you are using the ROM. We do not need any other control signals, so leave the other options unchecked. Click Next to go to the next panel (Fig. 6.11).

This next panel is why we first created the memory initialization file and put it in the directory where we are creating the ROM. Type in the file name, which in this example is **ROM.mif**. Click Next to go to the next panel (EDA), which you can skip so Click Next again (Fig. 6.12).

In the final panel, select both the VHDL component declaration file (*ROM.cmp*) and the instantiation template file (*ROM_inst.vhd*) so that both these files will be generated when ROM.vhd is generated. The file *ROM.cmp* (click here for file) is shown in Listing 6.21. The file *ROM_inst.vhd* (click here for file) is shown in Listing 6.22. This allows you to conveniently cut and paste both the component declaration and instantiation template into your design. Click Finish to generate the ROM IP.

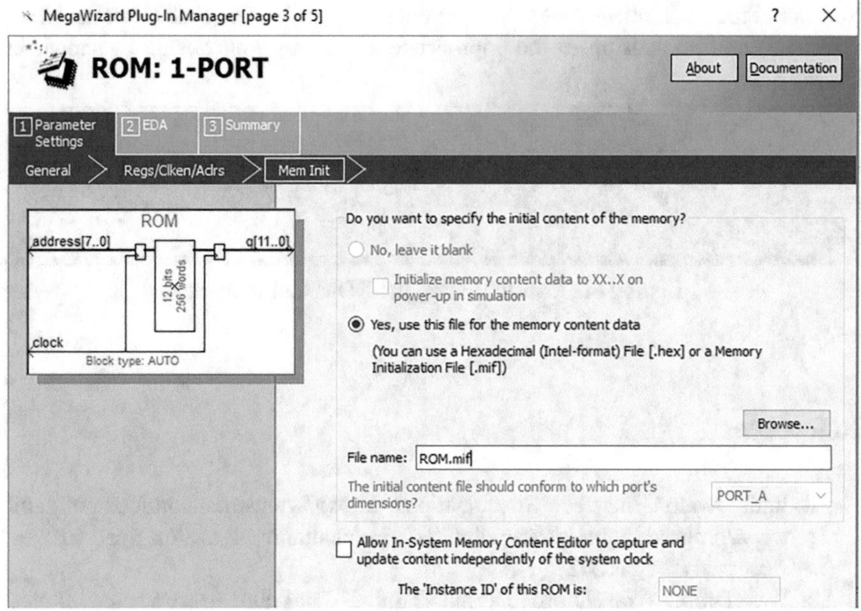

Fig. 6.11: Specifying the Memory Initialization File in the Quartus MegaWizard

File	Description
☑ ROM.vhd	Variation file
☐ ROM.inc	AHDL Include file
☑ ROM.cmp	VHDL component declaration file
☐ ROM.bsf	Quartus Prime symbol file
☑ ROM_inst.vhd	Instantiation template file

Fig. 6.12: Creating the Component Declaration and Instantiation Files in the Quartus MegaWizard

Note: The ROM.qip file that is generated is what you add to your Quartus project to be able to use the ROM IP. You can also just add the file *ROM.vhd* to your Quartus project. To use the ROM IP, you first need to declare the component in your VHDL code (before the *begin* statement) as shown in Listing 6.21.

```
component ROM
    PORT
    (
        address         : IN STD_LOGIC_VECTOR (7 DOWNTO 0);
        clock       : IN STD_LOGIC  := '1';
        q               : OUT STD_LOGIC_VECTOR (11 DOWNTO 0)
    );
end component;
```

Listing 6.21: Declaring the ROM Component

You would then instantiate the ROM component in your VHDL code (after the *begin* statement) and hook it up to the appropriate signals in your design as shown in Listing 6.22.

```
ROM_inst : ROM PORT MAP (
    address  => address_sig,
    clock    => clock_sig,
    q        => q_sig
);
```

Listing 6.22: Instantiating the ROM Component

References

1. Intel, Avalon Interface Specifications. https://www.intel.com/content/dam/www/programmable/us/en/pdfs/literature/manual/mnl_avalon_spec.pdf. Accessed 01 Mar 2022
2. WaveDrom, WaveDrom Digital Timing Diagram everywhere. https://wavedrom.com/. Accessed 01 Mar 2022

Chapter 7
Introduction to Memory Addressing

The ARM CPUs in the Cyclone V SoC FPGA are 32-bit CPUs and can only address memory in the range from 0x00000000 to 0xFFFFFFFF (4 GB). The SoC FPGA contains a number of memory-related devices and peripherals, and the view of memory in the SoC FPGA depends on the particular device vantage point. This is illustrated in Fig. 7.1.

Of particular interest to the designs in this book are the far right column that is labeled *MPU* and the annotated row labeled *Lightweight Bridge*. The *MPU* column is what the ARM CPUs see in terms of memory. Notice that all the peripherals and memory bridges are located above 0xC0000000, which places them under control of the Linux kernel that is running in the MPU. The *Lightweight Bridge* is the address range starting at 0xFF200000, and this is where the registers of our custom hardware are memory-mapped.

7.1 The View of Memory from the ARM CPUs and Platform Designer

When custom hardware is placed in the FPGA fabric, the control registers for the custom component are connected to the lightweight bus in Platform Designer as shown in Fig. 7.2. Platform Designer will assign a "base address" to the custom component, but in reality this is an offset to the lightweight bridge address of 0xFF200000. How the memory addressing is calculated for registers in custom hardware is given in Sect. 7.1.1.

The view of memory that the ARM CPUs see is shown in the right column of Fig. 7.1. Of particular interest is the location of the lightweight HPS-to-FPGA AXI bridge that is located at memory address 0xFF200000 and ends at 0xFF3FFFFF having a span of 2 MB where the bridge address width is 21 bits. It is within this address range that our control registers will exist when our custom hardware com-

R. K. Snider, *Advanced Digital System Design using SoC FPGAs*,
https://doi.org/10.1007/978-3-031-15416-4_7

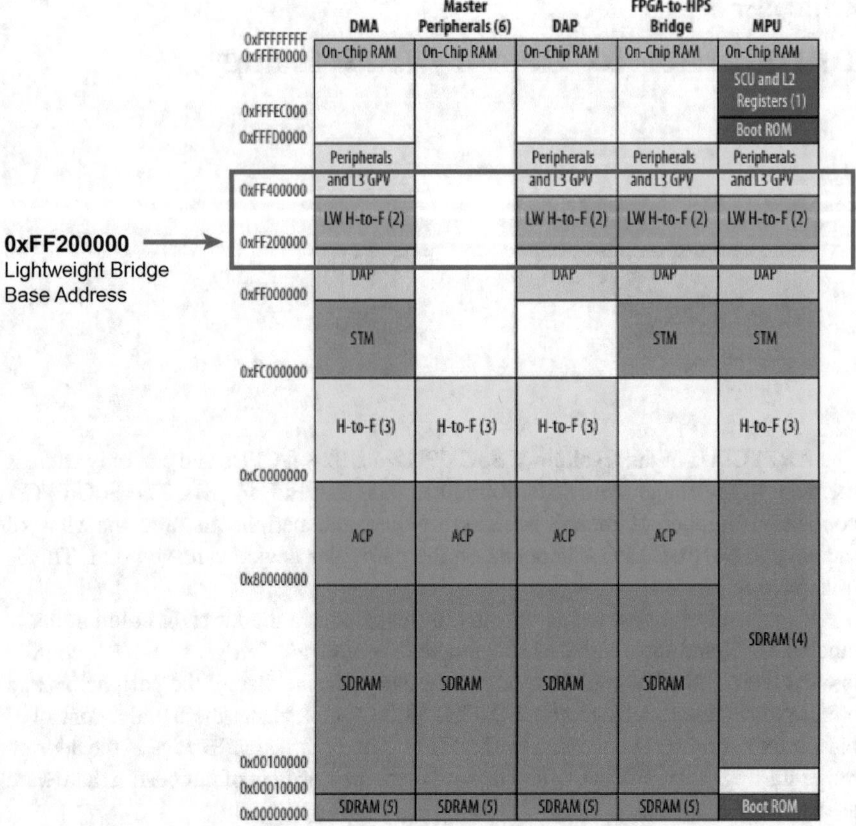

Fig. 7.1: SoC FPGA multiple views of memory. Custom hardware that is attached to the lightweight bus (LW H-to-F) is accessed through the lightweight bridge that has a base address of 0xFF200000. Figure adapted from [1]

ponent is attached to the HPS bus signal *h2f_lw_axi_master* in the HPS component in Platform Designer.

7.1.1 Memory Addressing for Registers on the HPS Lightweight Bus

In Platform Designer, the term "Base Address" for the custom component is misleading since it is the base address on the bus that it is attached to. In reality, it is an **offset** to the address of the lightweight HPS-to-FPGA AXI bridge that has an address of 0xFF200000.

The register address calculation for your custom component when using C on Linux running on the ARM CPUs is:

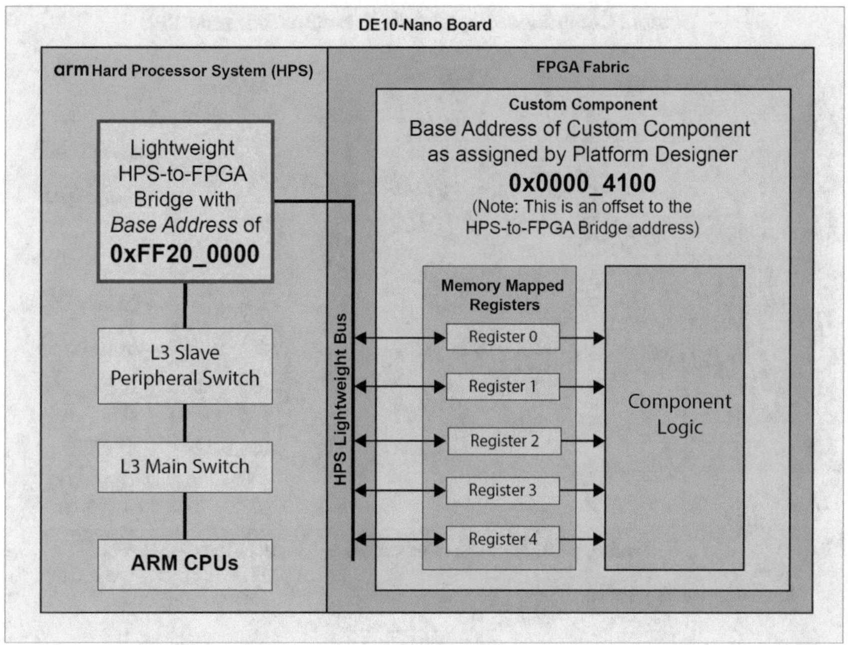

Fig. 7.2: Memory Addressing of Custom Component

$$\text{Register}[i] = \text{Lightweight Bridge Address} + \text{Component Address} + i * 4 \quad (7.1)$$
$$\text{where } i = [0 \ldots N-1] \quad (7.2)$$
$$N = \text{Number of Registers} \quad (7.3)$$

where you take the base address of the component that has been assigned by Platform Designer and add the appropriate number of register offsets that are 4 bytes since the sizes of the registers that have been created are assumed to be 32 bits. The number of registers N is determined by the components address width W in the entity, i.e., $N = 2^W$.

Thus Register[3] in the example shown in Fig. 7.3 has the physical memory address:

$$\text{Register}[3] = \text{Lightweight Bridge Address} + \text{Component Address} + i * 4 \quad (7.4)$$
$$= \text{0xFF200000} + \text{0x4100} + 3 * 4 \quad (7.5)$$
$$= \text{0xFF200000} + \text{0x4100} + \text{0xC} \quad (7.6)$$
$$= \text{0xFF20410C} \quad (7.7)$$

7.2 The View of Memory from System Console

The view of memory that System Console sees is taken from the perspective of the attached **JTAG to Avalon Master Bridge** component and is shown in Fig. 7.4. The

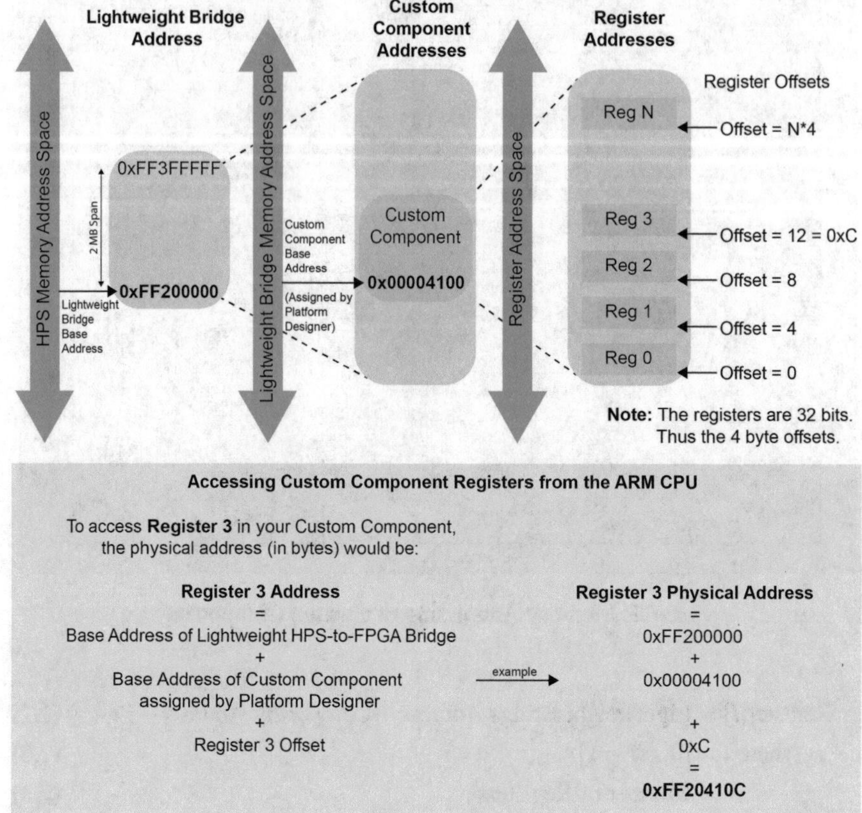

Fig. 7.3: Component Register Addressing from ARM CPUs

memory addressing is relative to this JTAG to Avalon Master Bridge. This means that the component that you are testing needs to be attached to the same bus as the JTAG to Avalon Master Bridge component. See Sect. 6.3.2 Modifying the Design in Platform Designer (page 71) on how to add this component to your Platform Designer system.

In Platform Designer, the term "Base Address" can be a bit misleading since it is the base address on the bus that it is attached to. In the System Console case, this is all we need to know since the JTAG to Avalon Master Bridge component will see the component under test at this address. This is different from the ARM CPU case as illustrated in Fig. 7.3.

The register address calculation for your custom component when using System Console is:

$$\text{Register}[i] = \text{Base Address given by Platform Designer} + i * 4 \tag{7.8}$$
$$\text{where } i = [0...N-1] \tag{7.9}$$
$$N = \text{Number of Registers} \tag{7.10}$$

where you take the base address of the component that has been assigned by Platform Designer and add the appropriate number of register offsets that are 4 bytes since the sizes of the registers that have been created are assumed to be 32 bits. The number of registers N is determined by the component address width W in the entity, i.e., $N = 2^W$.

Thus Register[3] in the example shown in Fig. 7.4 has the physical memory address:

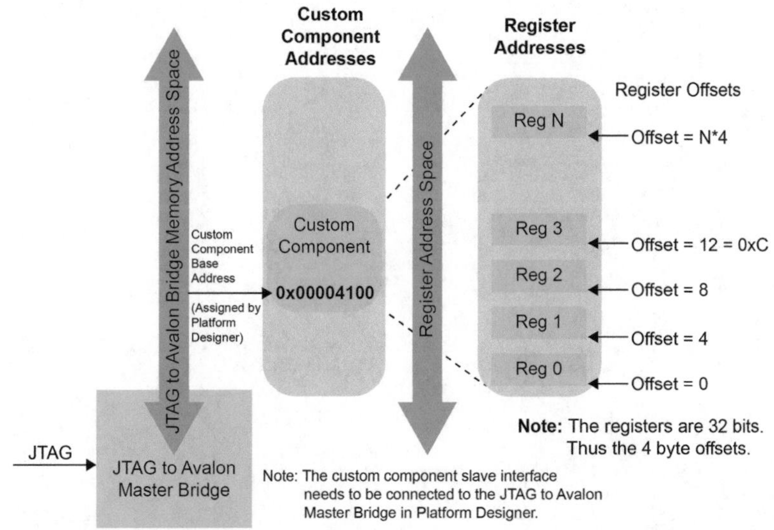

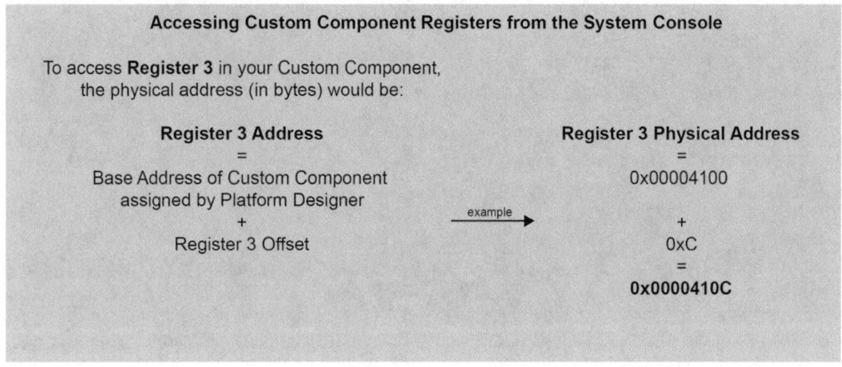

Fig. 7.4: Component Register Addressing from System Console

$$\begin{aligned}
\text{Register}[3] &= \text{Base Address given by Platform Designer} + i * 4 && (7.11)\\
&= 0\text{x}4100 + 3 * 4 && (7.12)\\
&= 0\text{x}4100 + 0\text{xC} && (7.13)\\
&= 0\text{x}410\text{C} && (7.14)
\end{aligned}$$

Reference

1. Intel, Cyclone V Hard Processor System Technical Reference Manual. https://www.intel.com/content/dam/www/programmable/us/en/pdfs/literature/hb/cyclone-v/cv_5v4.pdf. Figure 8-4, Address Maps for System Interconnect Masters, page 8-6, Accessed 23 June 2022

Chapter 8
Introduction to Verification

8.1 Design Assumptions

8.1.1 Synchronous Designs

Our working assumption is that the system we are creating is a *synchronous* digital system where all signals change in lockstep to the system clock. This is why you should always use the rising_edge() function in your VHDL designs when you create a *process* in VHDL. You should get into the habit of always writing *if rising_edge(clk)* right after *begin* in a *process(clk)* statement as shown in Listing 8.1 (unless you have a very good reason to do otherwise).

```
my_process_name : process(clk)
begin
    if rising_edge(clk) then
        <...VHDL code here...>
    end if;
end process;
```

Listing 8.1: Synchronous VHDL process using the rising_edge() function. You should always write *if rising_edge(clk)* right after *begin* in a *process(clk)* statement

8.1.2 Hierarchical Designs

We are assuming a hierarchical design process where larger components are comprised of smaller ones. If you waited till the very end and your complicated system did not work, where would you find the error? It would be hard to find. Systems using SoC FPGAs are even more complex and harder to debug than your typical computer system because you are developing both the underlying hardware and the software that uses your hardware. If the system breaks, the problem could exist in

R. K. Snider, *Advanced Digital System Design using SoC FPGAs*,
https://doi.org/10.1007/978-3-031-15416-4_8

your hardware or in the software using your hardware. Thus to reduce the time you spend debugging, it is important to verify that your building blocks are correctly created before you use them.

A systematic design approach is to design small components and make sure they are correct before moving on. It is tempting to jump in and starting using them right away, but this approach will in the end cause you to waste a lot of time debugging the system that could have been avoided if you tested and verified right after creating a building block. (Yes, I have been guilty of this myself.) If you verify the building blocks as you create them, then when you create a larger component, and it does not work, you know the error is unlikely to be coming from the smaller components, but rather in how you are using them in the larger component.

8.1.3 VHDL Code Formatting

When you write code, it is good to follow a style guide. This can help reduce coding mistakes. Here we present how to use Python to format your VHDL code using the VSG package.

There are many Python Integrated Development Environments (IDEs) (List of Python IDEs), so if you already have your own Python environment, then use that one. If not, then we give PyCharm as a suggested Windows environment and give the steps to set it up to format VHDL code. The steps are:

Step 1: Install the **PyCharm** Community Edition. (Download Link)

Step 2: In PyCharm, install the *VHDL Style Guide* (**VSG**) package. You can do this in PyCharm by opening a *terminal* window and entering the command pip install vsg.

```
> pip install vsg
```

Step 3: Create a project in PyCharm (suggested project name = **vsg**). Having a project ready to open and run on a file will make it convenient to use.

Step 4: Add the VHDL style guide adsd_vhdl_style.yaml to the project folder. Change any style rules in the file if so desired. It will work fine as is.

Step 5: Add the Python script format_vhdl_file.py to the project folder and make the following edits in the file:

- Edit 1: Add the name of your VHDL file (line 18) that you want to format or check.
- Edit 2: Add the path to your VHDL file (line 19).
- Edit 3: Add the path to the .yaml file you downloaded in step 4. (It will be the path to the vsg project directory.)

Step 6: Run the python script **format_vhdl_file.py**, and fix any errors and warnings in your VHDL file. Repeat until there are no errors or warnings. If you disagree on what a warning or error should be, modify the file *adsd_vhdl_style.yaml* accordingly and rename it something like *my_style.yaml*.

8.2 Verification

8.2.1 Why Verify?

Computer systems are arguably one of the most complicated systems to develop, and it is easy to introduce errors in the design process. A design goal is to minimize errors, while you are developing a system, and it is best (and cheaper) to find potential errors right away, rather than having your customers find them for you. This is why you should test and verify your code and digital logic building blocks as you develop them.

8.2.2 Verification Process

Here we assume that you have just finished creating a VHDL component that you wish to verify. The process is to send values, known as test vectors through the component, and check if the results are correct. But how do you know what the correct result should be? Well, in our case, we will use Matlab and create a Matlab function that should produce the same output as the VHDL component. Then if both the VHDL and Matlab code agree with each other, we are pretty certain that the component is correct. Theoretically, it is possible to make the same mistake in two different languages where the same mistake manifests itself over all the test cases. However, the probability of this actually happening is pretty close to zero. The verification process shown in the following examples assumes that you have both ModelSim and Matlab installed on your system.

Two examples are provided that illustrate the verification process. **Example 1** performs a simple VHDL component verification using the file *input.txt* for the input test vectors, and the results are written to the file *output.txt*. Matlab is used to generate the test vectors contained in the input file and then to verify that the output is correct.

Example 2 builds upon Example 1 in two ways. The first extension is that the number of I/O ports in the VHDL component has increased. There are two input ports to create test vectors for and two output ports to verify. The second extension is that there is an IP component created with the Quartus IP Wizard (a ROM memory). We will create this ROM, generate the memory initialization file, and setup ModelSim so that it can simulate this ROM IP component.

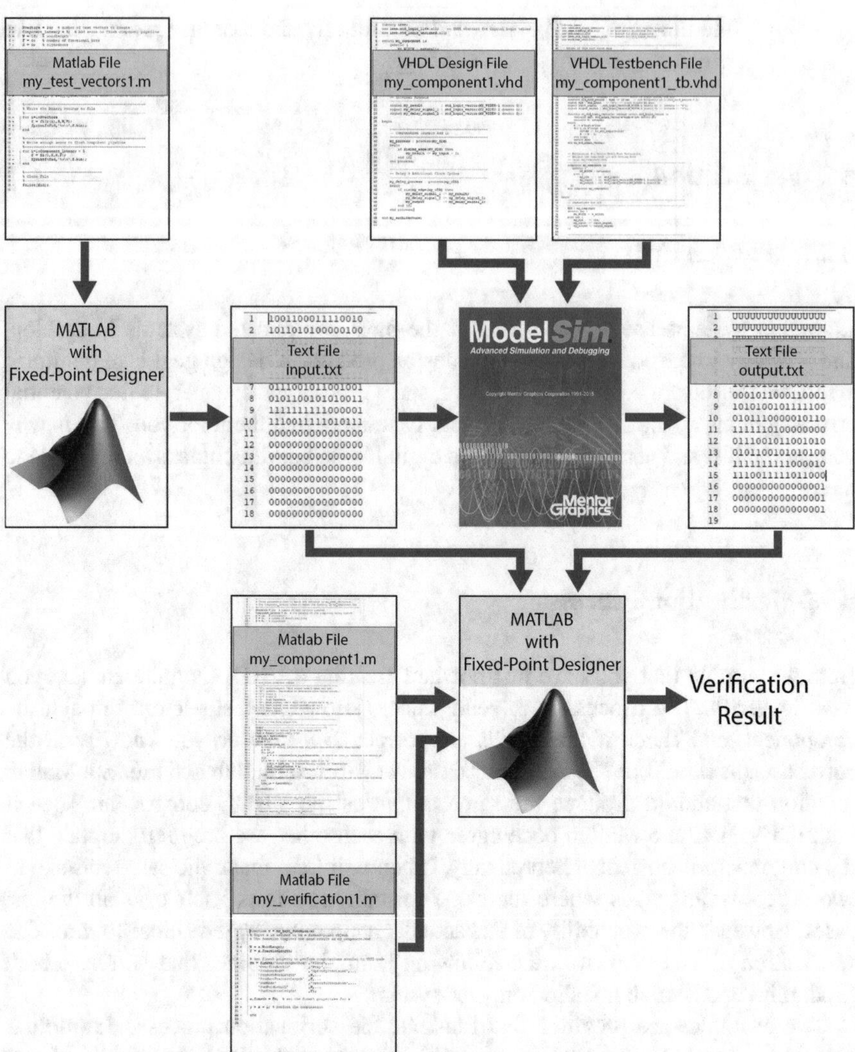

Fig. 8.1: Verification Flow Diagram. ModelSim verifies the VHDL component my_component1.vhd using the associated testbench my_component1_tb.vhd by reading the test vectors from file input.txt and writing the results to output.txt. The Matlab script my_test_vectors1.m is used to create the test vectors in the file input.txt. Verification is performed by the Matlab scrip my_verification1.m that reads input.txt and computes what the results should be using my_component1.m. It then compares this output to what is in output.txt

8.3 Verification Example 1: File I/O

Verification Example 1 performs verification of a simple VHDL component where the file input.txt contains that test vectors and the results are written to output.txt. Matlab is used to generate the test vectors and then to verify that the output is correct.

The starting point for verification is the VHDL design file that we wish to verify, which in this example is **my_component1.vhd**. The verification flow is illustrated in Fig. 8.1, which shows how all the files relate to each other. These files are listed in Table 8.1.

Table 8.1: Files used to verify **my_component1.vhd**

File	Description
my_component1.vhd	**VHDL component to be verified**
my_component1_tb.vhd	VHDL testbench file that verifies **my_component1.vhd**
my_test_vectors1.m	Matlab script that creates the input test vectors and writes them to **input.txt**
input.txt	File that contains the input test vectors. It is created by the Matlab script **my_test_vectors1.m**
output.txt	File that contains the simulation results when **my_component1_tb.vhd** is run in ModelSim
my_component1.m	Matlab function that computes the same results as **my_component1.vhd** and can generate the same results as is found in **output.txt**. Verification is complete when the Matlab function **my_component1.m** agrees with the VHDL component **my_component1.vhd**
my_verification1.m	Matlab script that reads **input.txt** and runs these test vectors through **my_component1.m** and compares these results to the results found in **output.txt**
Note: These files can be found on GitHub (click here)	

8.3.1 VHDL File to Verify: my_component1.vhd

The file **my_component1.vhd** (click here for the source file) will be the VHDL code that we wish to verify. The entity can be seen in Listing 8.2, which has a generic MY_WIDTH that defines the signal widths of the signals *my_input* and *my_output*.

```
entity my_component1 is
   generic (
      MY_WIDTH : natural);
   port (
      my_clk    : in  std_logic;
      my_input  : in  std_logic_vector(MY_WIDTH-1 downto 0);
      my_output : out std_logic_vector(MY_WIDTH-1 downto 0)
   );
end my_component1;
```

Listing 8.2: Entity of my_component1.vhd

The computation that my_component performs is to simply add 1 to the input, i.e., $my\_output = my\_input + 1$. This is done by the synchronous process my_add1_process that adds 1 to my_input on the rising edge of the clock. This process can be seen in Listing 8.3.

```
   my_add1_process : process(my_clk)
   begin
      if rising_edge(my_clk) then
         my_result <= my_input + 1;
      end if;
   end process;
```

Listing 8.3: my_add1_process

To simulate a component that has multiple clock cycles of latency, we arbitrarily delay the signal *my_result* three clock cycles before we send it out of the component. The *my_delay_process* is shown in Listing 8.4, and this coding style is only suitable for a couple of delays. We also use this style to illustrate the behavior of VHDL processes. Although processes allow sequential programming, they are still different from the usual serial programming languages. If we follow a value through this process, on rising edge 1 of the clock, the contents of *my_result* are assigned to *my_delay_signal_1*. On rising edge 2, *my_delay_signal_1* is assigned to *my_delay_signal_2*, and on rising edge 3, *my_delay_signal_2* is assigned to *my_output*. This ordering may look like it will not work if you are looking at it from the perspective of a typical serial programming language, where it looks like the value in *my_result* overwrites all the signals with the same value in successive order. In a VHDL process however, the assignments do not occur till you are at the *very end* of the process where they all occur "instantaneously" and in parallel (actually the simulator has delta cycles with zero time for these assignments). Thus the behavior of the process is that all these assignments occur simultaneously at the rising edge of the clock. This is why in a VHDL process a signal could have several different assignments occurring at different locations in the sequential code, but it is only the last assignment made in the process that actually occurs when you get to the end of the process. Since my_clock is in the sensitivity list, the process runs again when any change in my_clock occurs, and we condition our code to run on the rising edge of the clock to make our designs synchronous with this rising clock edge.

```
  my_delay_process : process(my_clk)
  begin
     if rising_edge(my_clk) then
        my_delay_signal_1 <= my_result;
        my_delay_signal_2 <= my_delay_signal_1;
        my_output         <= my_delay_signal_2;
     end if;
  end process;
```

Listing 8.4: my_delay_process

8.3.2 VHDL Testbench File: my_component1_tb.vhd

The file **my_component1_tb.vhd** (click here for the source file) is the VHDL testbench file that verifies my_component1.vhd, which is the *Device Under Test* (**DUT**). The signal width that is controlled by the generic MY_WIDTH is set by assigning the constant W_WIDTH to the desired value. In this example, we will be creating signals that are 16 bits wide.

```
   constant W_WIDTH : natural := 16;              -- width of ↲
   ↪input signal for DUT
```

Listing 8.5: Defining signal widths

Since our components are synchronous, we need to create a clock for our simulation. We do not really care what the frequency of the clock is, just that we have rising edges. We set the half period of the clock and create the clock signal as shown below.

```
   constant clk_half_period : time := 10 ns;   -- clk ↲
   ↪frequency is 1/(clk_half_period * 2)
   signal clk : std_logic := '0';                -- clock ↲
   ↪starts at zero
```

Listing 8.6: Clock definitions

Note that we need to initialize the clock to zero since we will be toggling the clock every half period.

```
   clk <= not clk after clk_half_period;
```

Listing 8.7: Clock creation

The main process in this testbench is reading the test vectors from the file *input.txt* and writing the results to *output.txt*. We start the process by first creating the local variables for this process as seen in Listing 8.8. The variables *read_file_pointer* and *write_file_pointer* are declared as file objects that will point to files of *text* (characters) where *text* is a VHDL type defined in the package textio. This is why we need to include this package at the beginning of the testbench (use std.textio.all).

Next, we create two variables *line_in* and *line_out* that are declared as *line* types, and a *line* type is a pointer to a string. This allows us to hold the address of a string (i.e., keep track of where the string is in memory), but it is not the string itself. To create a place to hold the string, we create a variable *input_string*, and we must tell it how many characters the string will contain. We also create the variable *input_vector* of std_logic_vector type because this is the type that we want to convert the input vectors to.

```
    process
        file read_file_pointer    : text;
        file write_file_pointer   : text;
        variable line_in          : line;
        variable line_out         : line;
        variable input_string     : string(W_WIDTH downto 1);
        variable input_vector     : std_logic_vector(W_WIDTH-1 downto 0);
    begin
```

Listing 8.8: Variables for file I/O

Using the *file_open()* function, we open the file *input.txt* in read mode, and *read_file_pointer* points to this opened file. In a similar fashion, we open the file *output.txt*, but in write mode, it is pointed to by our variable *write_file_pointer*.

```
        file_open(read_file_pointer,  "input.txt",  read_mode);
        file_open(write_file_pointer, "output.txt", write_mode);
```

Listing 8.9: Opening files

We want to read all the test vectors in the file *input.txt* so we create a while loop to keep reading until there are no more input test vectors. When that happens, the *endfile()* function returns true and the loop stops.

```
        while not endfile(read_file_pointer) loop -- Read input file until end of file
```

Listing 8.10: While loop

The function *readline()* moves a line of text from the file *input.txt* into an internal buffer, and it gives us the address of this buffer in the pointer *line_in*. We then use the *read()* function where we give it the *line_in* pointer that points to the string in the buffer, and it copies it into our string variable *input_string*. We want this string to be of type std_logic_vector so we use the conversion function *to_std_logic_vector()* to perform the conversion and place it in our *input_vector* variable. Now we can make the signal assignment to *input_signals* since they are the same type and same vector width. Thus a line of text in *input.txt* has been converted into a std_logic_vector type and placed into the input of the component under test. The line that starts with the keyword *report* is used to display the values as they are being read in. If you have a lot of test vectors, you will want to delete or comment this line out.

```
          readline(read_file_pointer, line_in); -- Read a line from input file
          read(line_in, input_string);          -- convert line to a string
          input_vector := to_std_logic_vector(input_string); -- convert string to std_logic_vector
          report "line in = " & line_in.all & " value in = " & integer'image(to_integer(unsigned(input_vector))); -- display what is being read in
          input_signal <= input_vector;   -- assign to the input_signal going into the DUT
```

Listing 8.11: Reading test vectors

In parallel with a test vector being read in, we have a result vector *output_signal* that we want to save to *output.txt*. The function *write()* takes our std_logic_vector *output_signal* and writes it to the buffer pointed to by *line_out*. It converts it into a string in the buffer, and we have to tell it how many characters to write and if it should be right or left justified. We then write the buffer to the file with the *writeline()* function.

```
          write(line_out, output_signal, right, W_WIDTH);
          writeline(write_file_pointer, line_out);
```

Listing 8.12: Writing results

Notice that this file I/O process does not have a sensitivity list. Instead we use wait statements within the process. All the statements in the process do not happen until a wait occurs, and since we want a line to be read in and converted into our input vector and a result to be written every clock cycle, we wait until we have a rising edge of the clock signal. If this wait was omitted, the process would run, looping and reading in all the test vectors until it saw the next wait statement. Then the *last* assignments would occur, and you would only see the last test vector in the input file being assigned and fed into the test component.

```
          wait until rising_edge(clk);
```

Listing 8.13: Synchronizing file I/O with clock

When done, the files are closed, and a final wait statement is written so that the file I/O process is not restarted again during simulation.

```
        file_close(read_file_pointer);
        file_close(write_file_pointer);
        wait; -- we are done so don't read the file again
    end process;
```

Listing 8.14: Closing files

8.3.3 Creating Test Vectors with Matlab Script my_test_vectors1.m

In our testbench, we want to verify a component that has an input that is 16 bits wide and where these 16-bit vectors will be read from the file *input.txt*. We will create this input.txt file using the Matlab script called **my_test_vectors1.m** (click here for the source file)

The first parameter is *Nvectors* that will be the number of test vectors created. Here it is set to 10 so the simulation will be short. Normally, you want as many test vectors as possible while still being practical given the simulation time. Ideally you would generate all possible input bit patterns as the test vectors, which is known as complete coverage. However, if you had an input vector that was 128 bits wide, complete coverage would require 2^{128} test vectors. If your simulator could process 2^{32} or 4.3 billion test vectors per second, it would take 2.5×10^{19} *centuries* to complete the simulation, clearly not practical. Rather you would need to sample the input range to reduce the number of test vectors to a more manageable simulation time.

The second parameter is *Component_latency* that should be set to the latency of the component being tested. This causes zeros to be written to *input.txt* to make sure that the last non-zero test vector gets completely through the component before the simulation stops.

The next parameter *W* controls the number of bits or word size in the test vectors. Even though we will be using fixed-point values, we only need to know the test vector's width to create the input bit patterns. We do not concern ourselves with what these bit patterns mean at this point, and so we set the number of fractional bits to zero and make the values unsigned so we can treat the vector as an unsigned integer when creating the test vector bit patterns.

```
Nvectors = 10;            % number of test vectors to create
Component_latency = 3;    % Add enough zeros to flush ↲
 ↪component pipeline
W = 16;   % wordlength of my_input
F = 0;    % we don't care about fraction bits so setting to ↲
 ↪zero
S = 0;    % we don't care about sign bit so setting to ↲
 ↪unsigned
```

Listing 8.15: Parameter settings

In Matlab, we use the *fopen()* function to open the file *input.txt* in write mode (W) where we will be writing the test vectors. The function returns the file identifier that we store in the variable *fid* (short for file ID).

```
fid = fopen('input.txt','w');
```

Listing 8.16: Opening a file

Since the number of test vectors being generated is likely much smaller than the number of possible test vectors, we sample a small subset from the much larger set of all input possibilities. We treat the possible input set as a uniform distribution and draw from this uniform distribution. We use the *rng()* function in Matlab to set up the random number generator. The *shuffle* argument seeds the generator with the current time, so that when each time Matlab is started up, the random numbers that are generated will be different. The *twister* argument specifies to use the Mersenne Twister algorithm. This algorithm is good enough for our purposes, and we will not concern ourselves with finding the "best" random number generator since this is a rabbit hole we could go down into and not come back from. We create a vector of random integers using the *randi()* function. The range of the uniform distribution of integers is specified by the interval [0 2^{W-1}]. This is the range of unsigned integers that gives us all possible bit patters for a word length of W. The last two parameters *1,Nvectors* in *randi()* tell it to create a matrix of random values that has 1 row and Nvectors columns, which is returned in the variable *r*. The vector *r* contains the values that we will use for the test vectors.

```
rng('shuffle','twister');          % 'shuffle' seeds the random number generator with the current time so each randi call is different;  'twister' uses the Mersenne Twister algorithm
r = randi([0 2^W-1],1,Nvectors);   % select from a uniform integer distribution over all possible integers
```

Listing 8.17: Random coverage

We then go through all the *Nvectors* random values in the variable *r* by using a *for* loop. We make use of Matlab's Fixed-Point Designer toolbox by creating a fixed-point object using the *fi()* function. The first argument in this function is the value that will be used for the fixed-point number, which we get by indexing into our *r* vector. The second argument is the sign bit S, which we set to zero for unsigned data types. The third argument is the word length in bits, which we set to W. The last argument is the number of fraction bits in the word, which we set to zero since we are assuming unsigned integers. The fixed-point object is returned as the variable *f*.

A convenient radix conversion that is associated with these fixed-point objects is the binary string conversion that is accessed by using the suffix *.bin* (*.hex* for hexadecimal and *.oct* for octal). We can then use this string directly in our fprintf() function to write the binary string representation directly to our test vector file *input.txt*. An example of this test vector input file can be seen in **input.txt** (click here for the source file) and is also shown in Listing 8.20.

```
for i=1:Nvectors
    f = fi(r(i),S,W,F);
    fprintf(fid,'%s\n',f.bin);
```

```
end
```

Listing 8.18: Writing binary strings

When we are done writing the test vectors to *input.txt*, we close the file.

```
fclose(fid);
```

Listing 8.19: Closing a file

```
1110100011011100
1100101100111000
1110011111010000
1101010111111101
0101000100001011
1101110001011110
1101111100100000
1011111010000101
1011100101001101
0011110111110110
0000000000000000
0000000000000000
0000000000000000
```

Listing 8.20: File contents : input.txt

8.3.4 Computing the Results with Matlab Function my_component1.m

When the ModelSim simulation is run, the testbench *my_component1_tb.vhd* produces the results in file *output.txt* (click here for the text file), which can also be seen in Listing 8.21. You can see the latency of *my_component1.vhd* as the lines comprised of the character "U," which means *Uninitialized*. This means that ModelSim is putting out values from internal signals that have not been initialized and do not know what these values should be. When the first input finally gets through the pipeline, we see a result showing up on line 6. If you compare *output.txt* with *input.txt*, you can see that one has been added to all the input binary values.

```
1  UUUUUUUUUUUUUUUU
2  UUUUUUUUUUUUUUUU
3  UUUUUUUUUUUUUUUU
4  UUUUUUUUUUUUUUUU
5  UUUUUUUUUUUUUUUU
6  1110100011011101
7  1100101100111001
8  1110011111010001
9  1101010111111110
10 0101000100001100
11 1101110001011111
12 1101111100100001
13 1011111010000110
14 1011100101001110
15 0011110111110111
16 0000000000000001
17 0000000000000001
18 0000000000000001
```

Listing 8.21: File contents: output.txt

In order to verify that these results are correct, we need a Matlab function that produces the same results. This function is called **my_component1.m** (click here for the source file). If this Matlab function agrees with the ModelSim output produced by *my_component1.vhd*, we can be confident that both the VHDL and Matlab codes are correct.

The Matlab function declaration of *my_component1.m* is shown below where we expect a fixed-point variable to be passed into the function.

```
function y = my_component1(x)
```

Listing 8.22: Function declaration

We want our fixed-point Matlab operations to reflect what our VHDL code is doing. This means that we need to change the default behavior of Matlab's Fixed-Point Toolbox. The normal behavior is to automatically extend the word length of the result, which is not always what we do in our VHDL code. In our VHLD code, when we add 1 to a std_logic_vector signal, the signal size stays the same (yes, overflow is possible). To make our fixed-point variable have the same behavior, we need to modify the fixed-point math settings for this variable. We can do this by using the *fimath()* function to create a fimath object with our desired settings and then apply this to our fixed-point variable. We can create multiple fimath objects, each with their own settings, and apply these to different variables. In this example, we just use a single fimath setting and apply it to the fixed-point variable *x*.

```
W = x.WordLength;          % Extract the word length W of input variable x
F = x.FractionLength;      % Extract the fraction length F of input variable x
Fm = fimath('RoundingMethod' ,'Floor',...
     'OverflowAction'         ,'Wrap',...
     'ProductMode'            ,'SpecifyPrecision',...
     'ProductWordLength'      ,W,...
     'ProductFractionLength'  ,F,...
     'SumMode'                ,'SpecifyPrecision',...
     'SumWordLength'          ,W,...
     'SumFractionLength'      ,F);
x.fimath = Fm;  % Apply these fimath properties to x
```

Listing 8.23: Setting fimath properties

Even though we really only care about the *SumWordLength* property in this example, the fimath settings shown can be used in many situations where you are adding and multiplying in VHDL, and you want to reflect this in your Matlab verification code. We will discuss only a few of the fimath options. A complete list can be seen in Matlab's fimath reference page.

The first property that is being set is *RoundingMethod*, which is set to *Floor*. Floor causes rounding to round toward negative infinity. This is useful if you are truncating a fixed-point value where you are taking a VHDL signal slice that is eliminating some of the least significant fractional bits.

The next property that is being set is *OverflowAction*, where it can be set to either *Saturate* or *Wrap*. Both cases can occur in your VHDL code. *Wrap* is more common where you have a counter in VHDL using a std_logic_vector signal and you add 1 to the largest value (all ones in the vector), which causes it to become zero (wraps to zero). Since this is desired behavior in many cases, you want your Matlab code to reflect this so you would use the *Wrap* option in this case.

Where wrapping behavior is *not* desirable is when you are operating on audio signals. A wrap will cause a large positive value to immediately turn negative (i.e., two's complement overflow), and this will sound like a "pop" in your audio. If the signal value is too large to fit into your data type, what you want to do is to clip it to the maximum value that can be represented by the data type (fixed-point vector). This clipping will cause harmonic distortions to occur, but this will sound much better than a loud pop. Thus in audio applications, you would use the *Saturate* option for overflow behavior of your audio signals.

The next property that is being set is *ProductMode*, which is set to *SpecifyPrecision*. The default behavior is *FullPrecision* where Matlab will automatically grow the data type size in order to keep all of the bits after a multiplication. Here we want to control both the word length and fraction length of the result. To control the word length, we set *ProductWordLength* to *W*, which means we keep the same word length as *x*, which we get from the fixed-point object field x.WordLength. To control the fraction length, we set *ProductFractionLength* to *F*, which means we keep the same fraction length as *x*, which we get from x.FractionLength.

You will find in audio hardware that the audio signal is kept as a fractional data type, i.e., between $[-1\ \ 1]$. The reason is that after a multiplication occurs in hardware, you can easily resize the vector just by throwing away the least significant bits that do not fit into the vector. For example, if you had a 24-bit fractional signal and multiply it by another 24-bit fractional signal, you could keep it as a 24-bit fractional signal by throwing away the least significant 24 fractional bits. It is unlikely that you could hear these bits anyway so there would be no point in keeping them around.

The next property that is being set is *SumMode*, which is set to *SpecifyPrecision*. The default behavior is *FullPrecision* where Matlab will automatically grow the data type size in order to keep all of the bits after an addition. Here we want to control both the word length and fraction length of the result. To control the word length, we set *SumWordLength* to *W*, which means we keep the same word length as *x* where we can get it from x.WordLength. To control the fraction length, we set *SumFractionLength* to *F*, which means we keep the same fraction length as *x* where we can get it from x.FractionLength. This is how we keep the output the same length as the input since this is the behavior of the add 1 operation in *my_component1.vhd* and why the vectors in *output.txt* are the same length as the vectors in *input.txt*.

Once a fimath object has been created with the properties that we want, which are now in the fimath variable *Fm*, we assign it to the *x.fimath* field that controls what happens when *x* is used.

Finally, we add one to x where we have now specified that x should keep the same length after addition. This results in the variable y with the same bit width as variable x, which reflects what our VHDL code is doing.

```
y = x + 1; % perform the simple add 1 computation
```

Listing 8.24: Adding one

8.3.5 Performing Verification with the Matlab Script my_verification1.m

When the ModelSim simulation is run, the testbench *my_component1_tb.vhd* produces the results in file *output.txt*, which can be seen in Listing 8.21. You can see the latency of *my_component1.vhd* as the lines comprised of the character "U," which means that ModelSim is putting out values from internal signals that have not been initialized and do not know what these values should be. When the first input finally gets through the pipeline, we see a result showing up on line 6. If you compare *output.txt* with *input.txt*, you can see that one has been added to all the input values.

We want to compare what is in *output.txt* with what *my_component1.m* produces, and we automate the comparison process using the Matlab script **my_verification1.m** (click here for the source file).

The one parameter setting in *my_verification1.m* that needs to match the parameter settings in *my_test_vectors1.m* is the test vector bit width *W*. The test vectors were created with no concern as to the interpretation of these binary patterns. Now we *do care* what these binary patterns mean. Thus the parameters *F* and *S* need to be set with how we really will be interpreting these input numbers. It just so happens that in this case we will interpret the numbers again as unsigned integers, keeping these parameters the same, but typically this is not the case when using fixed-point numbers.

```
Nvectors = 10;  % number of test vectors created
Component_latency = 3;  % the latency of the component being tested
W = 16;  % wordlength
F = 0;   % number of fractional bits
S = 0;   % signedness
```

Listing 8.25: Verification parameter settings

The test vectors are read in from the file *input.txt* as shown in the code in Listing 8.26. The file is opened in read mode by Matlab's *fopen()* function, and the returned value assigned to the file handle *fid1*, which is the pointer to the input file. The Matlab function *fgetl()* then reads the first line in the file and puts this in the variable line_in.

Since we want to interpret what the binary strings in *input.txt* mean in terms of numerical values, we create a fixed-point object where variable S has been set to zero, which means the bit strings will be interpreted as unsigned numbers. W has been set to 16, which must match the number of bits on each line as seen in Listing 8.20. The parameter F is set to zero since our numbers are unsigned integers and have no fractional bits. At this point, in the code, we do not care what the value of the fixed-point object is and we set it to zero (first argument). We will use this fixed-point object in a bit. We are going to create an array of fixed-point objects for each line in *input.txt* so we set the starting index of this array to one (index = 1;).

We create a while loop and test line_in with the function *ischar()*. Function *fgetl()* returns the lines of text until it reaches the end of the file. When this occurs, a value of −1 is returned, which is not a character, and this causes *ischar()* to return false, stopping the while loop.

We assign the binary bit string that was read in to the fixed-point variable *a*. By assigning the binary string to the .bin field of *a*, Matlab updates the value of *a* to be consistent with the binary string and the parameters (S,W,F) that we used when we first created this fixed-point variable.

We then assigned *a* to our test_vector array with the current index and display what this value is. We increment the index for the next round and read in a new line from the input file and the while loop starts again. When an end-of-file condition is encountered, the loop ends and we close the input file.

When we are done reading in the test vectors from *input.txt*, we end up with the Matlab array ***test_vectors*** that contains all of the test vectors where the interpretation of the numbers given the binary strings have been controlled by the parameters (S,W,F).

```
fid1 = fopen('input.txt','r');
line_in = fgetl(fid1);  % read the first line in the file
a = fi(0,S,W,F);   % interpret the bit string appropriately ↲
   ↪by creating a fixed-point object with appropriate ↲
   ↪parameters
index = 1;
while ischar(line_in)
    a.bin = line_in; % push the binary string into the fixed↲
       ↪-point object where it will be interpreted with the ↲
       ↪given S (sign), W (word length), and F (frational ↲
       ↪bits) values
    test_vectors(index) = a;  % save this fixed-point object
    disp([num2str(index) ' : ' line_in ' = ' num2str(a)]) % ↲
       ↪display what we are reading in (comment out if there ↲
       ↪a lot of test vectors)
    index = index + 1;
    line_in = fgetl(fid1);
end
fclose(fid1);
```

Listing 8.26: Reading in the test vectors

When ModelSim writes the test vectors to *output.txt*, it can write any of the characters associated with the VHDL std_logic type as seen in Listing 8.27. Since we only want to deal with numeric values, we are going to ignore any output that has characters other than "0" or "1." We create a list of these nonbinary characters in the string *stdchar*. We will check for these characters and ignore them, when reading in the result vectors from *output.txt*.

```
% 'U': uninitialized. This signal hasn't been set yet.
% 'X': unknown. Impossible to determine this value/result.
% '0': logic 0
% '1': logic 1
% 'Z': High Impedance
% 'W': Weak signal, can't tell if it should be 0 or 1.
% 'L': Weak signal that should probably go to 0
% 'H': Weak signal that should probably go to 1
% '-': Don't care.
stdchar = 'UXZWLH-'; % create a list of all non-binary std_logic characters we will ignore non-binary strings when reading output.txt
```

Listing 8.27: std_logic characters

The result vectors that ModelSim produced are read in from the file *output.txt* as shown in the code seen in Listing 8.28. This code is similar to reading in the test vectors in Listing 8.26 so we will only describe the differences here. In the while loop, we want to see if *line_in* contains any of the nonbinary std_logic characters. We start by setting the variable s to zero. Then, in a for loop, we add to s the output of the *contains()* function that will return 1 if *line_in* contains one of these nonbinary characters as we index through all of them. If there are not any, s will remain zero, and we assign the string to a.bin, which in turn gets saved into the array ***vhdl_vectors***. If the string does contain nonbinary characters, we ignore this line and print out message that we are doing so.

```
fid2 = fopen('output.txt','r');
line_in = fgetl(fid2);  % read first line in file
a = fi(0,S,W,F);  % interpret the bit string appropriately by creating a fixed-point object with appropriate parameters
index = 1;
while ischar(line_in)
    % check if the input string contains any std_logic characters other than the binary characters
    s = 0;
    for i=1:7
        s = s + contains(line_in,stdchar(i));  % check line_in for each non-binary std_logic value contains() will return 1 if it finds such a value
    end
    if s == 0  % s will be zero if line_in contains only 0s or 1s, which means we have a valid string that we can convert
        a.bin = line_in;  % convert binary string to fixed-point
        vhdl_vectors(index) = a;
        disp([num2str(index) ' : ' line_in ' = ' num2str(a)])
        index = index + 1;
    else
```

```
        disp([num2str(index) ' : ' line_in ' ~~ Ignoring line since it contains non-binary std_logic characters'])
    end
    line_in = fgetl(fid2);
end
fclose(fid2);
```

Listing 8.28: Reading in the result vectors

We have now read in the test vectors from *input.txt*, which have been saved in the array ***test_vectors***, and the result vectors from *output.txt*, which have been saved in the array ***vhdl_vectors***. We next run the ***test_vectors*** values through our Matlab function **my_component1.m**, producing the vector ***matlab_vectors*** as seen in Listing 8.29. If our VHDL and Matlab codes are correct, ***matlab_vectors*** should contain the same values as ***vhdl_vectors***.

```
matlab_vectors = my_fxpt_function(test_vectors);
```

Listing 8.29: Computing the Matlab version of my_component1

The comparison of ***matlab_vectors*** with ***vhdl_vectors*** is shown in Listing 8.30. We need to take care of one wrinkle before we can fully automate the comparison of these vector values. We need to know where the results start showing up in ***vhdl_vectors***; otherwise, we will always be comparing wrong values. This is because of two things. One, we have the latency through the VHDL component. Second, we may or may not be ignoring initial result vectors depending on if there are initialized results (i.e., "U" values) or if these are zero. This can be seen by inspection by comparing the starting values of the vectors and seeing what the offset is between these two vectors. This offset is then set in the Matlab verification code (line 105) so the verification can then be automated for all the values in the vectors. A for loop is used to go through all the ***matlab_vectors***, comparing them to their associated ***vhdl_vectors*** values. A comparison of the bit patterns is made since verification needs to be bit true, and if the bit strings do not match, a verification error is produced for this case with the values being printed out. If no errors are seen, the variable error_flag will remain zero, and this will result in the final message *Verification Succeeded!*.

```
105 index_offset = 0;  % set the alignment offset if needed (initial output.txt values might be valid and not ignored if they contain only binary characters)
106 error_flag = 0;
107 for i=1:Nvectors
108     x = matlab_vectors(i);
109     y = vhdl_vectors(i + index_offset);
110     if strcmp(x.bin,y.bin) == 0
111         a = test_vectors(i);
112         disp(['    ----------------------------------------'])
113         disp(['    Verification Error Occurred for test vector ' num2str(a) ' = ' a.bin '(index = ' num2str(i) ')'])
```

```
        disp(['     Matlab result = ' x.bin ' = ' num2str(x)])
        disp(['     VHDL result   = ' y.bin ' = ' num2str(y)])
        error_flag = 1;
    end
end

disp(' ')
if error_flag == 0
    disp('   Verification Succeeded!')
else
    disp('    ******* Verification Failed *******')
end
```

Listing 8.30: Comparing the VHDL and Matlab results

8.3.6 Running the Example 1 Verification

Here are the steps to take to run this verification example:

Step 1: Create a `\Ex1` directory in Windows and download all the Example 1 verification files from GitHub (click here for the files), with the exception of the *input.txt* and *output.txt* files (you will be creating these), and put them in this new directory.

Step 2: Open up **my_test_vectors1.m** in Matlab and run the script. It will create a new file ***input.txt***, which will be different from the one found on GitHub because the test vectors are randomly generated.

Step 3: Open ModelSim:

a. Create a Project by selecting *File* → *New* → *Project.*
   i. Name the project Ex1.
   ii. Under *Project Location*, browse to the `\Ex1` directory you just created.
   iii. Keep the Default Library Name as "work."
   iv. Keep the "Copy Settings From" as is (keep default setting).
   v. Click OK.
b. When it asks to *Add items to the Project*, click on *Add Existing File*, browse to `\Ex1`, and add the three VHDL files. Click OK and close the *Add items to the Project* window.
c. Modify the compilation order if it needs to be modified by going to *Compile* → *Compile Order...* and rearranging the compile order (select file and use up/down arrows to change the file order). The compile order should be:
   i. **text_util.vhd** (Order 0 in Project tab)
   ii. **my_component1.vhd** (Order 1 in Project tab)
   iii. **my_component1_tb.vhd** (Order 2 in Project tab)

d. Compile the files (*Compile → Compile All*).
   Note: If the compile fails (a red x will show up in the Status column), you can see the errors that have been logged in the compile summary (*Compile → Compile Summary*).
e. To run the simulation:
   i. Select the Library window in ModelSim (click Library tab above Transcript window).
   ii. Expand the work folder.
   iii. Double click on my_component_tb (this will open the Wave window).
   iv. To see the signals you want to see, select the component hierarchical level in the sim–Default window, click on a signal name in the Objects window, and either right click and select Add Wave or drag the signal into the first column of the Wave window. Add the following signals (from the my_component_tb instance):
      A. clk
      B. input_signal
      C. output_signal
   v. Set the simulation time to 500 ns. (This needs to be long enough to run all the stimulus vectors through the simulation, so if you change the number stimulus vectors, you will need to change this value.)
   vi. Run the Simulation (button to the right of the simulation time window). The file ***output.txt*** should be created that contains the result vectors.

Step 4: Open up **my_verification1.m** in Matlab and run the script. You should see the message *Verification Succeeded!* in the Matlab Command Window. **Note:** To see an error message, open the *output.txt*, change a bit in one of the result vectors, and re-run *my_verification1.m*.

8.4 Verification Example 2: Using a Quartus ROM IP Component

Example 2 builds upon Example 1 in two ways. The first extension is that the number of I/O ports in the VHDL component has increased. There are two input ports to create test vectors for and two output ports to verify. The second extension is that there is an IP component created with the Quartus IP Wizard (a ROM memory). The steps for creating the Quartus ROM IP can be found in Sect. 6.4.1 Creating a ROM IP Component (page 78). We will use this ROM IP and its memory initialization file when verifying **my_component2.vhd**. The files used in this Example 2 verification are listed in Table 8.2.

Table 8.2: Files used to verify **my_component2.vhd**

File	Description
my_component2.vhd	**VHDL component to be verified**
my_component2_tb.vhd	VHDL testbench file that verifies **my_component2.vhd**
my_test_vectors2.m	Matlab script that creates the input test vectors and writes them to the files **input1.txt** and **input2.txt**
input1.txt	File that contains the input test vectors for the *my_input* input signal. It is created by the Matlab script **my_test_vectors2.m**
input2.txt	File that contains the input test vectors for the *my_rom_address* input signal that contain the address values for the ROM. It is created by the Matlab script **my_test_vectors2.m**
output1.txt	File that contains the simulation results when **my_component2_tb.vhd** is run in ModelSim. It contains the output of the signal *my_rom_value*, which is the output from the ROM component
output2.txt	File that contains the simulation results when **my_component2_tb.vhd** is run in ModelSim. It contains the output of the signal *my_output*, which is the signal containing the computation results
my_component2.m	Matlab function that computes the same results as **my_component2.vhd**. Verification is complete when the Matlab function **my_component2.m** agrees with the VHDL component **my_component2.vhd**
my_verification2.m	Matlab script that reads **input1.txt** and **input2.txt** and runs these test vectors through **my_component2.m** and compares these results to the results found in **output1.txt** and **output2.txt**
ROM.vhd	ROM IP file that was generated by Quartus
ROM.mif	Memory initialization file (.mif) for the ROM
ROM.mat	Matlab file that contains the same information as the memory initialization file. It was generated for verification convenience so that the .mif file did not need to be parsed when read into Matlab
Note: The Matlab and VHDL files can be found on GitHub (click here), and similarly, the ROM files can be found on GitHub (click here)	

8.4.1 VHDL File to Verify: my_component2.vhd

The file **my_component2.vhd** (click here for the source file) will be the VHDL code that we wish to verify. The entity can be seen in Listing 8.31, which has generics that specify the signal widths for the ROM being used, the I/O signal widths for the component, and the number of clock cycles to delay the output.

```
entity my_component2 is
   generic (
        MY_ROM_A_W     : natural;    -- Width of ROM Address bus
        MY_ROM_Q_W     : natural;    -- Width of ROM output
        MY_ROM_Q_F     : natural;    -- Number of fractional bits in ROM output
        MY_WORD_W      : natural;    -- Width of input signal
        MY_WORD_F      : natural;    -- Number of fractional bits in input signal
        MY_DELAY       : natural);   -- The amount to delay the product before sending out of component
    port (
        my_clk          : in  std_logic;
        my_rom_address  : in  std_logic_vector(MY_ROM_A_W-1 downto 0);
        my_input        : in  std_logic_vector(MY_WORD_W-1  downto 0);
        my_rom_value    : out std_logic_vector(MY_ROM_Q_W-1 downto 0);
        my_output       : out std_logic_vector(MY_WORD_W-1  downto 0));
end my_component2;
```

Listing 8.31: Entity of my_component2.vhd

The computation that my_component2 performs is to multiply the input signal by a ROM value, i.e., $my\_output = my\_input\ x\ ROM\_signal$. The input fixed-point signal is *MY_WORD_W* bits wide and has *MY_WORD_F* fractional bits, and the ROM memory word is *MY_ROM_Q_W* bits wide and has *MY_ROM_Q_F* fractional bits. The ROM signal is specified by the input signal *my_rom_address*, which is *MY_ROM_A_W* bits wide and can address $2^{MY\_ROM\_A\_W}$ memory locations. The ROM takes two clock cycles for the output to appear after the ROM address is presented. Note that the ROM generic values need to be consistent with the generated ROM IP and ROM memory initialization file. The input signal *my_input* must be delayed by two clock cycles so that it time aligns with the specified ROM signal as seen in Listing 8.32. The resulting product *my_product* is a signal with $MY\_WORD\_W + MY\_ROM\_Q\_W$ bits and is trimmed so that it has the same fixed-point data type (same W and F) as the input signal *my_input*.

```
    my_process : process(my_clk)
    begin
        if rising_edge(my_clk) then
            my_input_delayed_1 <= my_input;
```

```
            my_input_delayed_2 <= my_input_delayed_1;
            my_product         <= my_input_delayed_2 * rom_value;
        end if;
    end process;

    my_product_trimmed <= my_product(MY_WORD_W + MY_ROM_Q_F - 1 downto MY_ROM_Q_F);  -- keep the output with the same W & F as the input.
```

Listing 8.32: Process that delays and multiplies

The product signal *my_product_trimmed* is then delayed by *MY_DELAY* clock cycles. We use VHDL generate statements to include the appropriate VHDL code depending on the delay value. If the delay is zero, we send the product out with no delay by connecting it directly to *my_output* as seen in Listing 8.33.

```
    gen_z0 : if MY_DELAY = 0 generate
        my_output <= my_product_trimmed;
    end generate;
```

Listing 8.33: Generated code when MY_DELAY is zero

If *MY_DELAY* is set to one, we generate a process where the output is assigned on the rising edge of the clock as seen in Listing 8.34.

```
    gen_z1 : if MY_DELAY = 1 generate
        my_delay_process : process(my_clk)
        begin
            if rising_edge(my_clk) then
                my_output <= my_product_trimmed;
            end if;
        end process;
    end generate;
```

Listing 8.34: Generated code when MY_DELAY is one

If *MY_DELAY* is set to two, we generate a process where the additional delay is made by assignment to an internal signal, which is declared before the generate begin statement as seen in Listing 8.35. Note that the generate *begin* statement is not needed if there are no signals to declare as seen in the previous generate statements.

```
    gen_z2 : if MY_DELAY = 2 generate
        signal my_output_delayed : std_logic_vector(MY_WORD_W-1 downto 0);
    begin
        my_delay_process : process(my_clk)
        begin
            if rising_edge(my_clk) then
                my_output_delayed <= my_product_trimmed;
                my_output         <= my_output_delayed;
            end if;
        end process;
```

```
    end generate;
```

Listing 8.35: Generated code when MY_DELAY is two

If *MY_DELAY* is greater than two, we generate a process where the additional delays are made by assignments to an array of signals, which can be seen in Listing 8.36. We do this by creating our own array type called *my_delay_array* and then specifying the size of this array when we create the signal *delay_vector*. We can then make the appropriate assignments between the *delay_vector*s inside the for loop. Note: There is a limit to the amount of delay that can be done by creating registers out of the logic elements in the FPGA fabric. If the delay needs to be large, then it is better to use dedicated memory in the fabric (e.g., FIFOs or circular buffers created out of the on-chip RAM blocks).

```
    gen_zg2 : if MY_DELAY > 2 generate
        -- delay array
        type my_delay_array is array (natural range <>) of std_logic_vector(MY_WORD_W-1 downto 0);
        signal delay_vector : my_delay_array(MY_DELAY-2 downto 0);
    begin
        my_delay_process : process(my_clk)
        begin
            if rising_edge(my_clk) then
                delay_vector(0) <= my_product_trimmed;
                for i in 0 to MY_DELAY-3 loop
                    delay_vector(i+1) <= delay_vector(i);
                end loop;
                my_output <= delay_vector(MY_DELAY-2);
            end if;
        end process;
    end generate;
```

Listing 8.36: Generated code when MY_DELAY is greater than two

8.4.2 VHDL Testbench File: my_component2_tb.vhd

The file **my_component2_tb.vhd** (click here for the source file) is the VHDL testbench file that verifies my_component2.vhd, which is the *Device Under Test* (**DUT**). This testbench is similar to *my_component1_tb.vhd*, which is described in Sect. 8.3.2 VHDL Testbench File: my_component1_tb.vhd (page 99). The difference is that two input files are read and two output files created. The input file *input1.txt* is fed to the input signal *my_input*, and the input file *input2.txt* is fed to the input signal *my_rom_address*. The output file *output1.txt* is created from the output signal *my_output*, and the output file *output2.txt* is created from the output signal *my_rom_value*.

8.4.3 Creating Test Vectors with Matlab Script my_test_vectors2.m

The file **my_test_vectors2.m** (click here for the source file) is the Matlab script that creates the two input test vector files. This script is similar to *my_test_vectors1.m*, which is described in Sect. 8.3.2 VHDL Testbench File: my_component1_tb.vhd (page 102). It creates the input file *input1.txt* with test vectors for the input signal *my_input* that are 16 bits wide as shown in Listing 8.37. It also creates the input file *input2.txt* with test vectors for the input signal *my_rom_address* that are 8-bit addresses as shown in Listing 8.38.

```
1011010000011111
0100100110000101
1010111001011101
0000010011010111
0101110001100000
0111001101011100
0010110000001010
0010110110011100
0101001011100010
1111110010010001
0000000000000000
0000000000000000
0000000000000000
```

Listing 8.37: File contents : input1.txt

```
00011011
11100010
10011100
00011100
00110001
01100011
01100011
11100111
01001101
00101110
00000000
00000000
00000000
```

Listing 8.38: File contents : input2.txt

8.4.4 Computing the Results with Matlab Function my_component2.m

When the ModelSim simulation is run, the testbench *my_component2_tb.vhd* produces two output files *output1.txt*, shown in Listing 8.39, and *output2.txt*, shown in Listing 8.40. **Output1.txt** (click here for the source file) contains the result of multiplying the fixed-point input signal (W=16, F=8) with the ROM word (W=12, F=11). You can see the latency of *my_component2.vhd* as the lines comprised of the "U" and "X" characters. The character "U" means *Uninitialized* and "X" means *Forcing Unknown*. This means that ModelSim is putting out values from internal signals that have not been initialized or do not know what these values should be yet. When the first result finally gets through the pipeline, we see it showing up on line 9.

The output file **Output2.txt** (click here for the source file) shows the output from the ROM component (Listing 8.40), which takes two clock cycles to appear. The latency is two clock cycles because it takes the first rising edge of the clock to capture the address and then a second rising edge to put out the value associated with this address.

```
1 UUUUUUUUUUUUUUUU
2 UUUUUUUUUUUUUUUU
3 UUUUUUUUUUUUUUUU
4 UUUUUUUUUUUUUUUU
5 UUUUUUUUUUUUUUUU
6 UUUUUUUUUUUUUUUU
7 XXXXXXXXXXXXXXXX
8 XXXXXXXXXXXXXXXX
9 1001101011110111
10 0001110001110111
11 0101010101100100
12 0000010000100100
13 0100011100001010
14 0100011010100010
15 0001101011110111
16 0001000101100100
17 0011011111011001
18 1100010100010010
```

Listing 8.39: File contents: output1.txt

```
UUUUUUUUUUUU
000000000000
000000000000
011011100010
001100011001
001111101011
011011011001
011000100111
010011100110
010011100110
001100001101
010101100100
011000111110
```

Listing 8.40: File contents: output2.txt

The Matlab function declaration of **my_component2.m** (click here for the source file) is shown below where three parameters are passed to the function. The parameter x is the input port that takes in the *my_input* values from the file *input1.txt*. The word length and the number of fractional bits of x are extracted in lines 39–40. The parameter *addresses* is the input port that takes in the *my_rom_address* values (W=8) from the file *input2.txt*. The parameter *rom* is the ROM variable *array* that was saved in *ROM.mat* when the ROM memory initialization file was created. This allows Matlab to easily access the ROM memory values just by loading in a .mat file rather than having to parse a .mif file. Note that the word length and number of fractional bits for the ROM memory need to be explicitly set in lines 43–44.

```
27 function z = my_component2(x,addresses,rom)
```

Listing 8.41: Function declaration

The address values are converted into the ROM memory words as seen in Listing 8.42. A value of one is added to the address values since the ROM hardware has a zero index offset and Matlab indexes into arrays with an index offset of one.

```
50 for i=1:length(addresses)
51     rom_values(i) = rom(addresses(i)+1);
52 end
```

Listing 8.42: ROM lookup

In order to account for the two clock cycle latency of the ROM, the array of ROM values is then shifted by two in the array as seen in Listing 8.43 where zeros of the same fixed-point data type are inserted as place holders. In a similar fashion, the input values are also delayed by two to match the VHDL component behavior (line 65).

```
57 f = fi(0,S3,W3,F3);
58 rom_values = [f f rom_values];  % align (shift) vectors to ↲
       ↪account for latency
```

Listing 8.43: ROM latency

The input values are then multiplied by their respective ROM value as seen in Listing 8.44. We allow Matlab to expand the fixed-point product to the full range, which is its default behavior and why we do not set any fimath properties as we did in my_component1.m. We extract the resulting binary string of the full product from the fixed-point object *result*. We then trim the binary string in a similar way to how the VHDL std_logic_vector was trimmed to get the same result. This string is then assigned to a fixed-point object so that we can interpret the results and save it in the z array. The verbose setting is to allow one to see the string results, which is useful during development to debug the indexing, but then is turned off once the indexing for the trimming is correct.

```
68 for i=1:length(x)
69     m1 = delayed_input(i);
70     m2 = rom_values(i);
71     result = m1 * m2;
72     result_bit_string = result.bin;
73     % perform the same signal slicing as in my_component2.↲
           ↪vhd to trim the result
74     left_trim_length = result.WordLength-result.↲
           ↪FractionLength-(W1-F1);
75     result_bit_string(1:left_trim_length) = [];
76     result_bit_string(end-F3+1:end) = [];
77     f.bin = result_bit_string; % put it back into the ↲
           ↪appropriate sized fixed-point object
```

```
    output = f;
    z(i)   = output;   % collect the results
    if verbose == 1
        disp(['i = ' num2str(i) ' ↲
          ↪-----------------------------------'])
        disp(['delayed_input = ' m1.hex ' = ' m1.bin ' = '↲
          ↪num2str(m1)])
        disp(['rom_value     = ' m2.hex ' = ' m2.bin ' = '↲
          ↪num2str(m2)])
        disp(['result        = ' result.hex ' = ' result.bin↲
          ↪ ' = ' num2str(result)])
        disp(['result        = ' output.hex ' = ' output.bin↲
          ↪' = ' num2str(output)])
    end
end
```

Listing 8.44: Computation of my_component2

8.4.5 Performing Verification with the Matlab Script my_verification2.m

When the ModelSim simulation is run, the testbench *my_component2_tb.vhd* produces the results found in the files *output1.txt* and *output2.txt*, which can be seen in Listings 8.39 and 8.40. We want to compare what is in these files with what *my_component2.m* produces, and we automate the comparison process using the Matlab script **my_verification2.m** (click here for the source file). This is similar to the process described in Sect. 8.3.5 Performing Verification with the Matlab Script my_verification1.m (page 107). The primary difference is reading in the additional input and output files and performing two separate verifications.

The first verification (lines 154–189) checks that the ROM values in *output2.txt* are consistent with the memory initialization file contents that were saved as a .mat file.

The second verification (lines 202–230) checks that the Matlab function *my_component2.m*, using the same inputs, *input1.txt* (test_vectors) and *input2.txt* (address_vectors), produces the same results as the VHDL component did in *output1* (vhdl_vectors).

When the Matlab code agrees with the VHDL ModelSim simulation over the entire input coverage, one is pretty certain that both the Matlab and VHDL codes are correct.

8.4.6 Running the Example 2 Verification

Here are the steps to take to run this verification example:

Step 1: Create a \Ex2 directory in Windows and download all the Example 2 verification files from GitHub (click here for the files), with the exception of the *input1.txt*, *input2.txt*, *output1.txt*, and *output2.txt* files (you will be creating these), and put them in this new directory.

Step 2: Download the files *ROM.vhd*, *ROM.mif*, and *ROM.mat* from GitHub (click here for the files) and place them in the \Ex2 directory.

Step 3: Open up **my_test_vectors2.m** in Matlab and run the script. It will create two new files ***input1.txt*** and ***input2.txt***, which will be different from the ones found on GitHub because the test vectors in these files are randomly generated each time.

Step 4: Compile the Quartus **altera_mf** Simulation Library:

a. Create a folder for the library: `\Ex2\simlib`.
b. Open Quartus and open the EDA Simulation Library Compiler (Tools → Launch Simulation Library Compiler).
c. Under *Tool Name*, select **ModelSim**.
d. Under *Executable Location*: browse to where you installed ModelSim, e.g., `C:\Modeltech_pe_edu_10.4a\win32pe_edu`.
e. Under *Library families*, add the **Cyclone V** device (using >). Note: The actual device family is not really important since we will just be simulating with ModelSim.
f. Under *Library Language*, select **VHDL**.
g. Under *Output Directory*, browse to the folder you created (i.e., `\Ex2\simlib`). **Note:** You will need this path information when compiling in ModelSim so save this path information. You also might want to put this library folder in a more general place since you will be using it for other simulations (e.g., in Lab 1).
h. Click the *Start Compilation* button.

Step 5: Open ModelSim:

a. Create a Project by selecting *File → New → Project.*
   i. Name the project Ex2.
   ii. Under *Project Location*, browse to the \Ex2 directory you just created.
   iii. Keep the Default Library Name as "work."
   iv. Keep the "Copy Settings From" as is (keep default setting).
   v. Click OK.
b. When it asks to *Add items to the Project*, click on *Add Existing File*, browse to \Ex2, and add the four VHDL files. Click OK and close the *Add items to the Project* window.
c. Modify the compilation order if it needs to be modified by going to *Compile → Compile Order...* and rearranging the compile order (select file and use up/down arrows to change the file order). The compile order should be:
   i. **text_util.vhd** (Order 0 in Project tab)

ii. **ROM.vhd** (Order 1 in Project tab)
iii. **my_component2.vhd** (Order 2 in Project tab)
iv. **my_component2_tb.vhd** (Order 3 in Project tab)

d. Tell ModelSim where the altera_mf library is. In ModelSim and in the Transcript (command) window, issue the following **vmap** command that defines a mapping between a logical library name (altera_mf) and the associated folder that contains the library.

```
%\myprompta% vmap altera_mf c:/..../simlib/
    vhdl_libs/altera_mf
```

Listing 8.45: ModelSim Tcl command: vmap altera_mf

Note 1: For the vmap command, the directories need to be separated by forward slashes ("/") rather than back slashes ("\"), since it follows a Linux path specification rather than Windows. If you cut and paste the path from Windows, you will need to change this; otherwise, ModelSim will complain.

Note 2: Replace /..../ with the appropriate path on your computer system to the location of the altera_mf simulation library.

Note 3: Create a ModelSim Do file to automate commands such as this vmap command. In ModelSim, go to File → New → Source → Do, and type the vmap command into the DO file. Save the file as **setup.do** To reissue this command, when you reopen ModelSim, you just type **do setup.do** in the command window. You can also add commands to add signals to the WAVE window. Notice that when you add a signal, that command was echoed to the command window. You can then just cut and paste this command into your .do file to automate any setup for the next ModelSim session.

```
%\myprompta% do setup.do
```

Listing 8.46: ModelSim Tcl command: do setup.do

e. Compile the files (*Compile → Compile All*).
Note: If the compile fails (a red x will show up in the Status column), you can see the errors that have been logged in the compile summary (*Compile → Compile Summary*).

f. To run the simulation,
   i. Select the Library window in ModelSim (click Library tab above Transcript window).
   ii. Expand the work folder.

iii. Double click on my_component2_tb (this will open the Wave window).

iv. To see the signals you want to see, select the component hierarchical level in the sim–Default window, then click on a signal name in the Objects window, and either right click and select *Add Wave* or drag the signal into the first column of the Wave window. Add the following signals from the *my_component_tb* instance. You can also add these to the setup.do file by copying the commands in the Transcript window after you add the signals:
   - clk
   - input1_signal
   - input2_signal
   - output1_signal
   - output2_signal

   Add the following signals from the *my_component_tb/DUT* instance so that you can see the results and the delay of the output:
   - my_product
   - my_product_trimmed
   - my_output

v. Set the simulation time to 500 ns. (This needs to be long enough to run all the stimulus vectors through the simulation, so if you change the number stimulus vectors, you will need to change this value.)

vi. Run the simulation (button to the right of the simulation time window). The files ***output1.txt*** and ***output2.txt*** should be created that contain the result vectors.

Step 6: Open up **my_verification2.m** in Matlab and run the script. You should see the message *Verification Succeeded!* in the Matlab Command Window.

8.5 Homework Problems

Problem 8.1

Perform the Example 1 verification as described in Sect. 8.3.6 but with **all** the modifications listed below where you operate on *signed fixed-point numbers* rather than on unsigned integers. The verification should result in no errors signifying that both your VHDL and Matlab codes are correct. **Note:** Specific values for **(Vadd, W, F, S)** may be individually assigned to you by the instructor.

Modification 1: Change the vector bit width **W** from 16 bits to 24 bits.
Modification 2: Change the number of fractional bits **F** from zero to 12 bits.

Modification 3: Change the signedness from unsigned to signed, i.e., change **S** from zero to one.

Modification 4: Instead of adding 1 to the input signal in my_component1.vhd, add the value **Vadd** = 4.125 to the input signal. This will require creating a constant VHDL signal that is initialized to 4.125 and where the binary point is aligned with the input signal in order to add properly. You will also need to modify my_component1.m to reflect this change in the Matlab function.

Problem 8.2

Perform the Example 2 verification as described in Sect. 8.4.6 but with **all** the modifications listed below. This will involve creating a new ROM IP component as described in Sect. 6.4.1 Creating a ROM IP Component (page 78). The verification should result in no errors signifying that both your VHDL and Matlab codes are correct. **Note:** Specific values for **(ROM_A_W, ROM_Q_W, ROM_Q_F, ROM_Q_S, WORD_W, WORD_F, DELAY)** may be individually assigned to you by the instructor:

Modification 1: Change the ROM **address** size **ROM_A_W** from 8 bits to 10 bits.

Modification 2: Change the ROM **word** size **ROM_Q_W** from 12 bits to 16 bits.

Modification 3: Change the number of **fractional** bits in the ROM word **ROM_Q_F** from 11 bits to 10 bits.

Modification 4: Change the **signedness** of the ROM word from unsigned to signed, i.e., **ROM_Q_S** from zero to one.

Modification 5: Create a new **memory initialization file** (**.mif** and associated **.mat** files) for the ROM that contains **random** values with the (ROM_Q_W, ROM_Q_F, ROM_Q_S) specifications given above.

Modification 6: Change the input word size **WORD_W** from 16 bits to 24 bits.

Modification 7: Change the number of fraction bits in the input word **WORD_F** from 8 bits to 16 bits.

Modification 8: Change the output **DELAY** from 4 clock cycles to 3 clock cycles.

Chapter 9
Introduction to Linux

9.1 The Linux View of Memory

Linux is a modern operating system that presents to each process, thread, and user space program an abstracted view of physical memory where each process and program has virtually unlimited uniformly contiguous memory. Thus the addresses that user space programs use are *virtual addresses* that are different from the physical address where the memory objects actually reside. Furthermore, there is a division between user space memory and kernel memory. Since Linux allows each process to pretend that it has unlimited contiguous memory, it needs to manage the following **Linux Address Types**:

- **Virtual Addresses** The addresses seen by user space programs. In 32-bit systems that have a maximum of 4 GB of memory, each user space process is limited to the lower 3 GB of virtual address space. The Linux kernel uses the top 1 GB of virtual address space.
- **Physical Addresses** The addresses used by the CPU to access the system's physical memory.
- **Kernel Logical Addresses** The normal address space of the kernel. Kernel logical addresses and the associated physical addresses differ only by a constant offset. Kernel address space is the area above *CONFIG_PAGE_OFFSET* = *0xC0000000* (3 GB), which is the default value for 32-bit CPUs but is configurable at kernel build time.
- **Kernel Virtual Addresses** Similar to logical addresses in that, they are a mapping from a kernel-space address to a physical address. Kernel virtual addresses do not necessarily have the linear, one-to-one mapping to physical addresses that characterize the logical address space, however.
- **Bus Addresses** The addresses used between peripheral buses and memory.

R. K. Snider, *Advanced Digital System Design using SoC FPGAs*,
https://doi.org/10.1007/978-3-031-15416-4_9

When each user space process needs to access memory, it uses the virtual address that the Linux kernel has given it. However, this virtual address needs to be translated to the physical address where the object being accessed actually resides in physical system memory. This translation is automatically performed by the *Memory Management Unit* (**MMU**) as shown in Fig. 9.1. Since this address translation occurs with *every* memory access, the MMU has its own cache called the *Translation Lookaside Buffer* (**TLB**) to speed up this translation process.

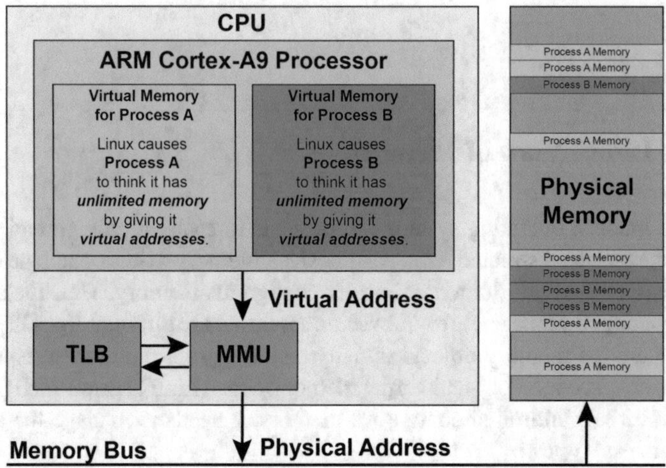

Fig. 9.1: The Linux operating system manages the virtual address spaces for all the processes (who each think they own all the memory in the world) and the allocation of real memory to virtual memory. Address translation hardware in the CPU, often referred to as a Memory Management Unit (MMU), automatically translates virtual addresses to physical addresses. This is done for each memory unit that is called a page, which is typically 4K bytes in size. The Translation Lookaside Buffer (TLB) is a cache that speeds up the page address translations

The address translation that occurs by the MMU is illustrated in Fig. 9.2. Linux keeps track of all the 4K memory pages in a Page Table that is indexed by the virtual page number. The virtual page number is used to find the associated page table entry that has the physical page number. If the virtual page number is not found in the TLB cache, then it is a cache miss, and the TLB is updated to include the new virtual page number. There are additional bits in the TLB that are associated with each page, and these bits contain the following information about the memory page:

- A **caching bit** is used to indicate that the processor should bypass the cache when accessing memory. This is important when the memory is a register associated with memory-mapped hardware. The register cannot be cached since it can change value at any time, and the change would not be reflected in the data cache.

- A **dirty (or modified) bit** indicates whether the page has been modified since it was brought into memory.
- A **present bit** indicates whether this page is in physical memory or on disk (i.e., it has been swapped out).
- A **reference bit (accessed bit)** is used to track whether a page has been accessed, and is useful in determining which pages are popular and which pages can be replaced in the cache.
- A **protection bit** is used to indicate whether the page can be read from, written to, or executed from. Accessing a page in a way not allowed by these bits will generate a trap to the OS.
- A **valid bit** is common to indicate whether the particular translation is valid; for example, when a program starts running, it will have code and heap at one end of its address space, and the stack at the other. All the unused space in between will be marked invalid, and if the process tries to access such memory, it will generate a trap to the OS that will likely terminate the process.

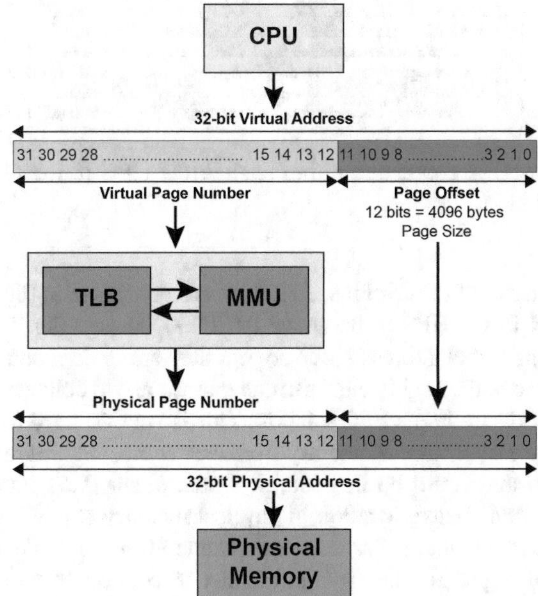

Fig. 9.2: The Memory Management Unit (MMU) along with the Translation Lookaside Buffer (TLB) automatically translates virtual addresses to physical addresses

There is a MMU for each CPU in the Cyclone V SoC FPGA as seen in Fig. 9.3. Microcontrollers typically do not have a MMU, so this is the primary reason why you do not see Linux running on microcontrollers, even 32-bit ones. Without a MMU and associated TLB, performance drastically suffers.

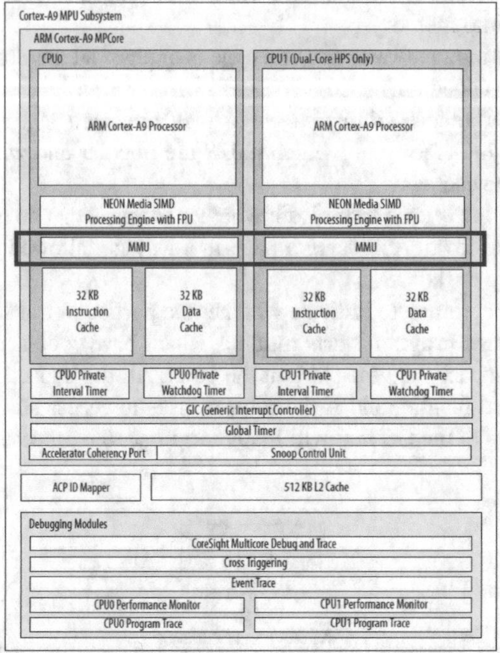

Fig. 9.3: The MMU is contained within each ARM CPU in the Cyclone V HPS. (figure from [1])

All the peripherals of the Cyclone V HPS, which are shown in Fig. 9.4, have been memory-mapped in the HPS to be above 0xC0000000 (see Fig. 7.1). This places them in the Linux kernel address space so that they are under control of the Linux kernel. This includes the Lightweight Bridge that we use to connect to the registers in our custom hardware in the FPGA fabric. This is why we need to create a Linux device driver. If we try to access memory from user space, the program will be using virtual addresses that could be mapped anywhere *below* 0xC0000000, and these memory pages could change location in physical memory at any time if the Linux kernel wishes to move them. We need a loadable kernel module for our custom hardware control registers where we tell Linux to connect to a specific physical memory address. Kernel modules are covered in Sect. 9.3 Kernel Modules (page 135).

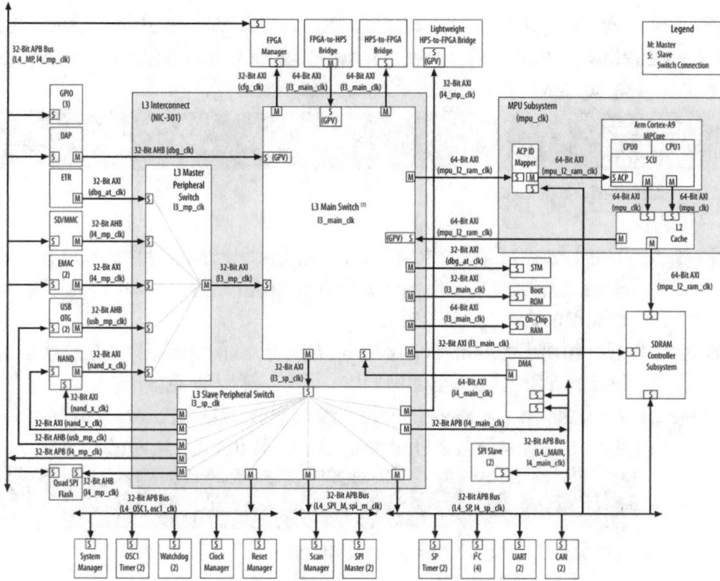

Fig. 9.4: The peripherals of the Cyclone V SoC FPGA are connected through the L3 Master Peripheral Switch and memory-mapped to be above 0xC0000000, putting them in the Linux kernel address space. (figure from [2])

9.2 Cross Compiling the Linux Kernel

Loadable kernel modules need to be built for the specific kernel version of Linux that they will be loaded into. This means that the Linux kernel version that is contained in **zImage** that gets loaded at boot time needs to match the kernel version that the kernel module was compiled for. To ensure that both the kernel version and your module version are the same, we will build both. This means we will simply ignore the issues that can arise when trying to insert a loadable kernel module into a kernel that has a different version number.

This section will give instructions on how to build a new kernel image (zImage) that will be loaded by U-boot at boot time. We will then know that when we compile our device drivers (see Sect. 9.6) that they will be built against the correct Linux kernel version that is running on the DE10-Nano board.

9.2.1 Packages Needed

The packages that need to be installed in order to cross compile the Linux kernel are listed below and some should already be installed. Remember, it is always good to check for updates in Ubuntu with the command ***sudo apt update*** before installing packages. **Note:** Running *apt install* again after a package has already been installed will not hurt anything. You will just get a message that it has already been installed (i.e., it will say <package> is already the newest version).

```
$ sudo apt update
```

Listing 9.1: sudo apt update

Package 1: The Linaro GCC tools. Instructions for installation can be found in Sect. 11.1.4.1 Configure and Set up the Linaro GCC ARM Tools in Ubuntu VM (page 238).

Package 2: The **build-essential** package. See item Step 5: Install and Configure Needed Software in the Ubuntu VM (page 205).

Package 3: The **bison** package, which is a general-purpose parser generator. Install bison in the Ubuntu VM with the command:

```
$ sudo apt install bison
```

Listing 9.2: sudo apt install bison

Package 4: The **flex** package, which is a fast lexical analyzer generator. Install flex in the Ubuntu VM with the command:

```
$ sudo apt install flex
```

Listing 9.3: sudo apt install flex

Package 5: The **ncurses-dev** package, which is a library and API for text-based user terminals Install ncurses-dev in the Ubuntu VM with the command:

```
$ sudo apt install ncurses-dev
```

Listing 9.4: sudo apt install ncurses-dev

9.2.2 Get the Linux Source Repository and Select the Git Branch

Step 1: Go to the directory where you want `/linux-socfpga` created, which will contain the Linux socfpga kernel source repository (e.g., you can put it in your home directory, so issue the command: cd ~).

Step 2: Using Git, download Intel's SoC FPGA kernel repository:

```
$ git clone https://github.com/altera-↲
  ↪opensource/linux-socfpga
```

Listing 9.5: git clone
https://github.com/altera-opensource/linux-socfpga

This will install the repository into the `/linux-socfpga` directory, which will be created where you issued the **git clone** command. If you have already cloned the repo before, you will want to update it by issuing **git pull** (issue the command in the `/linux-socfpga` directory):

```
$ git pull https://github.com/altera-↵
    ↪opensource/linux-socfpga
```

Listing 9.6: git pull https://github.com/altera-opensource/linux-socfpga

Step 3: Determine the git branches that exist by issuing the command:

```
$ git branch -a
```

Listing 9.7: git branch -a

In the output as shown in Fig. 9.5, you can see what the default branch version is by looking for the branch pointed to by the string `remotes/origin/HEAD ->`. As of this writing, this branch is `socfpga-5.4.124-lts`. However, we want to use the latest LTS (long-term support) branch, so let us change to this branch. In the list of branches, note the branch that has the latest version number and that also has "lts" at the end. In this example, it is *socfpga-5.10.60-lts*.

```
* socfpga-5.4.124-lts
  remotes/origin/HEAD -> origin/socfpga-5.4.124-lts
  remotes/origin/socfpga-5.10.50-lts
  remotes/origin/socfpga-5.10.60-lts
  remotes/origin/socfpga-5.11
  remotes/origin/socfpga-5.12
  remotes/origin/socfpga-5.13
  remotes/origin/socfpga-5.4.124-lts
```

Fig. 9.5: Git branches in the socfpga Linux source repository

Step 4: Change to the desired branch (socfpga-5.10.60-lts in this example) using the git checkout command:

```
$ git checkout socfpga-5.10.60-lts
```

Listing 9.8: git checkout socfpga-5.10.60-lts

The result (using the command: git branch -a) is shown in Fig. 9.6. Notice that socfpga-5.10.60-lts is now marked with a leading asterisk (*), signifying that it is the current branch.

```
* socfpga-5.10.60-lts
  socfpga-5.4.124-lts
  remotes/origin/HEAD -> origin/socfpga-5.4.124-lts
  remotes/origin/socfpga-5.10.50-lts
  remotes/origin/socfpga-5.10.60-lts
  remotes/origin/socfpga-5.11
  remotes/origin/socfpga-5.12
  remotes/origin/socfpga-5.13
  remotes/origin/socfpga-5.4.124-lts
```

Fig. 9.6: The newly selected git branch socfpga-5.10.60-lts

You can also see what branch you are using by issuing the command:

```
$ git branch
```

Listing 9.9: git branch

and the branch that you are using will be marked with a leading asterisk (*) as shown in Fig. 9.7.

```
* socfpga-5.10.60-lts
  socfpga-5.4.124-lts
```

Fig. 9.7: Checking what git branch you are using

9.2.3 Configuring the Kernel

Step 1: Navigate to the `/linux-socfpga` directory (the git repo you just downloaded).

Step 2: Configure your kernel with the default socfpga configuration options by issuing the command:

```
$ make ARCH=arm socfpga_defconfig
```

Listing 9.10: make ARCH

Note 1: ARCH=arm is an environment variable that tells `make` we are cross compiling for the ARM CPU.
Note 2: socfpga_defconfig contains all of the default `Kconfig` values for socfpga, as defined by Intel and the community. This file is located in `/linux-socfpga/arch/arm/configs/`.

Note 3: This creates the .config file that is used by `make` to configure the kernel when it compiles the kernel.

Step 3: You can see the contents of the .config file by issuing the command:

```
$ cat .config
```

Listing 9.11: cat .config

Since we will be creating kernel modules, we will want to add additional debugging support into the kernel. Configuration options for debugging that can be turned on in the kernel can be found (here). To make additional kernel configuration changes, navigate to the `/linux-socfpga` directory and type the command:

```
$ make ARCH=arm menuconfig
```

Listing 9.12: make ARCH

This pops up the Kernel Configuration panel as shown in Fig. 9.8.

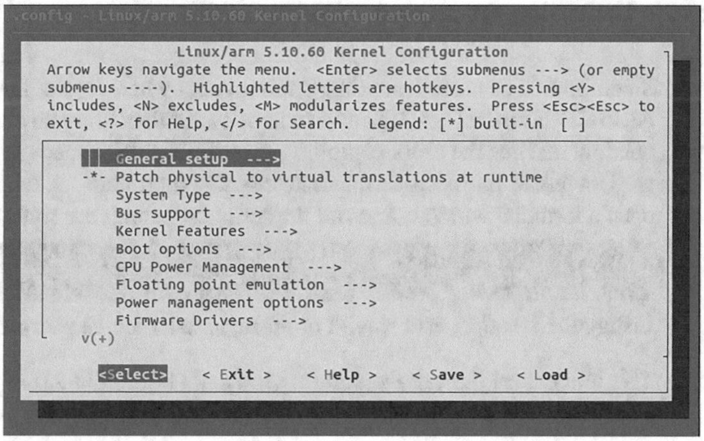

```
.config - Linux/arm 5.10.60 Kernel Configuration
                  Linux/arm 5.10.60 Kernel Configuration
 Arrow keys navigate the menu.  <Enter> selects submenus ---> (or empty
 submenus ----).  Highlighted letters are hotkeys.  Pressing <Y>
 includes, <N> excludes, <M> modularizes features.  Press <Esc><Esc> to
 exit, <?> for Help, </> for Search.  Legend: [*] built-in  [ ]

        General setup  --->
    -*- Patch physical to virtual translations at runtime
        System Type  --->
        Bus support  --->
        Kernel Features  --->
        Boot options  --->
        CPU Power Management  --->
        Floating point emulation  --->
        Power management options  --->
        Firmware Drivers  --->
    v(+)

     <Select>    < Exit >    < Help >    < Save >    < Load >
```

Fig. 9.8: Configuring the Linux kernel using menuconfig

Let us enable the following kernel debug options using menuconfig, which you just opened:

- CONFIG_DEBUG_KERNEL
- CONFIG_DEBUG_SLAB
- CONFIG_DEBUG_DRIVER
- CONFIG_MODULE_FORCE_UNLOAD

The easiest way to find where these options are located in menuconfig is to press the / key, which brings up the search window. Typing in

CONFIG_DEBUG_KERNEL (case does not matter, and you can type partial strings) shows that DEBUG_KERNEL is already enabled ([=y]).

Pressing / and typing in debug_slab show that it is not enabled ([=n]) and that it depends on SLAB as well, which is not enabled.

This means that we have to first enable SLAB, so type slab in the search window, which tells us that we have to enable it in menuconfig → General setup → Choose SLAB allocator → SLAB. Notice the (1) next to the choice. Pressing **1** takes us directly to this option. Once we have enabled SLAB, we press the / key and type debug_slab, which tells us that we enable it at menuconfig → Kernel hacking → Memory Debugging → Debug slab memory allocations. Pressing the number key associated with this choice will take us directly to this option, where we enable it.

Now enable the rest of the kernel debug options in the list above and any other kernel debugging options that you want to enable. Then save and exit from menuconfig.

9.2.4 Cross Compiling the Kernel

Step 1: Since we will be cross compiling a fair amount when developing for the ARM CPUs on the DE10_Nano, it will be convenient to create a terminal window that has the cross compiling export variables already set up for us. Download the bash script **arm_env.sh** (click here for the script), open a terminal window, and run the script by issuing the command:

```
$ source arm_env.sh
```

Listing 9.13: source arm_env.sh or . arm_env.sh (source shortcut)

This will export the variables ARCH=arm and CROSS_COMPILE= /usr/bin/arm-linux-gnueabihf-, and it will change the prompt as shown in Fig. 9.9 so that we know that this terminal window is set up for cross compiling since it has the prefix *arm|* followed by the directory name.

```
arm|~/linux-socfpga >>
```

Fig. 9.9: Terminal window prompt after running arm_env.sh that signifies that the terminal window is set up for cross compiling

Step 2: Change the directory to where the kernel source is located in the window that is set up for cross compiling:

```
arm|~» cd ~/linux-socfpga/
```

Listing 9.14: cd /linux-socfpga/

Step 3: Compile the Linux kernel by issuing the command:

```
arm|~/linux-socfpga/» make -j$(nproc)
```

Listing 9.15: make -j$(nproc)

Note 1: The **-j** flag specifies the number of simultaneous jobs, which we set to the number of logical processors on your machine with the **nproc** command, which results in faster compile times.
Note 2: Compiling the kernel can take a long time. On an i9-9900k (16 logical processors) with 32 GB of RAM, compilation took 1.5 minutes. It will likely take 10+ minutes on your virtual machine.
Note 3: If the shell scrip arm_env.sh was not run, you would need to enter the command:

```
$ make ARCH=arm CROSS_COMPILE=arm-linux-↵
    ↪gnueabihf- -j$(nproc)
```

Listing 9.16: Make command specifying cross compile export variables

Step 4: Copy the kernel image **zImage** from **~/linux-socfpga/arch/arm/boot** to **/srv/tftp/de10nano/kernel**. After compilation, the kernel image is located at ~/linux-socfpga/arch/arm/boot/zImage. The zImage is a self-extracting, compressed version of the kernel image that gets loaded by the U-boot bootloader. Since we are having U-boot to download the zImage over tftp, you need to copy the new zImage file to your tftp directory: /srv/tftp/de10nano/kernel/.

9.3 Kernel Modules

The Linux kernel is a modular operating system where kernel modules can be dynamically loaded into and unloaded from the kernel. This allows the capability of the kernel to be extended without rebooting the kernel. You can see what kernel modules exist by going to /lib/modules/<kernel_version>/kernel, which you can do by issuing the command:

```
$ cd /lib/modules/$(uname -r)/kernel
```

Listing 9.17: cd /lib/modules/$(uname -r)/kernel

Note 1: The command **uname -r** returns the kernel version number. This is then used in a command substitution **$(uname -r)**, which places the kernel version string into the path string.

Note 2: Kernel modules are specific to the Linux kernel version. This is why there is a directory under `/lib/modules` that has a specific kernel version. This is also why you cannot load kernel modules that have been compiled for a specific kernel into a kernel with a different version number.

Device driver modules that exist (they have been upstreamed and officially merged into the Linux source tree) can be seen by issuing the command:

```
$ cd /lib/modules/$(uname -r)/kernel/drivers
```

Listing 9.18: cd /lib/modules/$(uname -r)/kernel/drivers

You will notice that there are many types of device drivers in the Linux kernel! We will start the process of developing our own device driver, which will be a loadable kernel module. We will also keep our source code (and compiled module) separate from the Linux source tree. This means that our device driver will consider an **out-of-tree** build.

9.3.1 Loadable Kernel Modules

A **kernel module** is an object code file that extends the capability of the Linux kernel. These files have the extension **.ko** that stands for *kernel object*. We will used them to add device drivers for our new hardware when we create custom components in the FPGA fabric. Kernel modules that can be added at run time are known as **loadable kernel modules**. We will be taking this approach as compared to compiling our device driver code directly into the Linux kernel. Thus for our purposes, when we speak of developing device driver code, we will ultimately be creating a .ko file that will get loaded into the Linux kernel.

Adding a device driver kernel module is the way that user space programs can interact with hardware that have registers at very specific physical memory addresses and where the access to those memory locations is controlled by Linux. It is the portal between our custom hardware in the FPGA fabric and our software running in Linux user space where one does not have root access.

A simple kernel module is shown in Listing 9.19 (click here for the source file) that shows a minimal functioning kernel module, and yes, it is a "Hello World" kernel module. A kernel module needs to have at least two functions. An initialization function that is called when the kernel module is loaded and an exit function that is called when the kernel module is unloaded. The initialization function typically handles tasks such as setting default register values, registering sysfs entries, and enabling the device. The exit function typically frees any allocated memory and safely closes down any hardware it is controlling.

```
11 #include <linux/init.h>
12 #include <linux/module.h>
13
14 static int __init my_kernel_module_init(void) {
15     printk(KERN_ALERT "Hello, Linux Kernel World!\n");
16     return 0;
17 }
18 module_init(my_kernel_module_init);
19
20 static void __exit my_kernel_module_exit(void) {
21     printk(KERN_ALERT "Goodbye, Linux Kernel World!\n");
22 }
23 module_exit(my_kernel_module_exit);
24
25 MODULE_DESCRIPTION("Hello World Kernel Module");
26 MODULE_AUTHOR("myFirstName myLastName");
27 MODULE_LICENSE("Dual MIT/GPL");
28 MODULE_VERSION("1.0");
```

Listing 9.19: Hello world kernel module

Let us examine the initialization part of the kernel module shown in Listing 9.20 and parse Line 14.

```
14 static int __init my_kernel_module_init(void) {
15     printk(KERN_ALERT "Hello, Linux Kernel World!\n");
16     return 0;
17 }
18 module_init(my_kernel_module_init);
```

Listing 9.20: Initializing the kernel module

The `static` keyword means that the function is only visible to this file, i.e., other files will not be able to access this function since the scope of the function is local to this file. The `int` keyword means that the function returns an integer type. The `__init` string means the following: The double underscore __ means that this name has meaning for the compiler and is defined in init.h (click for reference), which is why the header `linux/init.h` is included on line 11.

The `__init` macro is placed before the function name **my_kernel_module_init**, and it tells the compiler that the function is used only during initialization so that it can free up memory once the function is finished. The keyword **void** in the function argument list means that the function takes no arguments. The opening brace { starts the beginning of the function body.

Let us now examine Line 15. You will notice that there is a **printk** rather than the typical *printf* that you typically use when programming in C. This is because we are programming for the Linux kernel, and we must make the distinction between **kernel space** and **user space**. *Kernel space* is where kernel code lives and runs, whereas *user space* is where normal applications run. There are important differences between kernel space and user space:

- Kernel space has its own memory space, which is not accessible from user space.
- Kernel code executes at the highest CPU privilege level. CPUs have multiple privilege levels, which are used to enforce protection rings. In short, privilege levels place restrictions on what operations/instructions processes have access to.
- User programs execute at the lowest CPU privilege level.
- The kernel does not have access to the C standard library; thus you cannot use any of those functions. However, the kernel reimplements some standard library functions. printk(), for example. Look through the Linux Kernel API, especially the *Basic C Library Functions* section, to see what is available.
- Floating point arithmetic is not allowed in kernel space. This is largely because of the overhead of having to save and restore the floating point unit's state on every transition between user space and kernel space.
- The kernel's stack is very small. Consequently, you should dynamically allocate any large data structures.

Now getting back to **printk**, it is for printing messages from the kernel and is similar to printf(), but with several notable differences, and has the form:

```
printk(KERN_INFO "Message: %s\n", arg);
```

First there is a log level *KERN_INFO* that determines if the message will be printed to the console or not. If it has higher priority (lower value) than the *console_loglevel*, it will be printed to the console. In any event, *all* messages are printed to the kernel log buffer `/dev/kmsg`, which can be read using the Linux command **dmesg**. We are using the KERN_INFO log level **KERN_ALERT**, which has high priority (only KERN_EMERG has higher priority). Click here for more information on printk() and log levels. To check what the current *console_loglevel* is, type the command:

```
$ cat /proc/sys/kernel/printk
```

Listing 9.21: cat /proc/sys/kernel/printk

The four numbers that are shown are the *current, default, minimum*, and *boot-time* log levels. The *current* level can be changed by the command:

```
$ echo 8 > /proc/sys/kernel/printk
```

Listing 9.22: echo 8 > /proc/sys/kernel/printk

where the *console_loglevel* value of 8 will cause all messages to be printed to the console.

A second difference from printf() is that printk() cannot use floating point specifiers (floating point is not supported in the Linux kernel). Format specifiers that can be used in printk are found here.

In line 18 of Listing 9.20, *module_init()* takes the function to be run when the kernel module is loaded into the Linux kernel, which in this case is the function *my_kernel_module_init*.

In a similar fashion, the exit function that runs when the kernel module is unloaded is shown in Listing 9.23.

```
static void __exit my_kernel_module_exit(void) {
    printk(KERN_ALERT "Goodbye, Linux Kernel World!\n");
}
module_exit(my_kernel_module_exit);
```

Listing 9.23: Exiting the kernel module

The kernel **module macros** are shown in Listing 9.24 and by convention are placed at the end of the module's source file. At a minimum, you are required to use **MODULE_AUTHOR()**.

```
MODULE_DESCRIPTION("Hello World Kernel Module");
MODULE_AUTHOR("myFirstName myLastName");
MODULE_LICENSE("Dual MIT/GPL");
MODULE_VERSION("1.0");
```

Listing 9.24: Module macros

These macros are defined in `<linux/module.h>` (click here). The descriptions of the modules used are:

- `MODULE_DESCRIPTION()`: Description of what the module does.
- `MODULE_AUTHOR()`: Who wrote the module.
- `MODULE_LICENSE()`: Specifies which license applies to your code. If a preferred free license is not used, the module is assumed to be proprietary, which will taint the kernel; we do not need to be particularly worried about these issues as proprietary kernel modules can be loaded, but your code has to be GPL-licensed if you want it to be merged into the mainline kernel. If you want a more permissive license, which we are suggesting, you can dual-license your module, e.g., `MODULE_LICENSE("Dual MIT/GPL")`. See the kernel license-rules for more info on the accepted license strings.
- `MODULE_VERSION()`: Code version number.

9.3.2 Cross Compiling the Kernel Module

Kernel modules must be built with the kernel's build system, known as kbuild. Modules can be either built in-tree (within the kernel's source tree) or out-of-tree (externally). We will be building our modules out-of-tree. kbuild uses a special build file syntax, described in the Linux Kernel Makefiles documentation. All we need at this point are "goal definitions" in the Makefile. For building a module that only has a single source file, our goal definition will look like this: `obj-m := <module-name>.o`, where `<module-name>` is the name of your source file (without an extension).

The **Makefile** we will use to compile hello_kernel_module.c is shown in Listing 9.25 (click here for source file).

```
1  ifneq ($(KERNELRELEASE),)
2  # kbuild part of makefile
3  obj-m  := hello_kernel_module.o
4
5  else
6  # normal makefile
7
8  # path to kernel directory
9  KDIR ?= ~/linux-socfpga
10
11 default:
12     $(MAKE) -C $(KDIR) ARCH=arm CROSS_COMPILE=arm-linux-gnueabihf- M=$$PWD
13
14 clean:
15     $(MAKE) -C $(KDIR) M=$$PWD clean
16 endif
```

Listing 9.25: Makefile for hello_kernel_module.c

The only line that needs modification is line 9 since this must use the correct path to where you installed `/linux-socfpga` (see Sect. 9.2.2 Get the Linux Source Repository and Select the Git Branch (page 130)).

In the directory that contains hello_kernel_module.c and the Makefile, run make:

```
$ make
```

Listing 9.26: make

This will create hello_kernel_module.**ko**, which is a kernel object file known as a *Loadable Kernel Module* (**LKM**) that now can be inserted into the Linux kernel.

To see information related to the kernel module, use the **modinfo** command:

```
$ modinfo hello_kernel_module.ko
```

Listing 9.27: modinfo hello_kernel_module.ko

In the output, you will see the information that you added using the module macros in the source file. In addition, you will see the **vermagic** string. If your kernel module did not load because it was compiled against a different kernel version, this is how you would check what kernel version the kernel module was compiled against. In our case, they are the same because we compiled both the kernel and kernel module. In the vermagic string, also notice **ARMv7**. This tells us that it has been cross compiled for the ARM CPUs on the DE10-Nano board.

Now we must copy the compiled kernel module (LKM or .ko file) to the rootfs seen by the ARM CPUs on the DE10-Nano board. This is located at `/srv/nfs/de10nano/ubuntu-rootfs/root/`. You need to be root to copy files to the root directory. To browse the `root` directory, you will need to be in a root shell (`sudo -i`).

9.3.3 Inserting the Kernel Module into the Linux Kernel

To insert the kernel module into Linux running on the DE10-Nano board, power up the board and log into Linux using the PuTTY terminal. Since you put the .ko file in the root directory, you should see the .ko file right after logging in as root when using the ls command.

To insert the LKM, use the **insmod** command:

```
$ sudo insmod hello_kernel_module.ko
```

Listing 9.28: sudo insmod hello_kernel_module.ko

To see the message printed out with printk that was in the initialization function, and that was logged in the kernel ring buffer, use the **dmesg** command:

```
$ dmesg
```

Listing 9.29: dmesg

To remove the LKM, use the **rmmod** command:

```
$ sudo rmmod hello_kernel_module
```

Listing 9.30: sudo rmmod hello_kernel_module

To see the exit message printed out with printk that was logged in the kernel ring buffer, use the **dmesg** command.

9.4 Device Trees

Device Trees tell Linux what hardware it is running on. When Linux boots on the DE10-Nano board, two Linux files are loaded by U-Boot. The kernel image (zImage) contains all the Linux code in a binary image and the device tree blob (.dtb) or binary file. The .dtb file is how Linux knows what hardware it is running on. When we create new custom hardware in the FPGA fabric, we have to tell Linux that this hardware exists by making an additional entry to the device tree for the new hardware. This is done in the device tree source (.dts) file that can be edited with a text editor, which is then compiled to the .dtb file. Then, to be able to actually talk to this hardware, we write a device driver for it. Thus the device tree tells Linux that our new hardware exists and the device driver allows us to use it from user space.

9.4.1 Device Tree Basics

The root of the device tree is designated as / (just as the root directory in the Linux file system is designated), and everything contained in the tree is enclosed by curly braces {} as shown in Listing 9.31, which shows the first part of *socfpga.dtsi*. Nodes in the tree describe hardware that the Linux kernel needs to be informed about. Nodes are identified by their names and contain associated node information within their curly braces {}. This node information can be other nodes (sub-nodes) in a hierarchically fashion. Notice that the node *cpus* on line 21 contains two nodes called *cpu0* (line 26) and *cpu1* (line 32) that describe the dual ARM core of the Intel Cyclone V SoC FPGA.

```
8  / {
9     #address-cells = <1>;
10    #size-cells = <1>;
11
12    aliases {
13       serial0 = &uart0;
14       serial1 = &uart1;
15       timer0 = &timer0;
16       timer1 = &timer1;
17       timer2 = &timer2;
18       timer3 = &timer3;
19    };
20
21    cpus {
22       #address-cells = <1>;
23       #size-cells = <0>;
24       enable-method = "altr,socfpga-smp";
25
26       cpu0: cpu@0 {
27          compatible = "arm,cortex-a9";
28          device_type = "cpu";
29          reg = <0>;
30          next-level-cache = <&L2>;
31       };
32       cpu1: cpu@1 {
33          compatible = "arm,cortex-a9";
34          device_type = "cpu";
35          reg = <1>;
36          next-level-cache = <&L2>;
37       };
38    };
```

Listing 9.31: Device tree root of socfpga.dtsi

Device tree nodes in general are identified with the following syntax:

node_label: node_unit@node_address

as can be seen in Listing 9.32. Let us take the interrupt controller on line 49 as an example. The **node_label:** is **intc:**. Labels are used to reference nodes, which we can see happening on line 42 where the *Performance Monitoring Unit* (**PMU**) (hardware unit that gathers operation statistics on the ARM CPUs) uses the label to reference the interrupt controller as the interrupt parent. The syntax **<&label>** or **<&intc>** in this case creates a *pointer handle* (**Phandle**), which is a pointer to the interrupt controller node on line 49. Labels are for human readable referencing.

After the label on line 49, we have the syntax **node_unit@node_address** given as **intc@fffed000**. The **node_unit** is a name for the hardware component, which can be the same or different from the label but typically has a more hardware labeling orientation. The **node_address** gives the base address of the node on the bus it is connected to. If it is connected to a specific bus, then it would be listed as a sub-node under the bus node. If it is sitting as a node under the root node, then the node address is the general memory address of the system. This is where we will put our custom hardware node.

```
40    pmu: pmu@ff111000 {
41        compatible = "arm,cortex-a9-pmu";
42        interrupt-parent = <&intc>;
43        interrupts = <0 176 4>, <0 177 4>;
44        interrupt-affinity = <&cpu0>, <&cpu1>;
45        reg = <0xff111000 0x1000>,
46              <0xff113000 0x1000>;
47    };
48
49    intc: intc@fffed000 {
50        compatible = "arm,cortex-a9-gic";
51        #interrupt-cells = <3>;
52        interrupt-controller;
53        reg = <0xfffed000 0x1000>,
54              <0xfffec100 0x100>;
55    };
```

Listing 9.32: Device tree node information

Node information (information contained within the node's curly braces {}) has the information presented as:

name = value;

where **name** is a string and where **value** can be an array of strings, numbers, or phandles.

There are several lines of node information that are of particular interest to us when creating custom hardware in the FPGA fabric. The first one has the form:

compatible = "<manufacturer>,<model>"[,"<manufacturer>,<model>"];

and we can see on Line 50 in Listing 9.32 the **compatible** string being given the value **"arm,cortex-a9-pmu"** where the <manufacturer> is **arm** and the <model> is **cortex-a9-gic** (gic stands for general interrupt controller). This is an important line for us because **the compatible string is used by the Linux kernel to "bind" a device driver to that hardware node.** The kernel is told that a hardware component exists by creating a node in the device tree, and it is told what device driver to use for the hardware by the *compatible* string. Thus this string needs to match the associated string in your device driver; otherwise, your custom hardware and associated device driver will not be connected by the kernel, and you will not be able to access the hardware from the device driver.

The next line of interest has the form:

reg = <address length>[,<address length>];

The address and length information contained in the **reg =** property can be variable in length, so we need two additional pieces of information so that we can interpret this information. These properties are **#address-cells** and **#size-cells**. If these are not explicitly given in the current node, this information is inherited from the parent node.

Let us take as an example lines 53–54 in Listing 9.32. This describes where the registers are located for the interrupt controller, and there are two register areas that exist for the controller since there are two <address length> entries. In order to interpret the address for the register locations, we first check to see if *#address-cells* and *#size-cells* have been defined in the intc node. They are not defined in the node, so we must check the parent, which is the root node, and they are defined in the root node on lines 9–10 in Listing 9.31 where they both have a value of 1. A value of 1 for *#address-cells* means that address is defined by a single number. However, we are not done yet because **numbers in the device tree are always 32-bit values** (big-endian). If we need to represent 64 bits, we must use two numbers (**#size-cells=<2>;**). Getting back to our example on Line 53, we know that *address* is defined as a single number (one address-cell that is specified by a size-cell of one) and similarly for *length*. Thus we know that bank1 of registers for the interrupt controller is located at memory address `0xfffed000` and has a span of `0x1000` bytes (4096 bytes). Similarly, we know that bank2 of registers is located at memory address `0xfffec100` and has a span of `0x100` bytes (256 bytes).

When you create a custom hardware component in Platform Designer, the device tree node that you create for your component will take the following form and naming convention:

```
my_label: platform_designer_name@component_address {
    compatible = "my_initials,platform_designer_name";
    reg = <component_address length>;
};
```

Listing 9.33: Device tree node for custom component in platform designer

and will be placed in the device tree root node. We also make the assumption that `#address-cells = <1>;` and `#size-cells = <1>;`, having been defined this way in the device tree root node. We take the *node_unit* as the name of the custom component as it has been entered into Platform Designer. We calculate the *component_address* of the *reg* property as explained in Sect. 7.1.1 Memory Addressing for Registers on the HPS Lightweight Bus (page 88). The length of the register span (in bytes) is determined by the size (the number of bits) of the address signal in the custom component in Platform Designer multiplied by the register size (32 bits):

$$\text{length} = 4x2^{\text{Number of address bits}} \tag{9.1}$$

The convention for the *compatible* string property is to use your initials for the manufacturer string entry and the name of the custom component in Platform Designer as the model string:

compatible = "my_initials,platform_designer_name"; (9.2)

9.4.2 Device Tree Hierarchy

Device tree sources can be monolithic, but they tend to be hierarchical, especially for SoC FPGA devices. If you look in the directory of the linux kernel source, located at `~/linux-socfpga/arch/arm/boot/dts`, you will see several thousand device tree associated files. We are particularly interested in the ones associated with SoC FPGAs. We can see these files using the command:

```
$ cd ~/linux-socfpga/arch/arm/boot/dts
$ ls socfpga*
```

Listing 9.34: Listing the SoC FPGA device tree files

There are three types of device tree files in this directory. The device tree include files have the **.dtsi** extension. The device tree board files have the **.dts** extension that then get compile and turn into binary (blob) device tree files with the **.dtb** extension.

These files are organized in a hierarchy as shown in Fig. 9.10 (left side). The Intel SoC FPGA devices have a base include file **socfpga.dtsi** that describes the hard processor system (HPS) that is common across these SoC FPGA devices. The Cyclone V SoC FPGA devices use the base socfpga.dtsi device tree hardware description by including this file (which is why it is called an include file and why it has the .dtsi extension). The information specific to the Cyclone V SoC FPGA is placed in the **socfpga_cyclone5.dtsi** file, which is also an include file. Thus include files can include other include files. When include files are organized this way, device tree information can be overlaid (added or overwritten) over the information that was pulled in earlier. Then there is specific board information, which in the case of Fig. 9.10 (left side) is the **socfpga_cyclone5_de0_nano_soc.dts** file that is the device tree file for the DE0-Nano-SoC or Atlas-SoC board.

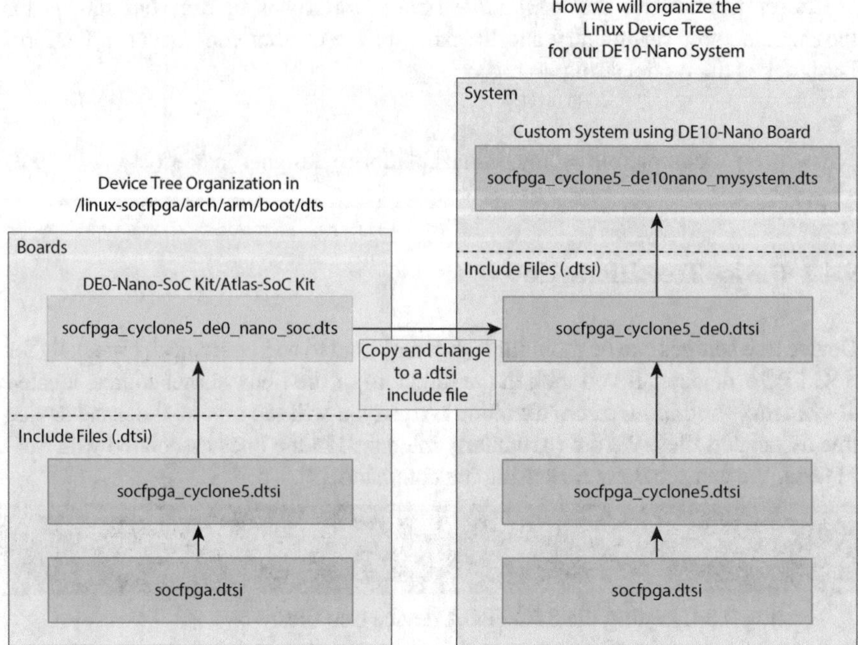

Fig. 9.10: Device tree source file hierarchy

Fortunately for us, the device tree for the DE0-Nano-SoC board also works for the DE10-Nano board since the boards are functionally very similar. All we have to do is just copy this file and change the extension to .dtsi (and shorten the name a bit) in order to turn it into an include file that we can pull into the device tree we will create.

Here we break from the Linux .dts convention since we need not only to have the board we are working on (DE10-Nano), but we also need a device tree for the custom system we are creating on top of the DE10-Nano board. Thus our final device tree (.dts) will have the following naming convention:

socfpga_cyclone5_de10nano_<system_name>.dts

and we will keep this file in the `~/linux-socfpga/arch/arm/boot/dts` directory in order to compile it into a .dtb file (see next section).

9.4.3 Creating a Device Tree for Our DE10-Nano System

When the kernel source was cloned using git (see Sect. 9.2.2 Get the Linux Source Repository and Select the Git Branch (page 130)), device trees for many ARM related boards were included in the `~/linux-socfpga/arch/arm/boot/dts` directory.

We will now create a device tree source file for our custom SoC FPGA system on the DE10-Nano board. The steps for creating this file and the associated final .dtb file are:

Step 1: Turn the de0_nano_soc board .dts file into a .dtsi file that we will include in our device tree source file (as illustrated in Fig. 9.10).

```
$ cd ~/linux-socfpga/arch/arm/boot/dts
$ cp socfpga_cyclone5_de0_nano_soc.dts ↵
    ↪socfpga_cyclone5_de0.dtsi
```

Listing 9.35: Copying the DE0-Nano-SoC .dts to .dtsi

Step 2: Perform any desired changes or node modifications in `socfpga_cyclone5_de0.dtsi`.

Step 3: Copy the example device tree source file for the DE10-Nano system *mysystem* that is named:
`socfpga_cyclone5_de10nano_mysystem.dts`
(click here for the source file) and put it in the Ubuntu VM in the linux source tree directory: `~/linux-socfpga/arch/arm/boot/dts`.

Step 4: Modify the Makefile in the directory `~/linux-socfpga/arch/arm/boot/dts`

Mod 1: Open the Makefile in an editor:

```
$ code Makefile
```

Listing 9.36: Editing the Makefile for creating .dtb's

Mod 2: In the editor, search for the section in the Makefile that has lines that start with `socfpga*`. (It will be lines ~1078-1091 for kernel version 5.15.60)

Mod 3: Insert the new line:
`socfpga_cyclone5_de10nano_mysystem.dtb \`
which can be done right after the line containing the string:
`socfpga_cyclone5_de0_nano_soc.dtb \`
and save the Makefile.

Step 5: Change to the top folder in the kernel source directories and compile all the .dts files:

```
$ cd ~/linux-socfpga
$ make ARCH=arm dtbs
```

Listing 9.37: Compile all the .dts files to .dtb files

Step 6: Copy `socfpga_cyclone5_de10nano_mysystem.dtb` and put it in the TFTP server so it will get loaded at boot time.

```
$ cp ~/linux-socfpga/arch/arm/boot/dts/↲
  ↪socfpga_cyclone5_de10nano_mysystem.dtb /↲
  ↪srv/tftp/de10nano/AudioMini_Passthrough/↲
  ↪soc_system.dtb
```

Listing 9.38: Copy the .dtb file to the TFTP server

Note 1: You can name it something different than soc_system.dtb (e.g., mysystem.dtb), but you will need to modify your u-bootscript accordingly.
Note 2: It is suggested that you backup your source .dts file socfpga_cyclone5_de10nano_mysystem.dts by putting it in your Quartus project folder.

9.5 Platform Device Driver

We treat the custom component *HPS_LED_patterns* (see Sect. 6.2.2 Creating a Custom Platform Designer Component (page 65) on how to create the HPS_LED_patterns component) as an independent device in Linux since we can directly address its registers in memory from the ARM CPUs. Since Linux associates each device with the bus it is attached to, we attach it to a *virtual bus* known as the **Platform Bus** as shown in Fig. 9.11. This is why the device driver is called a **Platform Driver**. In the figure, and in this section, we make the assumption that the HPS_LED_patterns component is located at memory address `0xff204100` as illustrated in Fig. 7.3, and the four register offsets are `0x0`, `0x4`, `0x8`, and `0xC` bytes, respectively, since they are 32-bit registers. When the platform driver is inserted into the kernel (it is a loadable kernel module), it creates a file for each register in the component. Typically, the filename created has the same name as the register signal in the component. In Fig. 9.11, in the **sysfs** virtual file system, we show only one register **LED_reg** in the locations that this register shows up due to space. However, there will be four registers associated with the HPS_LED_patterns component.

The platform device driver source **hps_led_patterns.c** (click here for the source file) is provided as an example along with the associated **Makefile** (click here for the Makefile). **Note:** Only two registers (HPS_LED_control and LED_reg) have been implemented in this source file. The two other registers need to be implemented as well.

At first glance, the platform driver code seems pretty scattered as shown in Fig. 9.12. The green arrows show the dependencies. We do not include green arrows for the data structure *hps_leds_patterns_dev* since most of the functions reference this data structure. The code looks scattered because we are just defining data structures and creating functions that need to be present and inserted into the kernel using the provided macros. The macros simplify writing the driver code so one only needs to

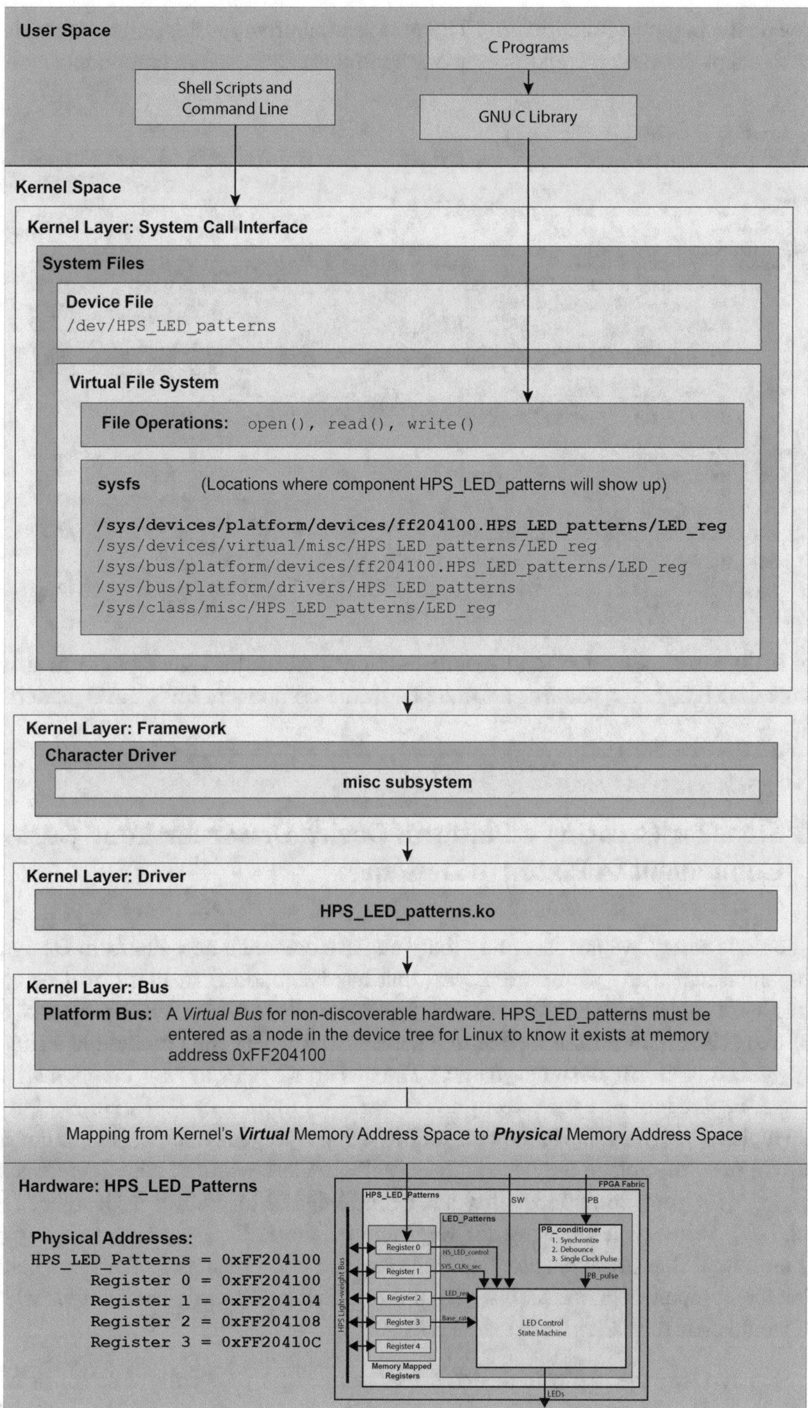

Fig. 9.11: Platform device driver hierarchy in the Linux Kernel for the hardware component HPS_LED_patterns

focus on the required functionality. The next section provides the steps one needs to take to adapt the example platform driver code for a new custom component.

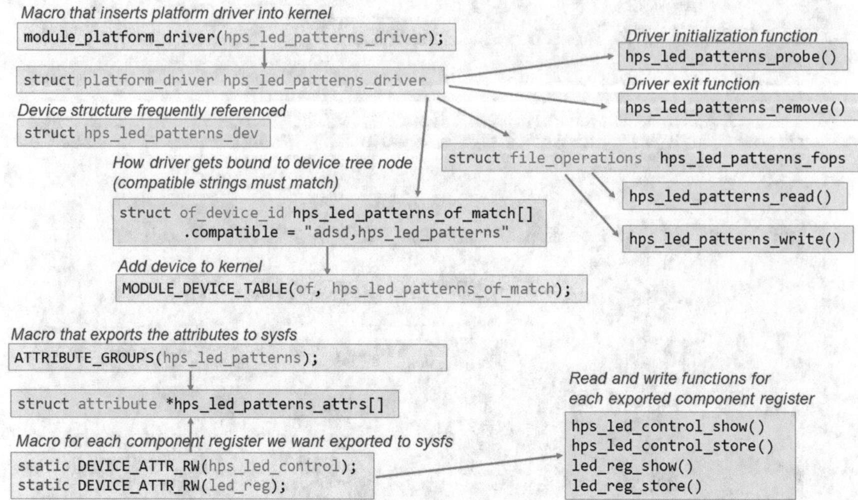

Fig. 9.12: Struct, function, and macro relationships in the Linux platform driver source file HPS_LED_patterns.c for the hardware component HPS_LED_patterns

9.6 Steps for Creating a Platform Device Driver for Your Custom Component in Platform Designer

In the steps below, we are assuming that you have created a new Platform Designer component called **my_component.vhd** that has been added in Platform Designer at the base address `0xFF204100`, and it has the device tree similar to this device tree (click here for the device tree source) and where the compatible string is "my_initials,platform_designer_name." Thus when we refer to *"my_component"* as a name, **replace it with your component name as named in Platform Designer**. Also replace *"my_initials"* in the compatible string with your initials. And of course, you will need to replace the base address `0xFF204100` with the actual address associated with your component that has been assigned by Platform Designer (see Sect. 7.1.1 Memory Addressing for Registers on the HPS Lightweight Bus (page 88) on how this address is calculated). **Note:** The code listings in the steps below show the component name as *hps_led_patterns* rather than *my_component*, which you would change in Step 2: below.

Step 1: Copy the platform device driver source **hps_led_patterns.c** (click here for the source file) and rename it **my_component.c**. Also copy the associated **Makefile** (click here for the Makefile).

Step 2: In both **my_component.c** and **Makefile**, **perform a find and replace string search** where the string segment "**hps_led_patterns**" is replaced by the string segment "**my_component**". (There will be something like 92 occurrences replaced if you do a *Replace All* operation with matching case. There will be some misses in the header strings due to case differences, but you can fix these later).

Step 3: In the #define section as shown in Listing 9.39 below, define the register offsets for all the registers in your component. The suggested naming convention is REG<number>_<register_name>_OFFSET. The offset values determine which register you are reading from (4 byte increments since there are 32-bit registers, see Sect. 7.1.1 Memory Addressing for Registers on the HPS Lightweight Bus (page 88)). The SPAN #define value should be the same value as the *length* value found in the *reg* property of the associated device tree node for my_component (see Sect. 9.1 Device Tree Basics (page 144)). **Note:** Make sure that your register offsets and ordering are consistent with your register read/write VHDL code (i.e., the case statement section operating on the address signal) in the associated hardware component (see Sect. 6.2.1.1 Creating Component Registers (page 64)).

```
/* Define the Component Register Offsets*/
#define REG0_HPS_LED_CONTROL_OFFSET 0x0
/* #define REG1 (Add offset for SYS_CLKs_sec) */
#define REG2_LED_REG_OFFSET 0x08
/* #define REG3 (Add offset for Base_rate) */

/* Memory span of all registers (used or not) in the                  */
/* component hps_led_patterns                                           */
#define SPAN 0x10
```

Listing 9.39: Defining the register offsets and component memory span

Step 4: Make sure the **compatible strings** of the driver and device tree node match. In the of_device_id structure shown in Listing 9.40, make sure that this *compatible* field entry matches the compatible string in the associated device tree node (see Sect. 9.4.1 Device Tree Basics (page 145)). This string matching operation is how the kernel connects the driver to the hardware. If the string does not match the compatible string of the device tree node (or one of the strings in the case of multiple string entries in the device tree node), the kernel will not "bind" the driver to the associated hardware when the driver is inserted.

```
static const struct of_device_id hps_led_patterns_of_match[] = {
    // ****Note:**** This .compatible string must be identical to the
```

```
    // .compatible string in the Device Tree Node for hps_led_patterns
    { .compatible = "adsd,hps_led_patterns", },
    { }
};
MODULE_DEVICE_TABLE(of, hps_led_patterns_of_match);
```

Listing 9.40: The .compatible string in the driver must match the compatible string in the device tree node

Step 5: Export your component registers to sysfs. Do this in the sysfs Attributes section shown in Listing 9.41. For each register that you want exported to sysfs, i.e., *for each register* in your Platform Designer component, there are two things that need to be done in this code section.

1: Use the macro `DEVICE_ATTR_RW()` to define the sysfs attributes where the filename you want to show up is used as the argument. This filename should be the same as what you used for the register name in Step 3:. Thus it should have the form:

```
static DEVICE_ATTR_RW(<register_name>)
```

2: Add an entry in the attribute group structure for the register. This should take the form:

```
&dev_attr_<register_name>.attr,
```

Note: This is where you define what the filenames will be when exported to sysfs. Thus, if you want a different filename from the component register name, you would make the appropriate change in the string <register_name>.

```
// Define sysfs attributes
static DEVICE_ATTR_RW(hps_led_control);     // Attribute for REG0
/* TODO: Add the attributes for REG1 and REG2 using register names        */
static DEVICE_ATTR_RW(led_reg);             // Attribute for REG3

// Create an atribute group so the device core can
// export the attributes for us.
static struct attribute *hps_led_patterns_attrs[] = {
    &dev_attr_hps_led_control.attr,
/* TODO: Add the attribute entries for REG1 and REG2 using register names*/
    &dev_attr_led_reg.attr,
    NULL,
};
ATTRIBUTE_GROUPS(hps_led_patterns);
```

Listing 9.41: Exporting the component registers to sysfs

Step 6: Define the `show/store` (i.e., read/write) functions for each register that has been exported to sysfs. *For each register* that was exported to sysfs in Step 5:, create the following two functions:

f 1: Create the register `show` function named **`<register_name>_show()`** using the same input arguments as in the example `show` function for the `LED_reg` register in the *HPS_LED_patterns* component as shown in Listing 9.42. The `show` function returns the value read by `ioread32` as an ascii text file where the value gets converted into a string by `scnprintf`. The conversion data type and string conversion function need to be consistent with the register data type. If your data type is different, see scnprintf and printk-specifiers. The register offset used in `ioread32` is what you defined in Step 3:.

```
static ssize_t led_reg_show(struct device *dev,
    struct device_attribute *attr, char *buf)
{
    u8 led_reg;
    struct hps_led_patterns_dev *priv = dev_get_drvdata(dev);

    led_reg = ioread32(priv->base_addr + REG3_LED_REG_OFFSET);

    return scnprintf(buf, PAGE_SIZE, "%u\n", led_reg);
```

Listing 9.42: Show function to read the component register exported to sysfs

f 2: Create the register `store` function named **`<register_name>_store()`** using the same input arguments as in the example `store` function for the `LED_reg` register in the *HPS_LED_patterns* component as shown in Listing 9.43. The `store` function converts the string written to a u8 value by using the kernel string conversion function `kstrtou8` and then writes it to your register memory location using `iowrite32`. The conversion data type and string conversion function need to be consistent with the register data type. If your data type is different, see kernel string conversion functions. The register offset used in `ioread32` is what you defined in Step 3:.

```
static ssize_t led_reg_store(struct device *dev,
    struct device_attribute *attr, const char *buf, size_t size)
{
    u8 led_reg;
    int ret;
    struct hps_led_patterns_dev *priv = dev_get_drvdata(dev);
```

```

    // Parse the string we received as a u8
    // See https://elixir.bootlin.com/linux/latest/source/lib/kstrtox.c#L289
    ret = kstrtou8(buf, 0, &led_reg);
    if (ret < 0) {
        // kstrtou16 returned an error
        return ret;
    }

    iowrite32(led_reg, priv->base_addr + REG3_LED_REG_OFFSET);

    // Write was succesful, so we return the number of bytes we wrote.
    return size;
```

Listing 9.43: `Store` function to write to the component register exported to sysfs

Step 7: Change the name in the MODULE_AUTHOR() macro to your name.
Step 8: Compile the driver using the Makefile.
Step 9: Copy the driver (.ko) to the DE10-Nano root file system in the Ubuntu VM.

9.6.1 Testing the Platform Device Driver

Power up the DE10-Nano board, load the Platform Driver using insmod, and change to the directory `/sys/devices/platform/<base-address>.hps_led_patterns`. In this directory, there should be files associated with your registers. Read and write values to these registers using echo and cat, i.e.,

```
$ echo 1 > hps_led_control
```

Listing 9.44: echo 1 > hps_led_control

```
$ cat hps_led_control
```

Listing 9.45: cat hps_led_control

```
$ echo 0x55 > led_reg
```

Listing 9.46: echo 0x55 > led_reg

```
$ cat led_reg
```

Listing 9.47: cat led_reg

9.7 List of Linux Commands

Below is a list of alphabetized Linux commands that are used in the book:

- **./hello** found in Step 7:
- **/usr/bin/arm-linux-gnueabihf-gcc –version** found in Step 2:
- **/usr/bin/arm-linux-gnueabihf-gcc -o hello hello.c** found in Step 5:
- **cat .config** found in Step 3:
- **cat /etc/exports** found in Step 3:
- **cat /proc/sys/kernel/printk** found in 9.3.1
- **cat hps_led_control** found in 9.6.1
- **cat led_reg** found in 9.6.1
- **cd /lib/modules/$(uname -r)/kernel** found in 9.3
- **cd /lib/modules/$(uname -r)/kernel/drivers** found in 9.3
- **cd ~/linux-socfpga/arch/arm/boot/dts** found in 9.4.2
- **code exports &** found in Step 2:
- **code hello.c** found in Step 3:
- **code tftpd-hpa &** found in Step 4:
- **dmesg** found in 9.3.3
- **dpkg -s <packagename>** found in Step 3:
- **echo 0x55 > led_reg** found in 9.6.1
- **echo 1 > hps_led_control** found in 9.6.1
- **echo 8 > /proc/sys/kernel/printk** found in 9.3.1
- **env | grep ARCH** found in 11.1.4.3
- **env | grep CROSS_COMPILE** found in 11.1.4.3
- **export -p** found in 11.1.4.3
- **file hello** found in Step 5:
- **gcc -o hello hello.c** found in Step 6:
- **getent group nfs** found in Step 5:
- **getent group tftp** found in Step 2:
- **getfacl /srv/tftp** found in Step 2:
- **git branch** found in Step 4:
- **git branch -a** found in Step 3:
- **git checkout socfpga-5.10.60-lts** found in Step 4:
- **git clone https://github.com/altera-opensource/linux-socfpga** found in Step 2:
- **git pull https://github.com/altera-opensource/linux-socfpga** found in Step 2:
- **ifconfig** found in Step 3:
- **ip addr** found in Step 3: and Step 6:
- **ip link** found in Step 3:
- **journalctl -u audio-mini-drivers.service** found in 1.5.4
- **ls /sys/class/net/** found in Step 3:
- **ls socfpga*** found in 9.4.2
- **make** found in 11.1.4.3
- **make -j$(nproc)** found in Step 3:

- **make ARCH=arm CROSS_COMPILE=arm-linux-gnueabihf- -j$(nproc)** found in Step 3:
- **make ARCH=arm dtbs** found in Step 5:
- **make ARCH=arm menuconfig** found in Step 3:
- **make ARCH=arm socfpga_defconfig** found in Step 2:
- **mkimage -A arm -O linux -T script -C none -a 0 -e 0 -n "My script" -d u-boot.txt u-boot.scr** found in Setting 5:
- **modinfo hello_kernel_module.ko** found in 9.3.2
- **nmap -sn 192.168.1.*** found in Step 4:a
- **route -n** found in Step 5:
- **source arm_env.sh** found in Step 1:
- **sudo -i** found in Step 4:
- **sudo adduser <user_account_name> nfs** found in Step 5:
- **sudo adduser <user_account_name> tftp** found in Step 2:
- **sudo adduser <user_account_name> vboxsf** found in Step 3:
- **sudo apt full-upgrade** found in Step 3:
- **sudo apt install bison** found in Package 3:
- **sudo apt install build-essential** found in Step 5:
- **sudo apt install flex** found in Package 4:
- **sudo apt install gcc-arm-linux-gnueabihf** found in Step 1:
- **sudo apt install git** found in Step 5:
- **sudo apt install iproute2** found in Step 2:
- **sudo apt install ncurses-dev** found in Package 5:
- **sudo apt install net-tools** found in Step 2:
- **sudo apt install nfs-kernel-server** found in Step 1:
- **sudo apt install nmap** found in Step 4:a
- **sudo apt install tftpd-hpa** found in Step 1:
- **sudo apt install u-boot-tools** found in Step 6:
- **sudo apt install vim nano** found in Step 4:
- **sudo apt install virtualbox-guest-utils** found in Step 6:
- **sudo apt update** found in Step 4: and 9.2.1
- **sudo chgrp -R tftp /srv/tftp** found in Step 2:
- **sudo chmod -R a+r /srv/tftp** found in 11.1.3.9
- **sudo chmod -R g+rwx /srv/tftp** found in Step 2:
- **sudo chmod g+s tftp** found in Step 2:
- **sudo chown :tftp tftp** found in Step 2:
- **sudo exportfs -v** found in Step 4:
- **sudo groupadd nfs** found in Step 5:
- **sudo groupadd tftp** found in Step 2:
- **sudo insmod hello_kernel_module.ko** found in 9.3.3
- **sudo mkdir -p <directory>** found in Step 3:, Step 3:, Step 3:, and Step 1:
- **sudo rmmod hello_kernel_module** found in 9.3.3
- **sudo service nfs-kernel-server restart** found in Step 4:
- **sudo setfacl -d –set u::rwx,g::rwx tftp** found in Step 2:
- **sudo snap install –classic code** found in Step 4:

- **sudo snap install atom –classic** found in Step 4:
- **sudo systemctl enable tftpd-hpa** found in Step 1:
- **sudo systemctl restart tftpd-hpa** found in Step 1:
- **sudo systemctl restart tftpd-hpa** found in Step 5:
- **sudo systemctl status tftpd-hpa** found in Step 1:
- **sudo tar -cpzf ubuntu-rootfs.tar.gz ubuntu-rootfs** found in 11.1.3.9
- **sudo tar -zxpvf ubuntu-rootfs.tar.gz** found in 11.1.3.9
- **systemctl enable audio-mini-drivers.service** found in Step 5:

References

1. Intel. *Cyclone V Hard Processor System Technical Reference Manual.* https://www.intel.com/content/dam/www/programmable/us/en/pdfs/literature/hb/cyclone-v/cv_5v4.pdf. Figure 10-2, Cortex-A9 MPU Subsystem Block Diagram, page 10-3, Accessed 24 Jun 2022
2. Intel. *Cyclone V Hard Processor System Technical Reference Manual.* https://www.intel.com/content/dam/www/programmable/us/en/pdfs/literature/hb/cyclone-v/cv_5v4.pdf. Figure 8-1, Interconnect Block Diagram, : L3 Interconnect and L4 Buses, page 8-2, Accessed 24 Jun 2022

Chapter 10
Introduction to Digital Signal Processing

This section will not be a comprehensive coverage of Digital Signal Processing that a student would learn as part of their electrical engineering curriculum, which often comes as a two-course sequence. The first course is a "Signals and Systems" course that would include the *Continuous Time Fourier Transform* (**CTFT**). Since computers are heavily involved in signal processing, the theory has been extended to signals that have been sampled in time. This would be the topic of the second course called Digital Signal Processing.

Although this section will be math heavy (yes, there is calculus involved), there will still be useful takeaways for those that use these signal processing algorithms but do not have the math background. I will highlight areas where you need to be careful and why certain choices are often made when using these algorithms.

10.1 Sampling

To process a signal with a computer, we first need to create a representation of this signal. As we will show in Sect. 10.2, we can represent signals using simple sinusoidal functions. However, this will only work correctly if we ensure that we have sampled the signal properly. These are the steps for sampling correctly, which is how we implement the **Sampling Theorem**.

Step 1: Determine the *maximum frequency* $f_{\max}$ contained in the continuous signal we wish to represent in a computer. If we do not know what this maximum frequency is, which is typically the case, then we need to limit the frequencies to a known $f_{\max}$ by *low pass filtering* the signal to remove all frequencies above $f_{\max}$. This filter is known as an anti-aliasing filter and this filtering needs to be performed *before* sampling. If there is no analog anti-aliasing circuitry before the analog-to-digital converter where the sampling occurs, then the sampled signal is suspect.

R. K. Snider, *Advanced Digital System Design using SoC FPGAs*,
https://doi.org/10.1007/978-3-031-15416-4_10

Step 2: Sample the continuous signal that has been conditioned to have no frequencies greater than $f_{\max}$ with a sampling rate f_s that is *greater* than twice $f_{\max}$

Sampling Theorem
$f_s > 2f_{\max}$

The reason we must sample greater than twice the maximum frequency is due to the periodicity of the sine and cosine functions. This identity is listed below for the cosine function.

$$\cos(\theta) = \cos(\theta + 2\pi) \tag{10.1}$$

What this means in practical terms is that if we do not ensure that the sampling theorem has been followed, then there will be high frequencies masquerading as low frequencies, which is known as *aliasing*. To illustrate this, let us create two signals where signal one is comprised of frequency f_1, which will be less than one-half f_s and thus properly sampled. We will create a second signal that will be comprised of frequency f_2, and this frequency will violate the sampling theorem. For convenience, we will create f_2 that is some multiple of the sampling rate higher than f_1:

$$f_2 = f_1 + kf_s \tag{10.2}$$

If we take the first signal with frequency f_1

$$x(t) = \cos(2\pi f_1 t + \phi) \tag{10.3}$$

and sample it at the sampling period $T_s = 1/f_s$, it becomes

$$\begin{aligned} x[n] &= x(nT_s) \\ &= \cos(2\pi f_1 nT_s + \phi) \end{aligned} \tag{10.4}$$

Now take the second signal with frequency f_2

$$y(t) = \cos(2\pi f_2 t + \phi) \tag{10.5}$$

and sample it at the same sampling rate:

$$\begin{aligned} y[n] &= y(nT_s) \\ &= \cos(2\pi f_2 nT_s + \phi) \\ &= \cos(2\pi(f_1 + kf_s)nT_s + \phi) \\ &= \cos(2\pi f_1 nT_s + 2\pi k f_s nT_s + \phi) \\ &= \cos(2\pi f_1 nT_s + 2\pi kn + \phi) \\ &= \cos(2\pi f_1 nT_s + \phi) \\ &= x[n] \end{aligned} \tag{10.6}$$

Thus, this higher frequency signal with frequency f_2 ends up looking identical to the signal with frequency f_1 after sampling, i.e., Eq. 10.4 equals Eq. 10.6. In a similar manner, all high-frequency terms greater than $f_{\max}$ would end up masquerading as lower frequencies less than $f_s/2$ when sampled at f_s, which is known as **aliasing**. Since we do not want this to happen, we need to ensure that the sampling theorem has been followed.

The illustration of sampling theorem in the frequency domain is shown in Fig. 10.1. The spectrum of the signal gets replicated in the frequency domain in multiples of $2\pi f_s$ and only one replica ($k = 1$) is shown in the figure. The sampling frequency f_s controls the spacing between the replicas, and for no overlap to occur, we can see from the figure that the following inequality needs to hold:

$$\begin{aligned} 2\pi f_{\max} &< 2\pi(f_s - f_{\max}) && \text{Radian Frequency} \\ \implies f_{\max} &< f_s - f_{\max} && \text{Cyclic Frequency (Hz)} \\ \implies 2f_{\max} &< f_s && \text{Sampling Theorem} \end{aligned}$$

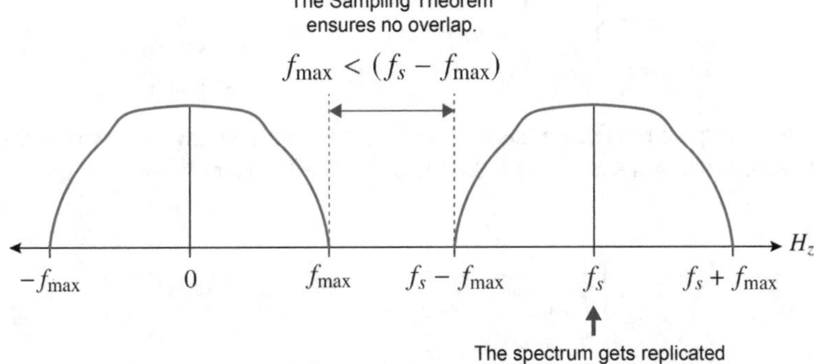
The Sampling Theorem
ensures no overlap.
$f_{\max} < (f_s - f_{\max})$
H_z
$-f_{\max}$
0
$f_{\max}$
$f_s - f_{\max}$
f_s
$f_s + f_{\max}$
The spectrum gets replicated
in the frequency domain
at multiples of the sample rate.

Fig. 10.1: Illustration of the sampling theorem in the frequency domain. The spectrum of the sampled signal gets replicated at multiples of the sample rate f_s (i.e., kf_s and only $k = 1$ is shown in the figure). The sampling theorem $f_s > 2f_{\max}$ ensures that there will be no spectral overlap, i.e., aliasing

10.2 Fourier Series

In signal processing, one of the fundamental ideas is that we can represent signals such as speech by simple sinusoids and it does not matter if the speech signal is acoustic where people are talking to each other across a room or if the speech signal is converted to a digital representation and people are talking to each other across the country using their cell phones. We can represent any signal just by adding the appropriate number of sines and cosines together, each with their own amplitude, frequency, and phase shift. This is known as a Fourier series, which has the following mathematical form:

$$s(t) = \frac{a_0}{2} + \sum_{k=1}^{N} \left(a_k \cos\left(\tfrac{2\pi}{T} kt\right) + b_k \sin\left(\tfrac{2\pi}{T} kt\right) \right) \tag{10.7}$$

This is typically written in a complex exponential form using Euler's formula:

$$s(t) = \sum_{k=-N}^{N} c_k e^{j\frac{2\pi}{T}kt} \tag{10.8}$$

since

$$e^{j\theta} = \cos\theta + j\sin\theta \tag{10.9}$$

Let us show how this works with the following fairly complicated arbitrary piecewise waveform that has a period of $T = 5$ seconds. The waveform has three segments given by the following function and is shown in Fig. 10.2.

$$s(t) = \begin{cases} 3\sin\left(2\pi\frac{1}{6}t\right) + \frac{1}{2}\sin(2\pi 8t) & 0 \le t < 3 \\ e^{(\ln(1)-\ln(4))t+(4\ln(4)-3\ln(1))} & 3 \le t < 4 \\ 3 & 4 \le t < 5 \end{cases} \tag{10.10}$$

Let us determine the Fourier series coefficients c_k that will allow us to reconstruct this waveform using a summation of sinusoids. The Fourier coefficients are defined as

$$c_k = \begin{cases} \frac{1}{T}\int_0^T s(t)dt & k = 0 \text{ (DC term)} \\ \frac{2}{T}\int_0^T s(t)e^{-j\frac{2\pi}{T}kt}dt & k > 0 \end{cases} \tag{10.11}$$

We will use the fact that integration is a linear operator, which means we can break the waveform into separate segments and integrate each segment separately and integrate the terms within each segment separately. The first segment is the time interval $0 \le t < 3$ and there are two terms (two sinusoids with different frequencies

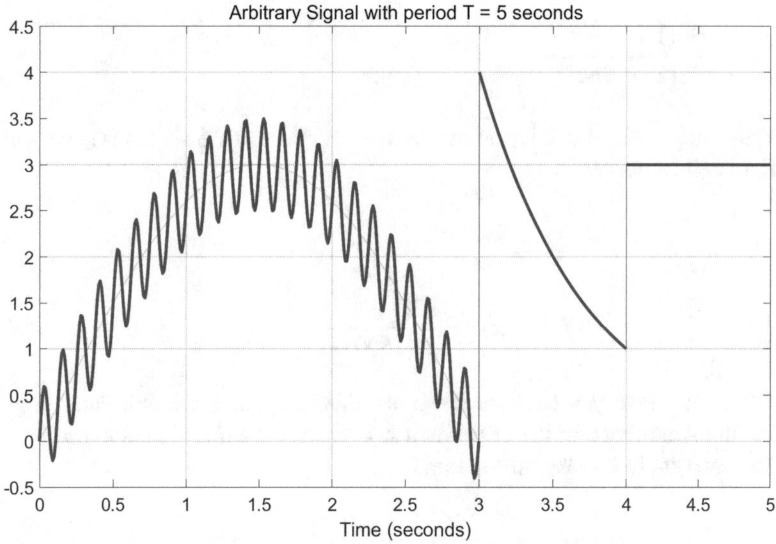

Fig. 10.2: An arbitrary signal given by Eq. 10.10

and amplitudes), so we will integrate each term separately over this interval. Thus the first term $3\sin\left(2\pi\frac{1}{6}t\right)$ in segment 1 has the DC coefficient:

$$c_0 = \frac{3}{5}\int_0^3 3\sin\left(\frac{\pi}{3}t\right)dt = \frac{3}{5}\left(\frac{-3}{\pi}\cos\left(\frac{\pi}{3}t\right)\Big|_0^3\right) \tag{10.12}$$

$$= \frac{-9}{5\pi}\left(\cos\left(\frac{3\pi}{3}\right) - \cos(0)\right) = \frac{-9}{5\pi}(-1-1) \tag{10.13}$$

$$= \frac{18}{5\pi} \tag{10.14}$$

Knowing that $\int e^{at}\sin(bt)dt = \frac{e^{at}}{a^2+b^2}\left[a\sin(bt) - b\cos(bt)\right]$, we get the rest of the coefficients for this first term:

$$c_k = \frac{6}{5}\int_0^3 3\sin\left(\frac{\pi}{3}t\right)e^{-j\frac{2\pi}{5}kt}\,dt \tag{10.15}$$

$$= \frac{6}{5}\left(\frac{1}{\frac{\pi^2}{9} + \left(\frac{-j2\pi k}{5}\right)^2}\right)\left[\left(\frac{-j2\pi k}{5}\sin\left(\frac{\pi}{3}t\right) - \frac{\pi}{3}\cos\left(\frac{\pi}{3}t\right)\right)e^{\frac{-j2\pi kt}{5}}\right]_0^3 \tag{10.16}$$

$$= \frac{6}{5}\left(\frac{1}{\frac{\pi^2}{9} - \frac{4\pi^2k^2}{25}}\right)\left[\frac{\pi}{3}e^{\frac{-j6\pi k}{5}} + \frac{\pi}{3}\right] \tag{10.17}$$

$$= \frac{2}{\frac{5\pi}{9} - \frac{4\pi^2k^2}{5}}\left(e^{\frac{-j6\pi k}{5}} + 1\right) \tag{10.18}$$

$$= \frac{90\left(1 + e^{-j\frac{6}{5}\pi k}\right)}{\pi\left(25 - 36k^2\right)} \tag{10.19}$$

Performing a similar integration for the second term $\frac{1}{2}\sin(2\pi 8t)$, we get the Fourier coefficients of

$$c_0 = 0 \tag{10.20}$$

$$c_k = \frac{20\left(1 - e^{-j\frac{6}{5}\pi k}\right)}{\pi\left(1600 - k^2\right)} \tag{10.21}$$

We are not done yet, because if we use this in Matlab, we will end up getting NANs (not a number) for the case when $k = 40$ because this results in $c_{40} = \frac{0}{0}$. For this case, we apply L'Hospital's rule:

$$\lim_{k\to 40} \frac{\frac{\partial}{\partial k}\left(20 - 20e^{-j\frac{6}{5}\pi k}\right)}{\frac{\partial}{\partial k}\left(1600\pi - \pi k^2\right)} = \lim_{k\to 40} \frac{24j\pi e^{-j\frac{6}{5}\pi k}}{-2\pi k} = \frac{-3j}{10} \tag{10.22}$$

Thus the coefficients that will give us the waveform in segment 1 ($0 \le t < 3$) are

$$c_k = \begin{cases} \dfrac{18}{5\pi} & k = 0\,(\text{DC term}) \\ \dfrac{90\left(1 + e^{-j\frac{6}{5}\pi k}\right)}{\pi\left(25 - 36k^2\right)} + \dfrac{20\left(1 - e^{-j\frac{6}{5}\pi k}\right)}{\pi\left(1600 - k^2\right)} & k > 0, k \neq 40 \\ \dfrac{90\left(1 + e^{-j\frac{6}{5}\pi k}\right)}{\pi\left(25 - 36k^2\right)} + \dfrac{-3j}{10} & k = 40 \end{cases} \tag{10.23}$$

We can check these coefficients by generating the waveform (blue curve) for segment 1 as shown in Fig. 10.3 where we use $N = 100$ coefficients and plot on top of the target waveform (green). Note that this curve is zero for the other two segments.

The waveform for segment 2 ($3 \le t < 4$) has the form $s(t) = e^{at+b}$, where a and b are chosen so that $s(t) = 4$ at $t = 3$ and $s(t) = 1$ at $t = 4$. This gives $a = \ln(1) - \ln(4)$ and $b = 4\ln(4) - 3\ln(1)$. The Fourier coefficients for segment 2 are calculated using Eq. 10.11 and are found to be

$$c_k = \begin{cases} \dfrac{e^{4a+b} - e^{3a+b}}{5a} & k = 0\,(\text{DC term}) \\ \dfrac{2}{5a - j2\pi k}\left(e^{4a+b-j\frac{8}{5}\pi k} - e^{3a+b-j\frac{6}{5}\pi k}\right) & k > 0 \end{cases} \tag{10.24}$$

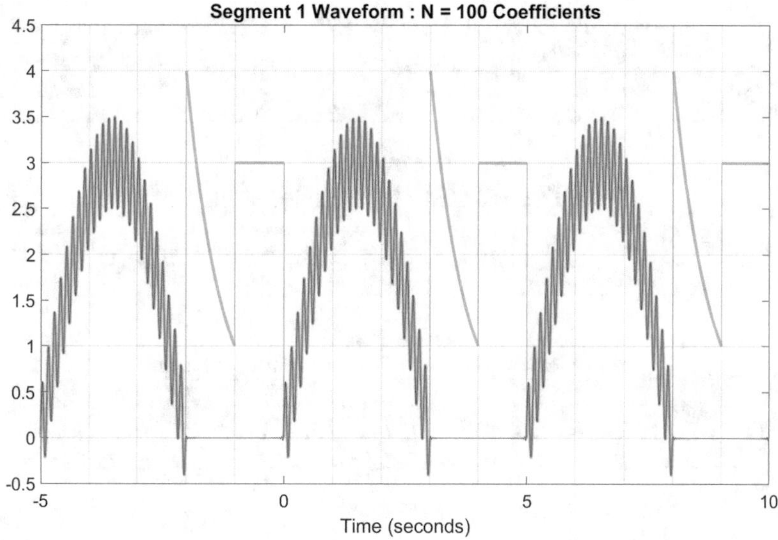

Fig. 10.3: Checking the coefficients for segment 1 ($0 \leq t < 3$). The Fourier series is linear, so we can treat each segment independently

The waveform for segment 3 ($4 \leq t < 5$) has a constant value of 3. This results in the following Fourier coefficients for segment 3:

$$c_k = \begin{cases} \dfrac{3}{5} & k = 0\,(\text{DC term}) \\ \dfrac{-6}{j2\pi k}\left(1 - e^{-j\frac{8}{5}\pi k}\right) & k > 0 \end{cases} \tag{10.25}$$

The final waveform uses all the coefficients from all the segments and is the sum of the waveforms for each segment. This reconstruction is shown in Fig. 10.4.

The Matlab files to plot this waveform are listed in Table 10.1 and can be downloaded to create the waveform with a different number of coefficients. In Fig. 10.4 and in the subplot with $N = 500$ coefficients, one can see overshoots and ringing still happening at the discontinuities. This is known as the Gibbs phenomenon [1]. To get rid of the overshoots and ringing, one has to include a large number of coefficients, which tells us that discontinuities and sharp corners contain very high frequencies.

The function *sumexp.m* in Table 10.1 is done with one line of Matlab code as shown in Listing 10.1 and illustrates vectorized Matlab code in contrast to using slower *for* loops.

```
s=real(C(:)'*(exp(1j*2*pi*f(:)*[0:(1/fs):dur])));
```

Listing 10.1: Matlab: Summation of complex exponentials

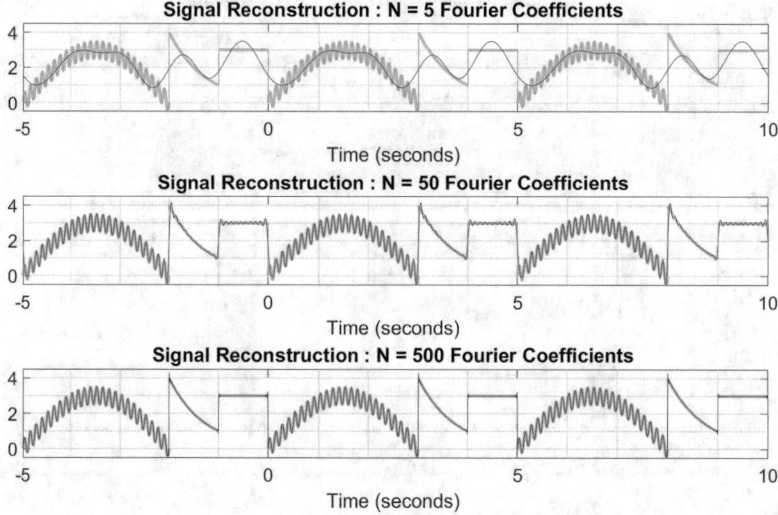

Fig. 10.4: Signal reconstruction using different numbers of Fourier coefficients ($N = 5, 50, 500$). As N increases, the signal converges to the "true" solution. The signal is periodic with period $T_0 = 5$ seconds and three periods are plotted in each case

Table 10.1: Matlab files used to plot arbitrary waveform in Fig. 10.4

File	Description	Link
fourier_series_Nvalues.m	Script that created Fig. 10.4	Click for file
fourier_series_waveform.m	Waveform function	Click for file
sumexp.m	Complex exponential summation	Click for file
fourier_series_target_template.m	Script to plot true waveform	Click for file

To interpret this line of code, we start with creating a row vector `[0:(1/fs):dur]` of time values that has a matrix dimension of $1 \times N_t$, where N_t is the number of samples (sample rate times time). We then take the vector of harmonic frequencies f and force it to be a column vector `f(:)` with dimension $N_f \times 1$, where N_f is the number of frequencies. Thus we do not care if f comes into the function as a row or column vector since we force it to be a column vector by using `f(:)`. We then multiple the column vector of frequencies by the row vector of time samples:

$$\underset{N_f \times 1}{f(:)} \; * \; \underset{1 \times N_t}{[0:(1/fs):dur]} = \underset{N_f \times N_t}{f(:)*[0:(1/fs):dur]} \tag{10.26}$$

This results in an outer product matrix $N_f \times N_t$ containing all combinations of frequencies and times. This matrix gets multiplied by the complex term $j * 2 * \pi$ and is the argument to Matlab's exp() function that results in a matrix of complex values of size $N_f \times N_t$.

We then take the vector of complex Fourier coefficients C that could be passed into the function as either a row or a column vector. It has the same number of elements as frequencies since this gives the amplitude and phase shift for each frequency (harmonic). To force it to be a row vector, we first force it to be a column vector $C(:)$ and then take its transpose $C(:)'$. We then have the product:

$$\underset{1\times N_f}{C(:)'} * \underset{N_f \times N_t}{\text{exp(1j*2*pi*f(:)*[0:(1/fs):dur])}} = \underset{1\times N_t}{\text{C(:)'*exp(1j*2*pi*f(:)*[0:(1/fs):dur])}} \tag{10.27}$$

which sums over all frequencies and we are left with the waveform as a function of time only. Finally, we use *real()* since we are only interested in the real part of the signal.

10.3 Geometric Interpretation of the Fourier Transform

If you came across the following equation for the *Fourier Transform* (**FT**), infrequently called the *Continuous Time Continuous Frequency Fourier Transform* (**CTCFFT**), could you picture in your head what this equation is doing?

$$\hat{s}(f) = \mathcal{F}\{s(t)\} = \int_{-\infty}^{\infty} s(t)e^{-j2\pi ft}\,dt \tag{10.28}$$

Let us break this apart, but first let us define what the variables are. The variable f stands for frequency with units of hertz (Hz) or cycles per second. The variable t stands for time with units of seconds. The grouping $2\pi f$ stands for angular frequency, often replaced by ω, and has units of radians per second. The variable j stands for the complex number $\sqrt{-1}$ (sometimes you will see the letter i instead of j).

The first step in breaking this apart is to use **Euler's formula**

$$e^{j\theta} = \cos(\theta) + j\sin(\theta) \tag{10.29}$$

which has the following geometric interpretation as seen in Fig. 10.5. The value $e^{j\theta}$ is a point on the unit circle that lives in the complex plane. The position of the point is determined by the angle θ and has the x coordinate value on the real axis of $\cos\theta$ and the y coordinate value on the imaginary axis of $\sin\theta$. We interpret the unit vector (radius one) that connects the origin with point $e^{j\theta}$ as being projected to both the real and imaginary axes using basic trigonometry. Note the circle is called the **Unit Circle** since it has a radius of 1 and is the range of $e^{j\theta}$. Thus Euler's formula maps all real values of θ to the unit circle and a point specified by θ has a projection to the real and imaginary axes. The real axis projection has the length of $\cos(\theta)$ and the imaginary axis projection has the length of $\sin(\theta)$.

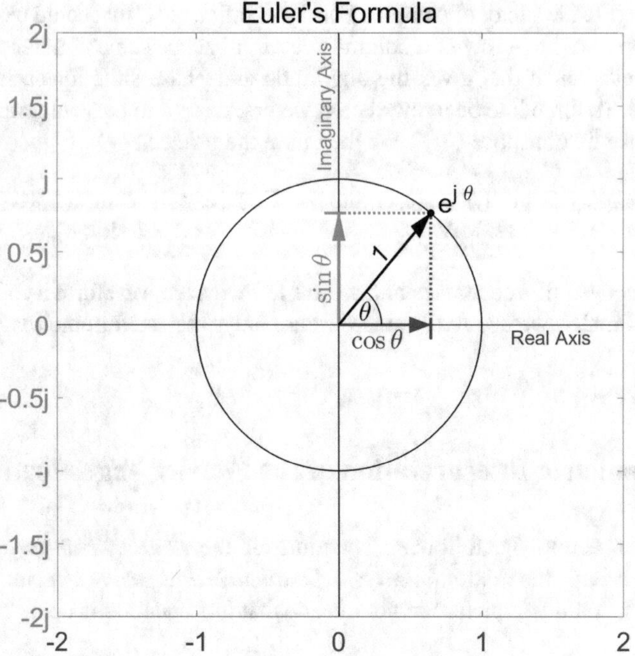

Fig. 10.5: Geometric interpretation of Euler's formula. We interpret the unit vector (radius one) that connects the origin with point $e^{j\theta}$ as being projected to both the real and imaginary axes using basic trigonometry

We can then interpret the term $s(t)e^{-j2\pi ft}$ as a vector with a time varying radius of $s(t)$ that is being projected to both the real and imaginary axes, while the Euler angle is also changing in time with the value of $\theta = 2\pi ft$. This causes the vector (whose radius is changing in time) to spin around the origin in the complex plane and this spinning vector is called a phasor. As we did for the unit vector in Euler's formula, we interpret the signal $s(t)$ as being projected to both the real and imaginary axes as illustrated in Fig. 10.6 where the projected signal length on the real axis is $s(t)\cos(2\pi ft)$ and the projected signal length on the imaginary axis is $s(t)\sin(2\pi ft)$.

Now, let us expand the Fourier Transform definition using Euler's formula:

$$\hat{s}(f) = \int_{-\infty}^{\infty} s(t)e^{-j2\pi ft}\, dt \tag{10.30}$$

$$= \int_{-\infty}^{\infty} s(t)[\cos(2\pi ft) + j\sin(2\pi ft)]\, dt \tag{10.31}$$

$$= \int_{-\infty}^{\infty} s(t)\cos(2\pi ft)\, dt + j\int_{-\infty}^{\infty} s(t)\sin(2\pi ft)\, dt \tag{10.32}$$

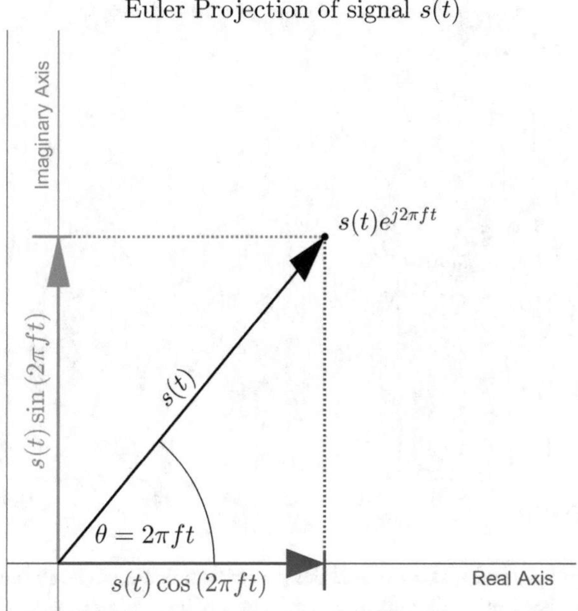

Fig. 10.6: Euler's formula projects the signal onto the real and imaginary axes. The signal $s(t)$ controls the vector length that varies in time along with the projection angle. The example shown here assumes that both $s(t)$ and cos() are positive. However, both terms can be both positive and negative and this will change the quadrant that the projection occurs in

Let us focus on the term $s(t)\cos(2\pi ft)$, which is the signal's projection to the real axis. For illustration purposes, we will use the speech signal shown as the blue line in Fig. 10.7. This will be the signal $s(t)$ being projected to the real axis by the cosine with time varying angle $\theta = 2\pi ft$ where $f = 112.51$ Hz (chosen to match the pitch of the speech signal well). This cosine with $f = 112.51$ Hz is shaded in a light green color that is behind the blue speech signal.

The product $s(t)\cos(2\pi ft)$, which is the projection to the real axis, is plotted in panel A of Fig. 10.8 as a function of time. The product $s(t)\sin(2\pi ft)$, which is the projection to the imaginary axis, is plotted in panel B as a function of time. The positive part of the product (as a function of time) is colored in blue and the negative part of the product is colored in red.

The first term in the Fourier integral

$$\int_{-\infty}^{\infty} s(t)\cos(2\pi ft)\,dt \tag{10.33}$$

is just the area under the curve in panel A of Fig. 10.8 where the positive areas (blue) add and the negative areas (red) subtract. In Matlab, this is just the summation

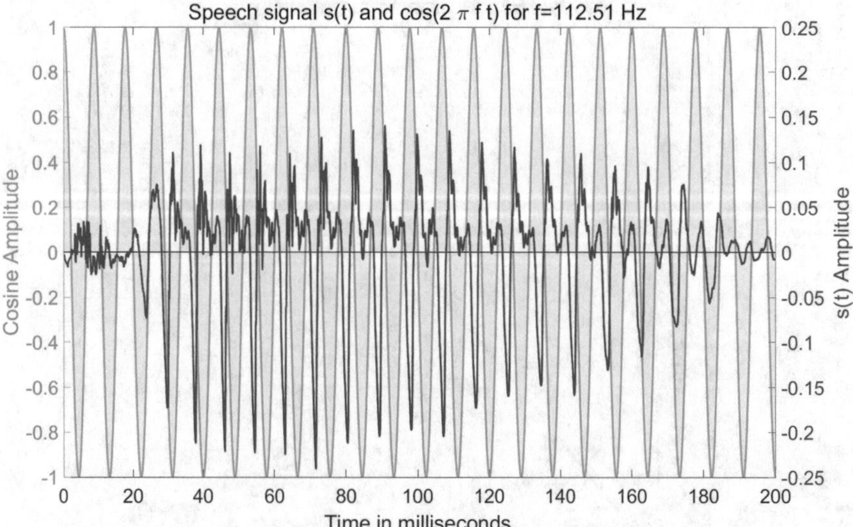

Fig. 10.7: Speech signal $s(t)$ used in Euler projection is the curve in blue. The cosine used for the projection is shaded in green. The cosine frequency of 112.51 Hz was chosen since it had the greatest Fourier Transform magnitude as can be seen in Fig. 10.10. The cosine frequency aligns well with the pitch frequency of the speech signal where most of the speech signal energy is contained

of the product vector $s(t)$.* $\cos(2\pi f t)$ over the interval shown, which results in the value -28.57. The signal goes significantly negative during the positive cycle of $\cos(2\pi 112.5t)$, so the area becomes significantly negative. This means that the signal matches this frequency well (the value would be positive if the negative part of the speech signal aligned with the negative cycle of the cosine). Similarly, the area for the imaginary axis project has a value of -7.88. Thus the Fourier Transform at the frequency $f = 112.5$ is $\hat{s}(f) = \hat{s}(112.5) = -28.57 - j7.88$, which has a magnitude of $m = \sqrt{(-28.57)^2 + (-7.88)^2} = 29.64$. This is the point of greatest magnitude marked in Fig. 10.10. Thus, when the signal oscillates at the same frequency as the cosine, the resulting area of the summation will be large (and can be either positive or negative), which tells us that there is a lot of signal energy at this frequency. Using both the sine and cosine tells us the phase shift of the signal relative to the sine and cosine functions, which is why the result is a complex number.

In contrast, an example of a frequency that does not match the speech signal well ($f = 636$ Hz) is shown in Fig. 10.9. At this projection frequency, the signal summations are close to zero (magnitude = 0.08), which is the low magnitude marked in Fig. 10.10. Notice that the positive and negative areas are similar and nearly symmetric. The cosine is going positive and negative, while the signal is not, which causes the product to be symmetric about zero. This tells us that the signal does not match this frequency well and thus does not have much signal energy at this frequency.

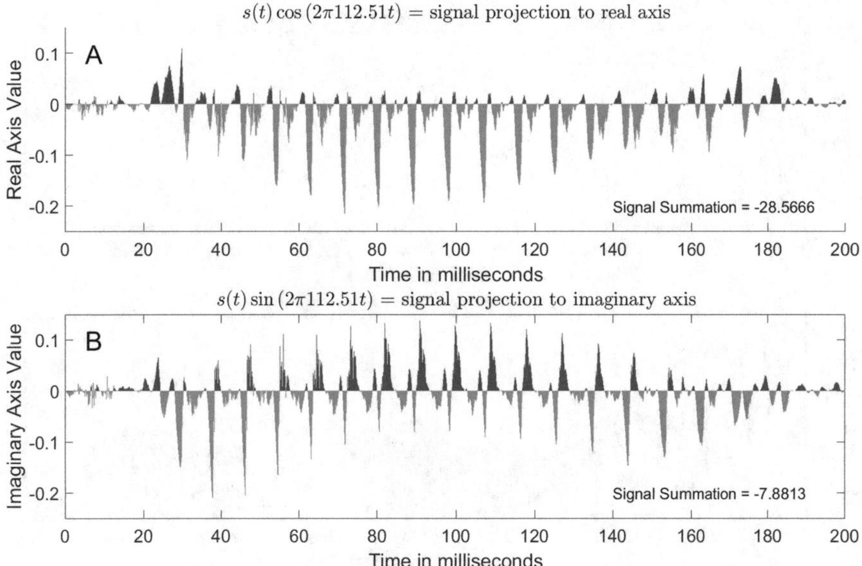

Fig. 10.8: In panel A, the speech signal $s(t)$ is projected to the real axis and the real axis value is plotted as a function of time. In panel B, the speech signal $s(t)$ is projected to the imaginary axis and the imaginary axis value is plotted as a function of time

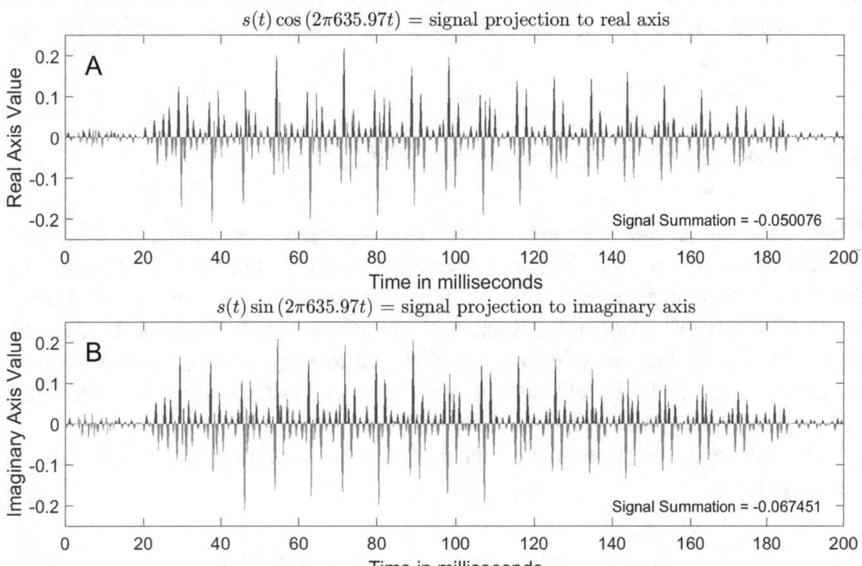

Fig. 10.9: Example where the frequency does not match the signal well. The positive area (blue) in this case is similar in area to the negative area (red)

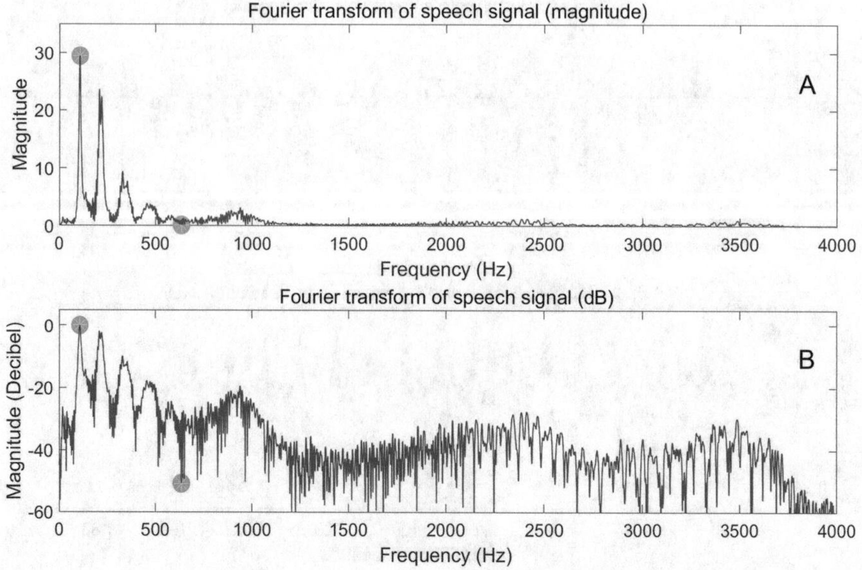

Fig. 10.10: Fourier transform of the speech signal from Fig. 10.7. Panel A shows the magnitude of the Fourier transform as a function of frequency. Plot B plots the normalized magnitude of the Fourier transform in $dB = 20\log_{10}(m/\max(m))$. The largest magnitude point in both A and B panels is at the frequency ($f = 112.51$ Hz), which is used in Figs. 10.7 and 10.8. The low magnitude point in both panels is at the frequency ($f = 636.97$ Hz), which is used in Fig. 10.9. Both these points are marked by green circles

10.4 The Fast Fourier Transform (FFT)

There are many reasons to use the Fourier Transform such as examining the frequency content of a signal or transforming convolution performed in the time domain to a much simpler multiply operation in the frequency domain. However we will ignore much of this mathematical infrastructure since there are many books devoted to this topic ([2, 3]). Rather, we will turn our attention to some practical considerations when using the FFT. We will use as our example a speech signal that has been sampled in time by an *analog-to-digital converter* (**ADC**).

Our mathematical starting point is the Fourier Transform (Eq. 10.28) that we restate here:

$$\hat{s}(f) = \mathcal{F}\{s(t)\} = \int_{-\infty}^{\infty} s(t)e^{-j2\pi ft}\,dt$$

Notice that the limits of integration are from $-\infty$ to $+\infty$. This is fine when we integrate continuous functions. However, when we use computers to compute the Fourier Transform, we immediately break this assumption since we do not have the time, memory, or patience to deal with infinities. Can we still use the Fourier Transform definition when we immediately modify the limits of integration? The general answer is that we cannot assume that it will give us valid answers. This would be like using an instrument outside of the manufacture's working specifications. Thus we need to know what this deviation from the definition is doing to our analysis. Modifying the limits of integration gets us into *windowing*, which is covered in Sect. 10.5.1 Windowing (page 178).

There are additional modifications that are made to the Fourier Transform as given in Eq. 10.28 to allow computers to implement the transform. These are:

Mod 1: Changing continuous time to discrete time. This is known as the *Discrete Time Continuous Frequency* (**DTCF**) **Fourier Transform** and is covered in Sect. 10.4.1 The Discrete Time Continuous Frequency Fourier Transform (page 173). This is useful when you want to determine the frequency content of a signal over an arbitrary frequency range (typically a small range) with arbitrary high precision.

Mod 2: Changing continuous frequency to discrete frequency (with discrete time). This is known as the *Discrete Time Discrete Frequency* (**DTDF**) Fourier Transform and is usually called the *Discrete Fourier Transform* (**DFT**), which is covered in Sect. 10.4.2 The Discrete Fourier Transform (page 174).

Mod 3: Speeding up the DFT, which is called the *Fast Fourier Transform* (**FFT**), which is covered in Sect. 10.4.3 FFT (page 176). This is the most commonly used form of the Fourier Transform when using computers.

10.4.1 The Discrete Time Continuous Frequency Fourier Transform

The first step toward using the Fourier Transform with computers is to use samples of a signal. This means that the time steps are discrete where they have been sampled at a particular time interval or sample rate. This changes the integral of the Fourier Transform definition to a summation and the signal samples are denoted $s[n]$. The Fourier Transform that uses discrete time samples is given as

$$\hat{s}(f) = \mathcal{F}_{dtcf}\{s[n]\} = \sum_{n=-\infty}^{\infty} s[n]e^{-j2\pi fn}$$

where $n \in \mathbb{Z}, f \in \mathbb{R}, \frac{-f_s}{2} \leq f \leq \frac{f_s}{2}$.

Notice that the limits of the summation are from $-\infty$ to $+\infty$, so we have not dealt with windowing yet (see Sect. 10.5.1 Windowing (page 178)). In this definition, we can use any real frequency value in the domain $\frac{-f_s}{2} \leq f \leq \frac{f_s}{2}$, which corresponds to the principal values of the sine/cosine functions ($-\pi \leq \omega \leq \pi$) where $\omega = 2\pi f$ and f_s is the sampling frequency.

This form is useful when you want to examine the frequency content of a signal within a specific frequency range and with arbitrary precision (limited by machine precision and data types used). This form was used to find the local frequency maximum and local frequency minimum with high precision (double precision) in Figs. 10.7, 10.8, 10.9 and 10.10. (click here for the source file of dtcfft.m)

10.4.2 The Discrete Fourier Transform

We converted time to discrete sample times in Sect. 10.4.1 The Discrete Time Continuous Frequency Fourier Transform (page 173) and we now need to convert the transform to discrete frequencies, so we can easily deal with them using computers. To do this, we sample the frequencies around the unit circle in the complex plane with uniform spacing. The full circle has 2π radians and we divide this into N intervals that gives us N frequencies (normalized). This gives the Discrete Fourier Transform:

$$\hat{s}[k] = \text{DFT}\{s[n]\} = \sum_{n=0}^{N-1} s[n]e^{-j\frac{2\pi}{N}kn} \quad \text{where} \quad k = 0, 1, 2, \ldots, N-1$$

The index n in the signal $s[n]$ is understood to represent samples in the signal that are spread apart in time by T_s seconds, where $Fs = 1/T_s$ is the sample rate in Hz. Furthermore, it is assumed that the maximum frequency content in the signal before sampling was less than $Fs/2$, which is why you always see anti-aliasing filters before analog-to-digital (ADC) converters (if you do not see them in the system, the hardware and signal are suspect). The index k in the spectrum $\hat{s}[k]$ represents the normalized radian frequency $\hat{\omega} = \frac{2\pi}{N}k$ (normalized to 2π), which means that regardless of FFT length N, the frequency sampling (evaluation) is done once around the unit circle. We can also normalize frequency f with respect to the sampling rate F_s, thus $\hat{\omega} = 2\pi\frac{f}{F_s}$. This means that $\frac{2\pi}{N}k = 2\pi\frac{f}{F_s}$ or $f = k\frac{F_s}{N}$, which is how you convert the DFT index k to frequency. **Note:** The DFT index k starts at zero, which is different from the Matlab FFT indexing that starts at one (see example in Table 10.3). A further wrinkle is that normalized radian frequencies $\pi < \hat{\omega} <= 2\pi$ (frequencies on the bottom half of the unit circle) actually represent negative radian frequencies since the principle argument for sinusoids must be in the interval $-\pi < \theta <= \pi$. Thus the normalized radian frequencies in the interval $\pi < \hat{\omega} <= 2\pi$ effectively have 2π subtracted from them resulting in negative frequencies.

In practice, the DFT is not used much because there is a much faster algorithm called the *Fast* Fourier Transform, or FFT (see Sect. 10.4.3). The computational cost for the DFT is $O(N^2)$ in contrast to the FFT's computational cost of $O(N \log N)$, which is significantly faster for larger FFT sizes as can be seen in Table 10.2.

Table 10.2: Computational cost of FFT vs DFT

Power of 2 v	FFT size $N = 2^v$	FFT Real ×'s $4N(\log_2(N) - 1)$	FFT Real +'s $2N\log_2(N)$	FFT Real ops	DFT Real ×'s $4N^2$	DFT Real +'s $4N^2 - 2N$	DFT Real ops	FFT Speedup
3	8	64	48	112	256	240	496	4.4
4	16	192	128	320	1024	992	2016	6.3
5	32	512	320	832	4096	4032	8128	9.8
6	64	1280	768	2048	16,384	16,256	32,640	15.9
7	128	3072	1792	4864	65,536	65,280	130,816	26.9
8	256	7168	4096	11,264	262,144	261,632	523,776	46.5
9	512	16,384	9216	25,600	1,048,576	1,047,552	2,096,128	81.9
10	1024	36,864	20,480	57,344	4,194,304	4,192,256	8,386,560	146.3
11	2048	81,920	45,056	126,976	16,777,216	16,773,120	33,550,336	264.2
12	4096	180,224	98,304	278,528	67,108,864	67,100,672	134,209,536	481.9
13	8192	393,216	212,992	606,208	268,435,456	268,419,072	536,854,528	885.6
14	16,384	851,968	458,752	1,310,720	1,073,741,824	1,073,709,056	2,147,450,880	1638.4
15	32,768	1,835,008	983,040	2,818,048	4,294,967,296	4,294,901,760	8,589,869,056	3048.2
16	65,536	3,932,160	2,097,152	6,029,312	17,179,869,184	17,179,738,112	34,359,607,296	5698.8

10.4.3 FFT

We will not get into the derivation of the FFT since there are good books on the subject ([4, 5]). Rather, we will look at how to use and interpret the FFT. The signal that we will examine is shown in Fig. 10.11 where the waveform is shown in the top plot. This signal (sampled at F_s = 2000 Hz) is zero for 0.5 seconds, a 10 Hz signal for 1 second, zero for 0.5 seconds, 100 Hz for 1 second, and then zero for 0.5 seconds. This results in a signal with 7002 samples. Since the FFT needs an input length that is a power of 2, we take an FFT of length N = 8192 where we add zeros to the end of the signal to make a signal with 8192 samples (known as zero padding). The output of the FFT is a vector of 8192 complex values, which is hard to plot. Since we are interested in the frequency content of the signal, we plot the magnitude, which is the middle plot of Fig. 10.11 where it has been plotted in decibels (i.e., Matlab command: m1 = 20*log10(abs(f1));) and where the peak dB value has been set to zero (i.e., Matlab command: m1 = m1-max(m1);). Setting the peak dB value to zero is typically done since we are usually more interested in the relative magnitudes of the frequencies in the signal than their absolute magnitudes. As an example in audio, we are typically more interested in how the audio sounds (harmonics, etc.) rather than how loud it is when it comes to viewing the audio spectrum.

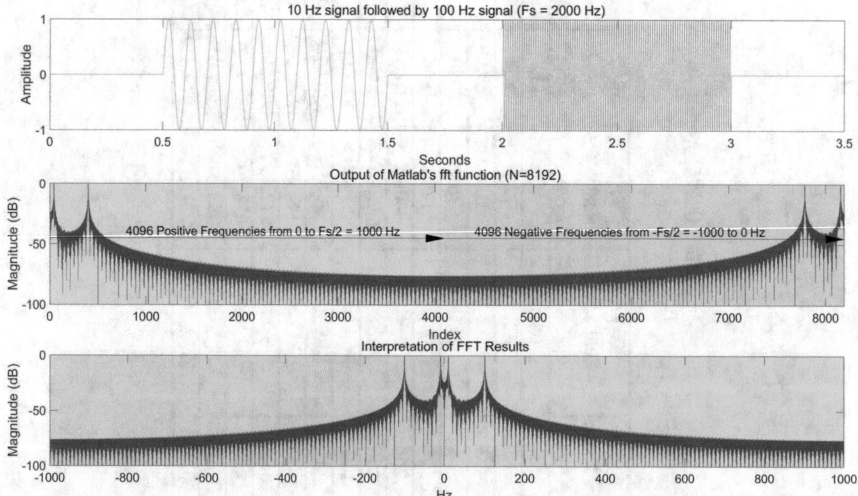

Fig. 10.11: Top figure: signal with 10 and 100 Hz components sample at Fs = 2000 Hz; middle figure: Output of Matlab's fft() function that puts the negative frequencies in the last half of the output vector; bottom figure: spectrum as expected with frequencies ordered on the real axis

One aspect to notice about the FFT result that is shown in the middle plot of Fig. 10.11 is the symmetry about the midpoint (DFT index k = 4096 or Matlab's

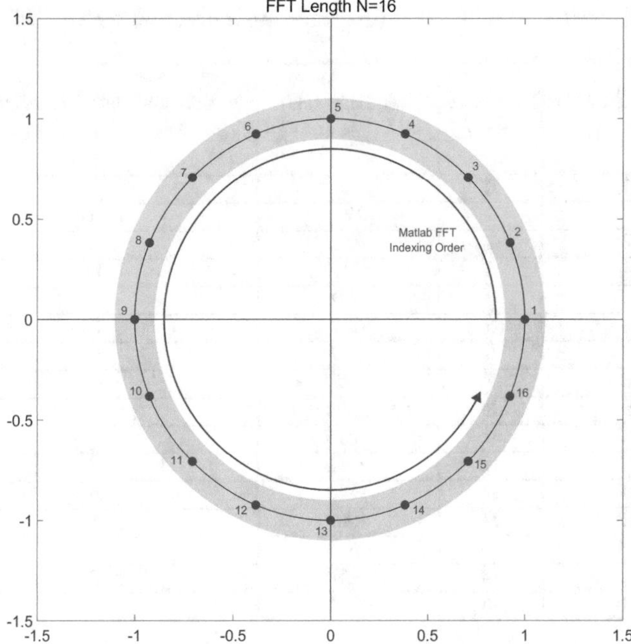

Fig. 10.12: Matlab's FFT indexing

index i = 4097), assuming that the FFT was taken of a signal with real (not complex) values. What can be confusing is that all the FFT values past the midpoint are associated with negative frequencies and you would expect to see the plot shown at the bottom of Fig. 10.11 with negative frequencies ordered as expected on the abscissa axis. Thus just plotting the FFT result will place the negative frequencies on the right side of the plot (red section) past the positive frequencies (green section). The reason this occurs is because the FFT normalized frequencies are evaluated around the unit circle from 0 to 2π as shown in Fig. 10.12 and the frequencies in the range $\pi < \hat{\omega} <= 2\pi$ (bottom half of the circle) are converted by the sinusoid functions to $-\pi < \hat{\omega} <= 0$ by effectively subtracting 2π due to the principle arguments of sinusoids being $-\pi <= \hat{\omega} <= \pi$. The associated DFT indexing and frequencies (assuming $Fs = 2000$) are listed in Table 10.3 for a 16-point FFT.

Due to the symmetry as seen in the middle plot of Fig. 10.11 when taking the FFT of real signals, typically only the first half of the FFT vector is plotted since it contains the positive frequencies. **Note:** If you are performing frequency domain processing of a real signal that involves taking the inverse FFT and you modify a positive frequency value by modifying either the magnitude or the phase, you also need to modify the associated negative frequency in the same manner, i.e., if you modify a Matlab FFT value at index i (DFT index *k=i-1*), you also need to modify the Matlab FFT value at index $j = N - i + 2$ (DFT index $j = N - k + 1$), where N is the FFT length.

Table 10.3: N = 16-point FFT indexing translations (Fs = 2000)

Matlab FFT index $i = 1 : N_{\text{FFT}}$	DFT index $k = i - 1$	Frequency (Hz) $f = k\frac{Fs}{N_{\text{FFT}}} = k\frac{2000}{16}$	Matlab conjugate frequency index $i_{\text{conj}} = N_{\text{FFT}} - k + 1 = N_{\text{FFT}} - i + 2$
1	0	0 (DC)	
2	1	125	16
3	2	250	15
4	3	375	14
5	4	500	13
6	5	625	12
7	6	750	11
8	7	875	10
9	8	1000 (Nyquist)	
10	9	−875	8
11	10	−750	7
12	11	−625	6
13	12	−500	5
14	13	−375	4
15	14	−250	3
16	15	−125	2

10.5 Practical Considerations When Using the FFT

10.5.1 Windowing

In the real-time analysis and synthesis FPGA example system in Sect. 3.1, one can see a "window" being applied in the Simulink model in Fig. 3.7 before being sent to the FFT engine. What window should be applied here? What would happen if you eliminated this windowing step? There are several reasons for performing this step.

To answer these questions, we first need to go back to the original definition of the Fourier transform, which we show here again:

$$\hat{s}(f) = \mathcal{F}\{s(t)\} = \int_{-\infty}^{\infty} s(t)e^{-j2\pi ft}\, dt \tag{10.34}$$

Notice the limits of integration in this definition. When we use computers, we do not have the time to start at time $t = -\infty$ and then wait until $t = +\infty$. Even if we could, we would not have the memory to be able to store a signal this long. So what happens if we have a signal that only lasts from time $t = t_1$ to time $t = t_2$ and is zero outside this time interval? Let us rewrite this as

$$
\begin{aligned}
\hat{s}(f) &= \int_{-\infty}^{\infty} s(t)e^{-j2\pi ft}\,dt && (10.35)\\
&= \int_{-\infty}^{t_1} s(t)e^{-j2\pi ft}\,dt + \int_{t_1}^{t_2} s(t)e^{-j2\pi ft}\,dt + \int_{t_2}^{\infty} s(t)e^{-j2\pi ft}\,dt && (10.36)\\
&= 0 + \int_{t_1}^{t_2} s(t)e^{-j2\pi ft}\,dt + 0 && (10.37)\\
&= \int_{t_1}^{t_2} s(t)e^{-j2\pi ft}\,dt && (10.38)
\end{aligned}
$$

However, we should not take this approach since we are getting away from using the definition of the Fourier Transform. So instead of playing with the integration limits, let us leave them alone but modify our signal instead. To do this, let us define a new signal $w(t)$ as

$$
w(t) = \begin{cases} 1 & \text{if } t_1 \leqslant t \leqslant t_2 \\ 0 & \text{otherwise} \end{cases} \tag{10.39}
$$

and we can rewrite the Fourier Transform where we multiple by this windowing function to time limit our signal and not have to mess with the limits of integration. The signal is now zero outside the time interval $t_1 \leqslant t \leqslant t_2$.

$$
\hat{s}(f) = \int_{-\infty}^{\infty} s(t)w(t)e^{-j2\pi ft}\,dt \tag{10.40}
$$

This means that when we take the FFT of a signal using a computer, we are ***always*** applying a window function, even if we do not think we are. **Note: If you do not explicitly apply a window to your signal, you are in effect using a *rectangular window* as defined in Eq. 10.40.**

By using a finite signal, which we have to do when using computers, we have essentially applied a rectangular window to an infinite signal. So what does this do to our FFT results? The mathematical answer is the following where we generalize to any window function $w(t)$. If $\hat{s}(f)$ is the Fourier Transform of $s(t)$ as seen in Eq. 10.34 and $\hat{w}(f)$ is the Fourier Transform of $w(t)$, we take the inverse Fourier Transform as defined by

$$
s(t) = \mathcal{F}^{-1}\{\hat{s}(f)\} = \frac{1}{2\pi}\int_{-\infty}^{\infty} \hat{s}(f)e^{j2\pi ft}\,df \tag{10.41}
$$

Then the Fourier Transform of the windowed signal is

$$\mathcal{F}\{s(t)w(t)\} = \int_{-\infty}^{\infty} [s(t)w(t)]\, e^{-j2\pi ft}\, dt \tag{10.42}$$

$$= \int_{-\infty}^{\infty} \left[\frac{1}{2\pi}\int_{-\infty}^{\infty} \hat{s}(\zeta)e^{j2\pi\zeta t}\, d\zeta\right] w(t)e^{-j2\pi ft}\, dt \tag{10.43}$$

$$= \frac{1}{2\pi}\int_{-\infty}^{\infty} \hat{s}(\zeta)\left[\int_{-\infty}^{\infty} w(t)e^{-j2\pi ft}e^{j2\pi\zeta t}\, dt\right] d\zeta \tag{10.44}$$

$$= \frac{1}{2\pi}\int_{-\infty}^{\infty} \hat{s}(\zeta)\left[\int_{-\infty}^{\infty} w(t)e^{-j2\pi(f-\zeta)t}\, dt\right] d\zeta \tag{10.45}$$

$$= \frac{1}{2\pi}\int_{-\infty}^{\infty} \hat{s}(\zeta)\hat{w}(f-\zeta)\, d\zeta \tag{10.46}$$

$$= \frac{1}{2\pi}[\hat{s}(f) * \hat{w}(f)] \tag{10.47}$$

Thus multiplication of the signal and the window function in the time domain is equivalent to **convolution** in the frequency domain.

What does this mean in practice? Before we answer this, we first need to know what the Fourier transform of our rectangular window is. For convenience, we will define $t_1 = -\frac{T}{2}$ and $t_2 = -\frac{T}{2}$ for our rectangular window in Eq. 10.39.

$$\mathcal{F}\{w(t)\} = \hat{w}(f) = \int_{-\infty}^{\infty} w(t)e^{-j2\pi ft}\, dt \tag{10.48}$$

$$= \int_{-\frac{T}{2}}^{\frac{T}{2}} e^{-j2\pi ft}\, dt \tag{10.49}$$

$$= \frac{1}{-j2\pi f}\left[e^{-j2\pi ft}\Big|_{-\frac{T}{2}}^{\frac{T}{2}}\right] \tag{10.50}$$

$$= \frac{1}{-j2\pi f}\left[e^{-j\pi fT} - e^{j\pi fT}\right] \tag{10.51}$$

$$= \frac{T}{\pi fT}\left[\frac{e^{j\pi fT} - e^{-j\pi fT}}{2j}\right] \tag{10.52}$$

$$= \frac{T}{\pi fT}\sin(\pi fT) \tag{10.53}$$

$$= T\text{sinc}(fT) \tag{10.54}$$

We can use the Matlab sinc() function to plot this function, which we do in Fig. 10.13 in the right column for two different values of time width T. The associated rectangular windows are shown on the left side and the associated Fourier transform on the right side. Note that the longer rectangular window in time (bottom left) results in narrower sinc function in frequency (bottom right). This means that to get a better frequency resolution in the frequency domain, a longer window of the signal in time needs to be taken.

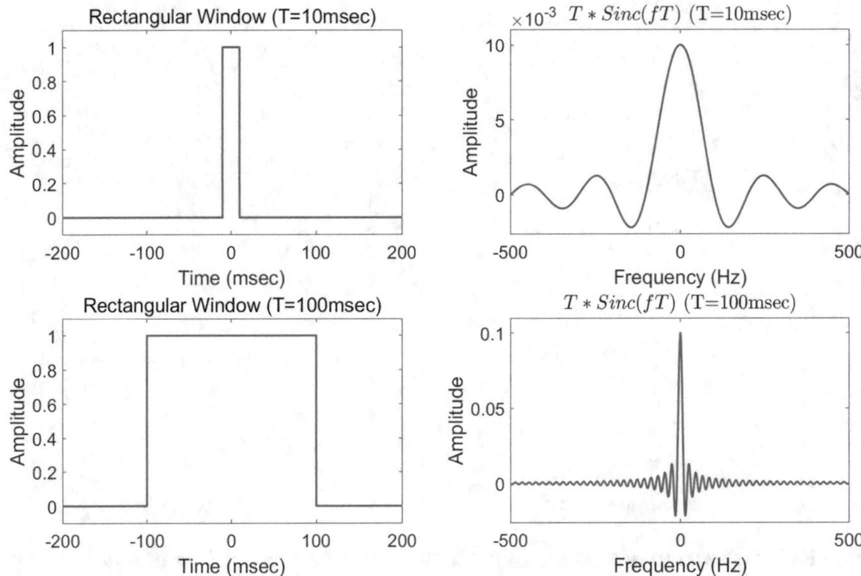

Fig. 10.13: The Fourier transform of a rectangular window is a sinc function. Longer windows in time result in narrower sinc functions in frequency

10.5.2 Window Length

In practice we can view the effect of the rectangular window (and other windows) as a blurring operation in the frequency domain where longer windows in time blur less in the frequency domain. Shorter windows blur more. To illustrate this, let us look at the spectrum of a cosine signal with frequency $f = 1000$ Hz where the Fourier transform has been compute of the cosine signal using rectangular windows of sizes [5 10 50 100 1000] msec as shown in Fig. 10.14. As we take longer rectangular windows of the cosine function in time, the resulting spectrum looks more and more like the line spectrum we expect for the cosine with $f = 1000$ Hz (shown as the green vertical line).

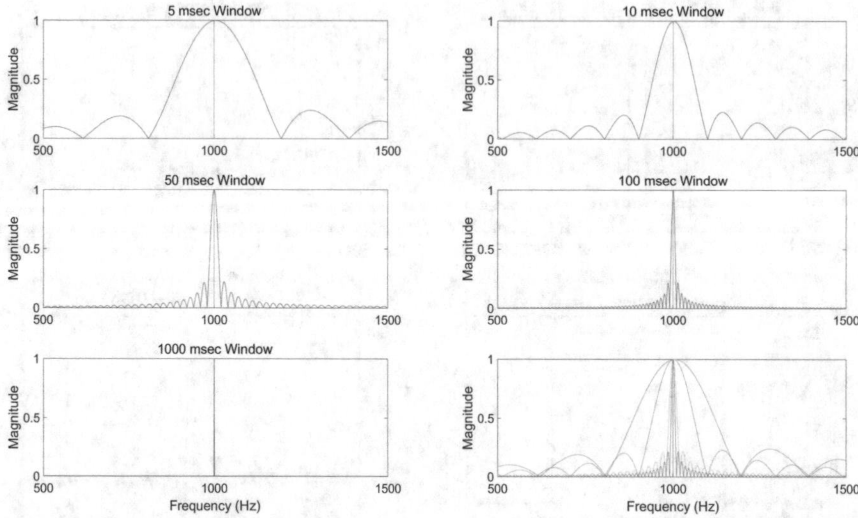

Fig. 10.14: Spectrum of a cosine signal with frequency $f = 1000$ Hz evaluated using windows of sizes [5 10 50 100 1000] msec ($Fs = 10{,}000$ Hz). The green vertical line is located at $f = 1000$ Hz. The last plot (bottom right) gives all the plots together for comparison purposes

Window length becomes important if we are trying to resolve frequencies that are close together. As an example, let us create a signal that is the sum of two cosines, one at $f1 = 1000$ Hz and the other at $f1 = 1050$ Hz. Now let us look at the resulting spectrum using windows of different lengths as shown in Fig. 10.15. With the 10 msec window, we do not even know that there are two frequencies. This is because the spectrum of the 10 msec window (shown in the upper right of Fig. 10.14) has been convolved with the two close frequencies and has blurred them together. As the windows get longer, the blurring from the associated sinc functions (convolution in the frequency domain) becomes less noticeable. Using a 1000 msec window results in hardly any blurring since the sinc function is very narrow relative to the separation (50 Hz) of the two frequencies.

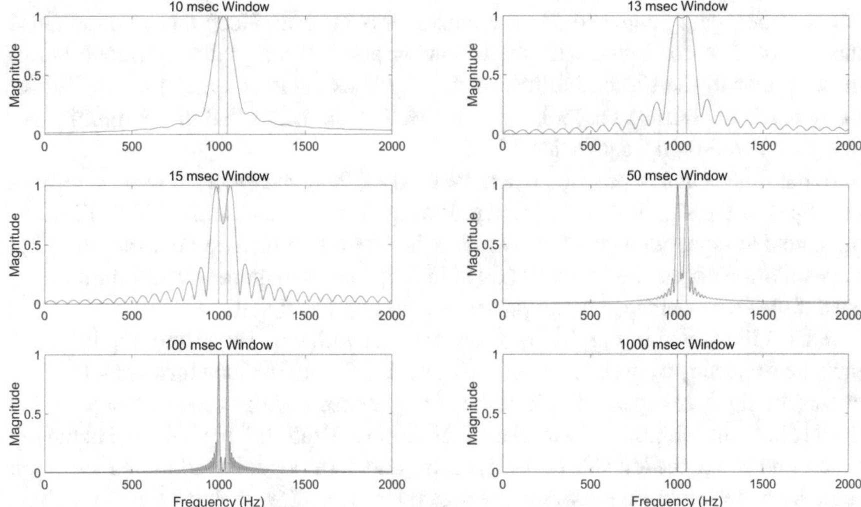

Fig. 10.15: Spectrum of a signal with two cosines with frequencies $f = 1000\,\text{Hz}$ and $f = 1050\,\text{Hz}$ evaluated using windows of sizes [10 13 15 50 100 1000] msec ($Fs = 10{,}000\,\text{Hz}$). If the window is too short (e.g., 10 msec window top left), the signal frequencies get blurred together and you cannot resolve them at all. The green vertical lines are located at $f = 1000\,\text{Hz}$ and $f = 1050\,\text{Hz}$

In summary, the longer the window is in time, the narrower the associated peak is in frequency. Thus if you are interested in good frequency resolution, use a longer window, which means using the FFT with more points. The trade-off is lower time resolution, since you do not know where the frequency occurs within a window, and longer latency for the FFT computation.

10.5.3 Window Edge Effects

If you look at the Fourier series example in Fig. 10.4, we determined what the Fourier series coefficients should be over the interval $0 \leq t \leq T = 5$ seconds. What happened outside of this interval? Since the sinusoids in the Fourier series are periodic, the signal represented by the Fourier coefficients becomes periodic with period T, as can be seen in the figure that shows three periods. Thus anytime we represent a signal in the frequency domain by taking the FFT, we have turned it into a periodic signal in time where it continuously repeats outside the window for all time. This is a subtle point since we typically expect the signal to remain the same when we extract a portion of it and do not realize that by extracting it (i.e., windowing) and taking the FFT, we have caused it to repeat continuously in time. This is because we are only paying attention to the time interval where we extracted the signal and do not consider what happened in time outside this interval. Keep this in mind since this needs to be combined with the following point. If a signal has discontinuities in it, similar to the discontinuities in the signal example in Fig. 10.4, it will take many harmonic terms to model the signal well. This means that discontinuities are associated with high frequencies.

What this means in practice when we used FFTs is that we need to be careful at the edges of the signal that we are windowing. This is because the FFT will cause our signal to repeat outside of our window in time and if there are discontinuities at the windowed edges, the FFT will add high frequencies to model the discontinuities, even if these frequencies are not present in the original signal.

This is illustrated in Fig. 10.16 where we start with a 64 Hz cosine (top left). We pick the frequency of the cosine and FFT length ($N = 1024$) such that when the FFT causes the signal to repeat outside of the $N = 1024$ rectangular window, the period of the 64 Hz cosine aligns well with the effective periodization. This is shown in the top center plot where the left side (in red) is at the end of the signal window and the green segment on the right side is where the signal has been effectively replicated in time. The FFT spectrum in this case is shown (top right). The rectangular window used had $N = 1024$ points, which resulted in a very narrow sinc spectrum in the frequency domain. The noise floor of the FFT spectrum is very low (−314 dB) because of the cosine that aligned perfectly with the window edges as it was replicated in time. This perfect alignment rarely happens in practice except in contrived cases like this one shown.

What typically happens is that the frequency components do not align nicely with the window replication. This is modeled in the example shown in the middle row of Fig. 10.16. Here we illustrate a typical practice of zero padding in order to get a power of two length that we want when applying the FFT. In this example we just set the last three points to zero. This causes a discontinuity to occur when the FFT replicates the signal (middle plot). The left side (in red) is at the end of the signal window and the green segment on the right side is where the signal has been effectively replicated in time. The FFT spectrum is now significantly different where the noise floor changed from −314 dB (shown as the green signal) to −55 dB (in blue). The discontinuity at the window edge has significantly changed the spectrum.

To get rid of discontinuities at the window edges, we can no longer use a rectangular window. We need a window that squashes the ends down to zero so that when the signal is replicated by the FFT, there are no discontinuities. A commonly used window that does this is is the **Hanning** (or Hann) window. A Hanning window ($N = 1024$) applied to the 64 Hz cosine signal with the edge discontinuity is shown in the bottom left plot in Fig. 10.16. We can see that there are no discontinuities at the window edges (middle plot). This improves the spectrum representation where the FFT floor drops from −55 dB (green) to −140 dB (blue) as seen in the bottom right plot. The −314 dB noise floor is also plotted in green. Note that the spectrum of the Hanning window has been convolved with the cosine line spectrum, so the Hanning spectrum can be seen in the blue spectrum curve. The Hanning window is only one of many windows that we can use, which gets us into the next section on window trade-offs.

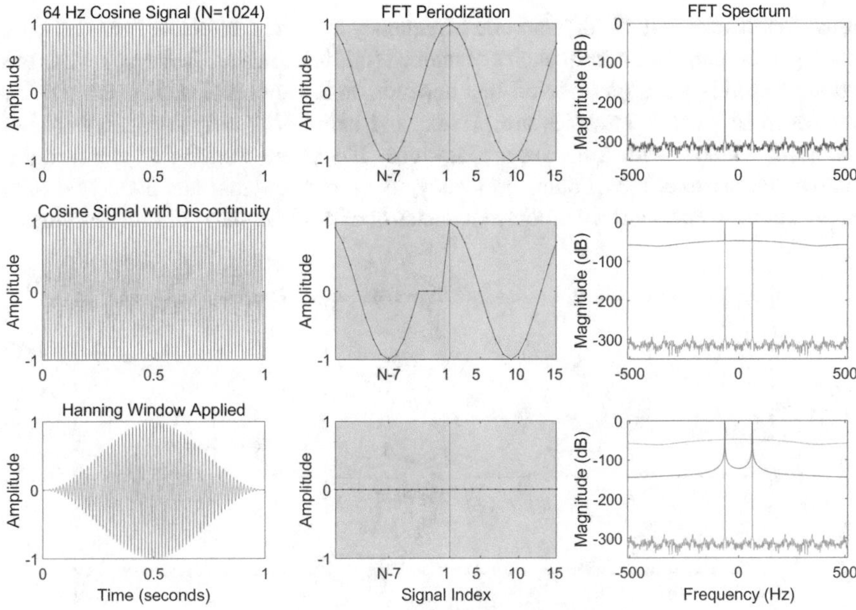

Fig. 10.16: Window edge effects

10.5.4 Window Trade-Offs

When you use the FFT, you are always using a window. If you have not explicitly applied a window, you have used the rectangular window, which means that you have convolved the spectrum of the rectangular window (sinc function) with the signal's spectrum (a blurring operation). In the previous section we discussed why using a Hanning window is better than using a rectangular window to eliminate edge discontinuities. A question you probably have is why a Hanning window? Are there other windows that could be used? The answer is that there are many other windows that can be used and this gets us into window trade-offs.

Let us look at the spectrum of the rectangular window (sinc function) shown in Fig. 10.17. Note that there are two parameters associated with this window spectrum (and all window spectrums). The width of the main-lobe (width at half-height or width at the −3 dB point) and how high the side-lobes are relative the height of the main-lobe. If we want good frequency resolution, we want the main-lobe to be as narrow as possible. An easy way to control the width of the main-lobe is to use longer windows in time. However, we can also affect the main-lobe width just by the choice of window where the windows have the same length in time. This is where the window trade-offs come into play. Choosing a window that has a narrower main-lobe width typically causes the side-lobe attenuation to be reduced (side-lobe peaks become more pronounced). Why is this an issue? Remember that this window spectrum gets convolved with the signal's spectrum. This means that energies associated with other frequencies will "leak" into other frequency locations, affecting the fidelity of energy measurement for frequencies of interest (this is known as "leakage"). Thus the window used is a design trade-off that depends on the application. If you are more interested in what frequencies are present and being able to resolve frequencies, choose a window with a narrower main-lobe. If you are wanting to measure the power that exists at a particular frequency, use a window that has good side-lobe attenuation, so energy from other frequencies does not leak into your measurement.

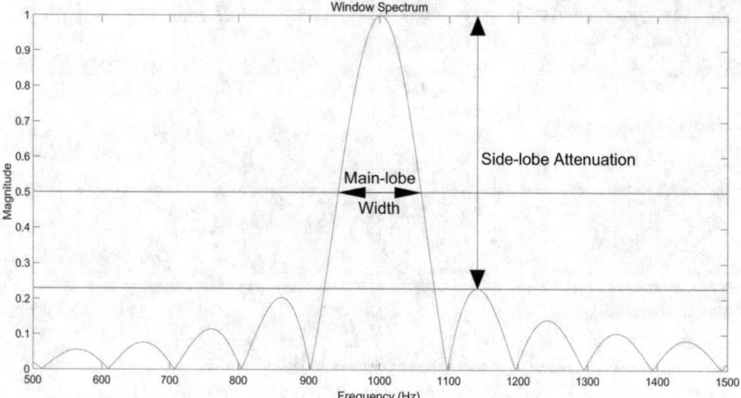

Fig. 10.17: Window trade-offs

The trade-off between main-lobe width and side-lobe attenuation can be seen in Fig. 10.18 for all the windows listed in Tables 10.4 and 10.5 . All the windows except for the rectangular window squash the window edges to zero in time to eliminate discontinuities. This is why the rectangular window is rarely used (except if you forget to apply a window). The Hanning (or Hann) window (blue circle #18) is typically chosen if you do not know what window to apply since it is a good balance between main-lobe width and side-lobe attenuation.

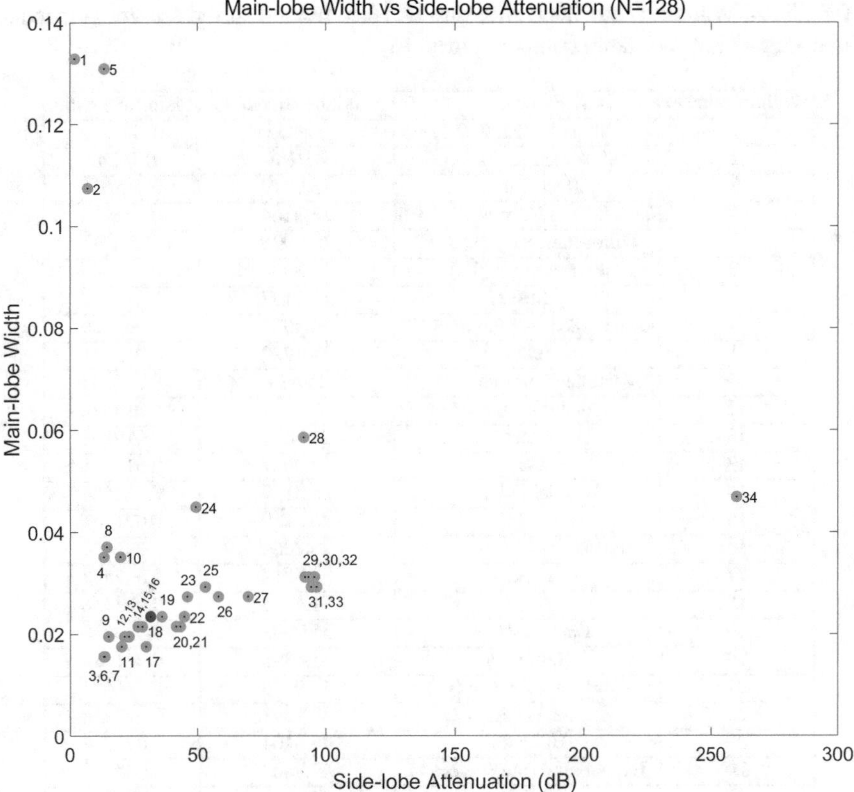

Fig. 10.18: Main-lobe width vs side-lobe attenuation for the windows listed in Tables 10.4 and 10.5. The commonly used Hann (or Hanning) window (number 18 in figure and colored blue) is a good compromise between main-lobe width and side-lobe attenuation. Windows were computed from Joe Henning's window utilities [6]

Table 10.4: Windows associated with Fig. 10.18. Windows sorted according to side-lobe attenuation. Windows computed from [6]

Window number	Window name	Side-lobe attenuation	Main-lobe width
1	Ultraspherical (0,0.5)	1.8255	0.13281
2	Ultraspherical (0.5,0.5)	6.9263	0.10742
3	Rectangular	13.2619	0.015625
4	Generalized Normal (P = 50)	13.3164	0.035156
5	Ultraspherical (−0.5,0.5)	13.3978	0.13086
6	Planck-taper (0.1)	13.3994	0.015625
7	Kaiser	13.6187	0.015625
8	Generalized Normal (P = 10)	14.4567	0.037109
9	Tukey	15.1229	0.019531
10	Generalized Normal (P = 4)	19.724	0.035156
11	Exponential (tau = 64)	20.158	0.017578
12	Welch	21.3084	0.019531
13	Sine	23.0021	0.019531
14	Triangular	26.5129	0.021484
15	Bartlett	26.5129	0.021484
16	Planck-Bessel (0.1, 4.45)	28.2264	0.021484
17	Taylor	29.6914	0.017578
18	Hann	31.4755	0.023438
19	Bartlett–Hann	35.9391	0.023438
20	Hamming	41.6395	0.021484
21	Gaussian (sigma = 0.4)	43.2658	0.021484
22	Slepian (alpha = 2)	44.7997	0.023438
23	Bohman	46.0164	0.027344
24	Exponential (tau = 9.2693)	49.3683	0.044922
25	Parzen	53.0471	0.029297
26	Blackman	58.1236	0.027344
27	Slepian (alpha = 3)	69.7603	0.027344
28	Flat Top	91.5097	0.058594
29	Blackman–Harris	92.0377	0.03125
30	Nuttall	93.3288	0.03125
31	Dolph–Chebyshev	94.4534	0.029297
32	Slepian (alpha = 4)	95.6357	0.03125
33	Blackman–Nuttall	96.5229	0.029297
34	Slepian (alpha = 10)	260.1093	0.046875

Table 10.5: Windows associated with Fig. 10.18. Windows sorted according to main-lobe width. Windows computed from [6]

Window number	Window name	Side-lobe attenuation	Main-lobe width
3	Rectangular	13.2619	0.015625
6	Planck-taper (0.1)	13.3994	0.015625
7	Kaiser	13.6187	0.015625
11	Exponential (tau = 64)	20.158	0.017578
17	Taylor	29.6914	0.017578
9	Tukey	15.1229	0.019531
12	Welch	21.3084	0.019531
13	Sine	23.0021	0.019531
14	Triangular	26.5129	0.021484
15	Bartlett	26.5129	0.021484
16	Planck-Bessel (0.1, 4.45)	28.2264	0.021484
20	Hamming	41.6395	0.021484
21	Gaussian (sigma = 0.4)	43.2658	0.021484
18	Hann	31.4755	0.023438
19	Bartlett–Hann	35.9391	0.023438
22	Slepian (alpha = 2)	44.7997	0.023438
23	Bohman	46.0164	0.027344
26	Blackman	58.1236	0.027344
27	Slepian (alpha = 3)	69.7603	0.027344
25	Parzen	53.0471	0.029297
31	Dolph–Chebyshev	94.4534	0.029297
33	Blackman–Nuttall	96.5229	0.029297
29	Blackman–Harris	92.0377	0.03125
30	Nuttall	93.3288	0.03125
32	Slepian (alpha = 4)	95.6357	0.03125
4	Generalized Normal (P = 50)	13.3164	0.035156
10	Generalized Normal (P = 4)	19.724	0.035156
8	Generalized Normal (P = 10)	14.4567	0.037109
24	Exponential (tau = 9.2693)	49.3683	0.044922
34	Slepian (alpha = 10)	260.1093	0.046875
28	Flat Top	91.5097	0.058594
2	Ultraspherical (0.5,0.5)	6.9263	0.10742
5	Ultraspherical (−0.5,0.5)	13.3978	0.13086
1	Ultraspherical (0,0.5)	1.8255	0.13281

References

1. D. Gottlieb, C.-W. Shu, On the Gibbs phenomenon and its resolution. SIAM Rev. **39**(4), 644–668 (1997)
2. A.V. Oppenheim, R.W. Schafer, *Digital Signal Processing* (Prentice-Hall, Hoboken, 1975)

3. J.G. Proakis, *Digital Signal Processing: Principles Algorithms and Applications* (Pearson, London, 2001)
4. R.N. Bracewell, *The Fourier Transform and Its Applications* (McGraw-Hill, New York, 1986)
5. E.O. Brigham. *The Fast Fourier Transform and Its Applications* (Prentice-Hall, Hoboken, 1988)
6. J. Henning, Window Utilities. MATLAB Central File Exchange https://www.mathworks.com/matlabcentral/fileexchange/46092-windowutilities. Accessed 18 Jun 2021

Part II
SoC FPGA System Development

SoC FPGAs allow you to create custom hardware and software designs, which gives great flexibility. The downside of developing a highly complex SoC system is that if the system does not work, where is the problem? The problem could be a bug in your VHDL code, a misconfiguration of the HPS, or a bug in your C code. How can you debug such a system? This is much more difficult than developing software for desktop computers since programmers do not expect that they will need to debug a PC's CPU or motherboard if their programs do not work.

There are two strategies that minimize SoC FPGA hardware/software debugging issues and we will use both.

Strategy 1: Start with a SoC FPGA hardware/software system design that you know already works. Ideally this will be close to the system you are targeting. Start with the working system and then slowly modify it into the system you want. After each modification, test the system to make sure that it is still working, fixing what broke at each modification step.

Strategy 2: When you add a hardware component to Platform Designer, test the component (e.g., with System Console, which is the topic of Lab 7), so you know that the hardware is functional before you start layering software on top of it. If you do not and the component does not respond to your software, you will not know if the problem is with your VHDL code, a Platform Designer connection, or with your C code.

Chapter 11
Development Environment Setup

11.1 Software Setup

11.1.1 Setting up a PuTTY Terminal Window in Windows to Communicate with Linux on the DE10-Nano Board

The DE10-Nano board uses the FT232R USB UART chip from Future Technology Devices International Limited or (FTDI) to convert the USB interface on the right side of the board as shown in Fig. 11.1 into a serial UART interface for the HPS in the Cyclone V SoC FPGA (schematic shown in Fig. 11.2). Once set up, this connection appears as a *Virtual Com Port* (**VCP**) in Windows. This connection is used to communicate with Linux running on the DE10-Nano board from the PuTTY terminal window from within Windows. The physical connection requires the use of a USB Mini-B cable (Fig. 11.3).

The UART connection is located above the Ethernet connector on the right side of the board, which is marked UART-to-USB in the figure. Do not use the USB-Blaster II port on the left side, which has the same USB Mini-B connector.

Fig. 11.1: The USB UART connection is on the right side, above the Ethernet port. It uses a USB Mini-B Connector. Figure is from the DE10-Nano getting started guide

R. K. Snider, *Advanced Digital System Design using SoC FPGAs*,
https://doi.org/10.1007/978-3-031-15416-4_11

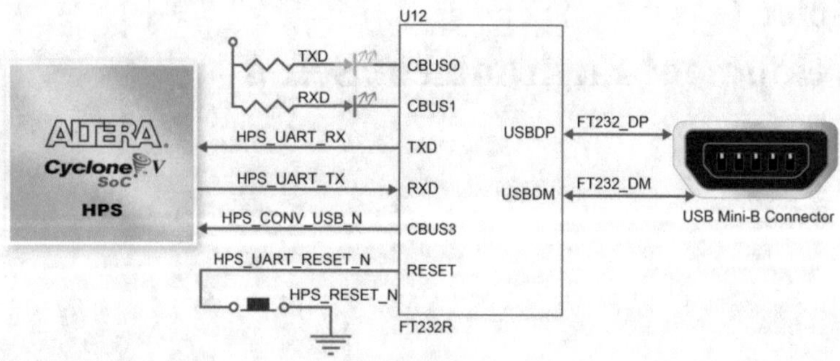

Fig. 11.2: The USB UART schematic connection. Figure is from the DE10-Nano schematic

Fig. 11.3: The USB Mini-B connector

To run the PuTTY UART Terminal in Windows, you will need the following software items:

Item 1: PuTTY, which can be downloaded from one of the two sources:

Source 1: (click here)
Source 2: (click here)

Item 2: FT232R USB UART Driver. The driver should automatically be installed by Windows. If not, you can download the driver from the (FTDI Drivers page).

11.1.1.1 Finding the COM Port Assigned by Windows

You need to know the communications port (e.g., COM3) that was assigned by Windows to the USB serial port when you run PuTTY. To find this out, check the Windows Device Manager using the following steps:

Step 1: Type *Device Manager* in the search window that is right next to the Windows icon that is at the lower left of the screen.

Step 2: Open Device Manager and expand the Ports (COM & LPT) section.

Step 3: Find entries that say *USB Serial Port*. Figure 11.4 shows an example where the COM port number is COM3. Note: The DE10-Nano board needs to be plugged into the Mini-B USB cable for the COM port to show up in Windows.

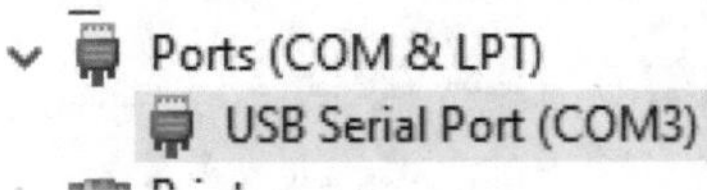

Fig. 11.4: Getting the COM port number from the windows device manager

Step 4: Verify that the COM port number is the one connected to the DE10-Nano board by unplugging the board. The COM port should disappear in the Windows device manager. Note: COM port numbers will change if you plug in another board but will stay the same if it is the same board.

11.1.1.2 PuTTY Setup

Configure the PuTTY Serial settings by selecting `\Connection\Serial` in the **Category** panel (panel on the left side). On the right side, set the following patterns to the given values as shown in Fig. 11.5:

Setting 1: **COM Port** (Serial line to connect to) = The COM port number determined in Sect. 11.1.1.1

Setting 2: **Speed (baud)** = **115200**

Setting 3: **Data bits** = **8**

Setting 4: **Stop bits** = **1**

Setting 5: **Parity** = **None**

Setting 6: **Flow control** = **None**

Next, select `\Session` in the **Category** panel (left side). Make sure that the **Connection type** is **Serial** as shown in Fig. 11.6.

Open the terminal window by clicking the Open button at the bottom. If PuTTY only beeps rather than opening a terminal window, you are probably trying to open the terminal window in the serial configuration setting panel `\Connection\Serial` rather than being in the `\Sessions` panel. Hit return if you do not see the Linux login prompt or power cycle the DE10-Nano board to see the boot process.

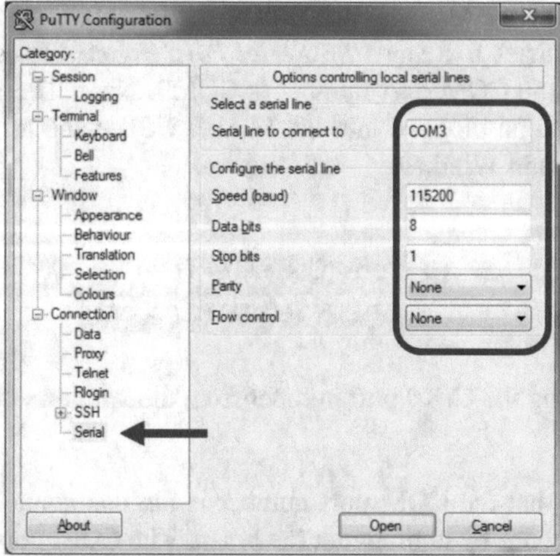

Fig. 11.5: Setting up the serial settings in PuTTY

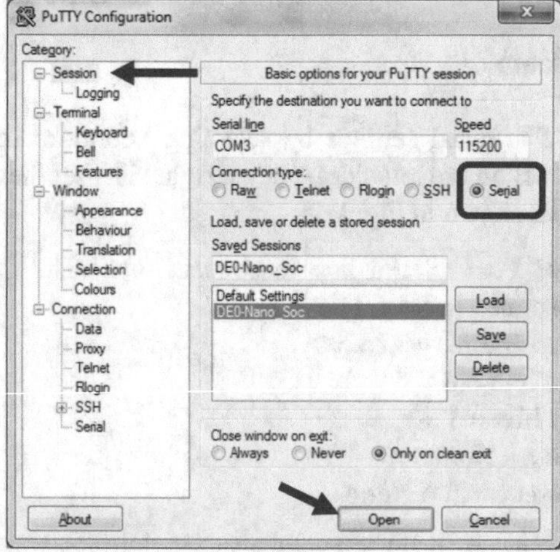

Fig. 11.6: Opening a terminal window in PuTTY

Finally, save the configuration settings in the Session window by clicking the Save button after entering an appropriate session name. This is so that you can load the settings next time rather than entering all the information again.

11.1.2 VirtualBox Ubuntu Virtual Machine Setup

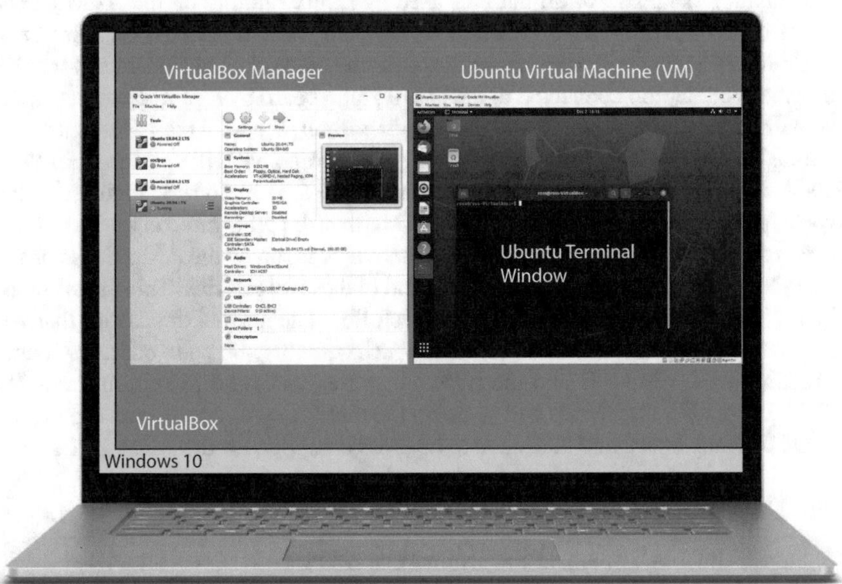

Fig. 11.7: The Ubuntu Virtual Machine (VM) running on a Windows 10 Laptop (illustration). The VirtualBox Manager is what opens when you run VirtualBox. The Ubuntu Virtual Machine (VM) is installed and opened from within the VirtualBox Manager. The Ubuntu Terminal Window is opened from within the Ubuntu VM and is where Ubuntu VM Linux commands are entered

11.1.2.1 Overview

We need a Linux development environment in order to compile Linux code that targets the ARM CPU inside the Cyclone V SoC FPGA. We will do this by creating a Linux virtual machine on Windows by using VirtualBox, which we will call the **Ubuntu VM**. Note that we will be compiling ARM CPU code in the Ubuntu VM, which is running on Windows 10 using a x86 CPU. Thus the compiler itself will be running and using x86 instructions, but creating machine code for the ARM CPU. This process of compiling code for a different CPU than what the compiler is using is called *cross compiling*. If you try to run this ARM code in the Ubuntu VM, it will not work since the Ubuntu VM requires programs to be compiled for the x86 CPU. The ARM code needs to be transferred to the DE10-Nano board before it can run (which cannot run x86 code) (Fig. 11.7).

In addition, we want a Linux environment from which to boot the DE10-Nano board during development, rather than booting from the microSD card that is shipped with the board. We will do this by booting from the Ubuntu VM using Ethernet where the Ubuntu VM will serve all the files used by Linux running on the ARM CPUs. This is called a **Network File System (NFS)**, and we will create a NFS server in the Ubuntu VM that will serve these files over the network. This local network will consist of your laptop connected to the DE10-Nano board by a short Ethernet cable. We will call booting via NFS the **Developer's Boot Mode**. This is in contrast to booting the DE10-Nano from the microSD card, which we will call the **Ship Boot Mode** since booting from the microSD card is the typical boot configuration that gets shipped when development is done. Note that the NFS directories and all the code in them that is served by the NFS server, which is run by the ARM CPUs on the DE10-Nano board, need to be cross compiled in these directories. This means in the Ubuntu VM, there will be two types of directories. The normal directories that you see when the Ubuntu VM is first created, which contains x86 code because Ubuntu is running on a x86 CPU and the NFS directories that need to run on the ARM CPUs. These NFS directories end up being the root file system for Linux running on the DE10-Nano board and thus need to be cross compiled for the ARM CPUs.

Setting up the Ubuntu virtual machine requires the following steps:

Step 1: Enable Virtualization in your PC's BIOS.
Step 2: Download VirtualBox.
Step 3: Download Ubuntu.
Step 4: Install VirtualBox.
Step 5: Install Ubuntu in VirtualBox.
Step 6: Install the VirtualBox Guest Additions.
Step 7: Install desired Linux software in the Ubuntu virtual machine (VM).

11.1.2.2 Enable Virtualization in Your PC's BIOS

VirtualBox's 64-bit Virtual Machine (VM) guests will not install unless hardware virtualization has been enabled in your PC's BIOS, so check and make sure your PC/Laptop's BIOS has virtualization enabled before you proceed with the VirtualBox installation. If you do not, and virtualization has not been enabled, you will probably get the error shown in Fig. 11.8.

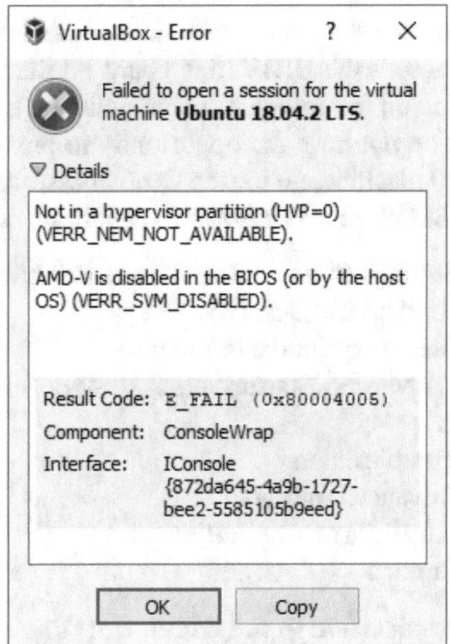

Fig. 11.8: Bad Bios setting for virtualization

First check if virtualization has been enabled on your PC. To do this in Windows 10, open up the **Task Manager**, click "More details" if it opens in a minimal window, and click on the **Performance Tab**. The task manager can be opened by typing *Task Manager* in the lower left search bar in Windows 10. The *Performance* tab is the tab just to the right of the *Processes* tab, which opens by default. Figure 11.9 shows an example where virtualization has not yet been enabled.

Utilization 2%
Speed 3.39 GHz
Processes 205
Threads 2583
Handles 85393
Up time 3:04:05:38

Base speed:	3.60 GHz
Sockets:	1
Cores:	8
Logical processors:	16
Virtualization:	Disabled
L1 cache:	768 KB
L2 cache:	4.0 MB
L3 cache:	16.0 MB

Fig. 11.9: Example showing where virtualization has been disabled and needs to be enabled

If virtualization is disabled, perform the following steps to enable virtualization in your PC's BIOS. **Before starting, make sure that you save any work and close down open programs because you will be rebooting your computer**:

Step 1: Reboot your computer, and when the computer starts booting up, press the key to access the BIOS (this could be the Delete, Esc, F1, F2, or F4 depending on the computer manufacturer). If you cannot see a BIOS message or do not have the opportunity to press a key, especially on a Windows 10 machine, go to step 2; otherwise, skip to step 3.

Step 2: **To set the BIOS on a Windows 10 machine**, do the following:

a. Open Settings (type "settings" in search bar).
b. Select Update and Security.
c. Select Recovery (in the left menu).
d. Under Advanced Startup, click Restart Now (your computer will reboot).
e. Click Troubleshoot.
f. Click Advanced options.
g. Select UEFI Firmware Settings.
h. Click Restart.

Step 3: **Enable virtualization in the BIOS**. How this is done depends on your motherboard's BIOS, so Google "enabling virtualization" and the motherboard or laptop model number. For example, I googled "enabling virtualization ax370" and found the following steps for my AMD Ryzen 7 machine, which was not at all obvious when I first examined the Gigabyte BIOS (which has a terrible user interface for finding and setting this parameter). Your steps are likely to be different depending on the make and model of your motherboard or laptop. However, my steps were:

a. Set Enabled: M.I.T. → Advanced Frequency Settings → Advanced CPU Core Settings → SVM Mode.
b. Set Enabled: Chipset → IOMMU.

Step 4: Check the performance tab in Windows 10 Task Manager to see that virtualization has now been enabled. You can see in Fig. 11.10 that Virtualization has now been enabled in the Task Manager.

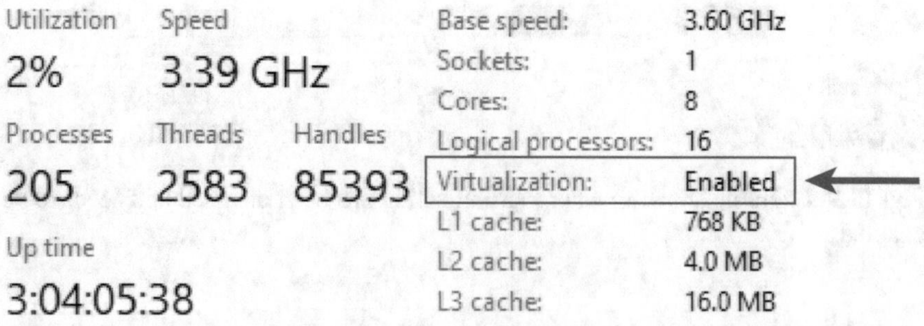

Utilization 2% Speed 3.39 GHz
Processes 205 Threads 2583 Handles 85393
Up time 3:04:05:38
Base speed: 3.60 GHz
Sockets: 1
Cores: 8
Logical processors: 16
Virtualization: Enabled
L1 cache: 768 KB
L2 cache: 4.0 MB
L3 cache: 16.0 MB

Fig. 11.10: Performance tab in task manager showing that virtualization has been enabled

11.1.2.3 Download VirtualBox

Download VirtualBox and its associated extension pack. We will be installing VirtualBox 6.1 in this example. In general, you should download the version specified since version dependencies can exist across different software packages. In the case of VirtualBox, the latest version should work, which can be found at:

https://www.virtualbox.org/

Step 1: Click the Download VirtualBox link. On the download page, you will see two headings: **VirtualBox 6.x.y platform packages** and **VirtualBox 6.x.y Oracle VM VirtualBox Extension Pack**. We will need both downloads. The version 6.x.y stands for the latest version you see on the download page (which is VirtualBox 6.1.12 at the time of this writing).

Step 1A: Under the **VirtualBox 6.1.12 platform packages** heading, click on *Windows hosts* to download VirtualBox for Windows 10.

Step 1B: Under the **VirtualBox 6.1.12 Oracle VM VirtualBox Extension Pack** heading, click on *All platforms* to download the extension pack. We need the extension pack so that the Ubuntu VM can access USB Flash Drives and share folders with Windows 10.

11.1.2.4 Download Ubuntu

We will typically use that latest LTS release of Ubuntu, which currently is Ubuntu 20.04 LTS . LTS stands for Long-Term Support, and Ubuntu 20.04 LTS will be supported by Canonical until 2030 [1, 2]. Standard Support typically lasts for 5 years, and the End of Life is 10 years after initial release. Ubuntu can be found at:

http://releases.ubuntu.com/

Select Ubuntu 20.04 LTS or the latest LTS release and download the 64-bit PC (AMD64) desktop image (ubuntu-20.04.1-desktop-amd64.iso). **Note:** This download is 2.6 GB, which will take a while to download.

11.1.2.5 VirtualBox Installation

Step 1: **Install Oracle's VM VirtualBox** by following the installation instructions. Default settings are OK (i.e., desired install location, etc.). If you have a previous version of VirtualBox, you can install over the previous installation (your VMs will be kept). Update by accepting default values.

When the new version of VirtualBox runs, if you are updating, it will prompt you to update the Extension Pack as well.

Step 2: **Install the VirtualBox Extension Pack** by following the installation instructions.

11.1.2.6 Setting up Ubuntu in VirtualBox

In this section, we will create the Ubuntu virtual machine (VM) that runs in VirtualBox. We will call this Ubuntu instance running on Windows 10 and x86 CPUs, **Ubuntu VM**, to distinguish it from **Ubuntu ARM** that we will be running on the DE10-Nano board using the ARM CPUs.

Note: Any time settings are changed in VirtualBox for a virtual machine, the virtual machine must be completely shut down and restarted in VirtualBox in order for the change to take effect.

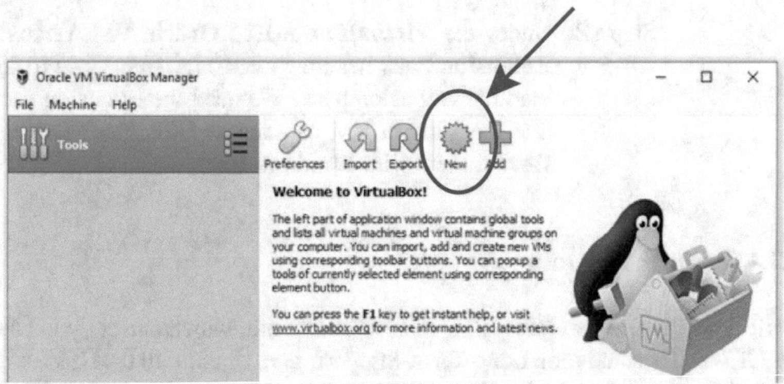

Fig. 11.11: Creating a new virtual machine in VirtualBox

Step 1: **Create a New Virtual Machine (VM)** by clicking on the New icon as shown in Fig. 11.11.

a. In the **Name and operating system** *panel*, add/select the following:
   i. Name: Ubuntu 20.04 LTS .
   ii. Machine Folder: `F:\Oracle\VMs`. (Your installation path will be different.)
   iii. Type: Linux.
   iv. Version: Ubuntu (64 bit).
   v. Click Next.

b. In the **Memory Size** *panel*:
   i. Enter: 8192 MB (Note: If your laptop only has 8 GB, you need to change this to 4 GB, which will work OK but likely slower).
   ii. Click Next.

c. In the **Hard Disk** *panel*:
   i. Select: Create a virtual hard disk now.
   ii. Click Create.
   iii. In the **Hard disk file type** *panel*:
      A. Select: VDI for a VirtualBox Disk Image.
      B. Click Next.
   iv. In the **Storage on Physical hard disk** *panel*:
      A. Select: Fixed Size. (Note: If your laptop's hard disk is not large, you will want to select dynamically allocated.)
      B. Click Next.
   v. In the **File location and size** *panel*:
      A. Accept the default file name and path.
      B. Enter: 100 GB for hard disk size. (Note: This is the recommended size, but if your laptop does not have much disk space, you might want to change this to 40 or 50 GB.)
      C. Click Create. (It will take a few minutes to create the virtual hard disk.)

Step 2: **Install Ubuntu** as the virtual machine operating system. With Ubuntu 20.04 LTS selected in the VirtualBox Manager (in left column), right click on [Optical Drive] as shown in Fig. 11.12 in the *Storage section* and "Choose a disk file. . . ."

a. Browse to where you downloaded the ubuntu-20.04.1-desktop-amd64.iso file and select it.
b. Click Open.
c. Double click the Ubuntu 20.04 LTS VM in the VirtualBox Manager to start up the Ubuntu VM (or right click and select Start). It will take a bit of time to start up, and the screen will be black for some time so wait until the install welcome screen shows up.
d. In the **Welcome screen**, click install Ubuntu:
   i. Select keyboard and click Continue.
   ii. In the **Updates and Other software** page:
      A. Select Normal Installation when asked what apps you would like to install.
      B. Select Download updates while installing Ubuntu in Other options.
      C. Select Install third-party software for graphics. . . in Other options.
      D. Click Continue.
   iii. In the **Installation type** page:
      A. Select Erase disk and install Ubuntu.
      B. Click Install Now.
      C. When it asks "Write the changes to disks," click Continue.
   iv. Select your time zone and click Continue.
   v. In the **"Who are you?"** *panel*:

A. Add your name. **Note:** A short name will make typing commands more convenient.
B. Change the computer and username if you wish from the default names generated from the Your name box.
C. Choose your password. **Note:** You will be using this password a lot in Ubuntu, so make it short and easy to remember. We are not going to worry too much about security while in Developers Mode since we will assume that the majority of work will be done in the Ubuntu VM where security will be handled by VirtualBox and Windows 10, and this VM will not be deployed where security is an issue.
D. Click continue. Installation will take several minutes.

vi. Reboot when it asks you by clicking Restart Now. Hit enter to continue and ignore the message about the optical drive.
vii. Power off the VM by closing the Ubuntu VM window and selecting "Power off the machine."

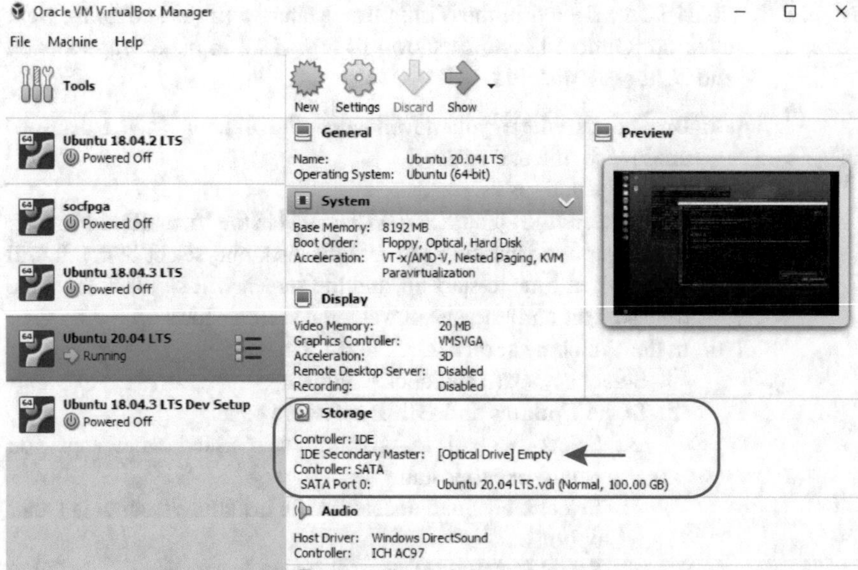

Fig. 11.12: Installing Ubuntu using the Ubuntu .iso image that is loaded into [Optical Drive]

11.1.2.7 Install and Configure Needed Software in the Ubuntu VM

We will now use a command line terminal window in the Ubuntu VM. The **apt** command is a command line utility for managing Debian packages in Ubuntu and related Linux distributions. It is a newer and a higher level package management command than apt-get and has most of the commonly used command options from apt-get and apt-cache. This means that you can just use apt, rather than apt-get for all our common use cases.

Step 1: Startup the Ubuntu VM in VirtualBox and log in.
Step 2: If Ubuntu asks about updating software, go ahead and update/install.
Step 3: Open a Terminal window by right clicking on the desktop and selecting Open Terminal. Note: The window resizing is unlikely to work at this point, which will be fixed when the Virualbox guest utilities are installed later in the VM.
Step 4: Update the Ubuntu package lists so that Ubuntu knows what all the latest update versions are. Do this by typing the Linux command (do not type the $ symbol, which is the last character of the command line prompt):

```
$ sudo apt update
```

Listing 11.1: Linux Command: sudo apt update

a. The **sudo** command stands for "superuser do", which allows you to run commands that require root privileges. It will ask for the root password. If you run sudo within a timeout window (default 15 min, which can be changed), it will not ask for a password again.
b. The **apt** command is a command line tool to manage software packages and works on a database of available packages (which needs to be updated to see if newer packages are available). APT stands for Advanced Package Tool, and apt is typically used to install new packages.
c. The **update** option tells apt to update the package list from a server on the Internet.

Step 5: Get the **GCC compiler** and related packages by typing the command:

```
$ sudo apt install build-essential
```

Listing 11.2: Linux Command: sudo apt install build-essential

a. The **install** option for apt installs the package specified, which in this case is build-essential.
b. The **build-essential** package is a reference for all the packages needed to compile a Debian package (Ubuntu is derived from Debian Linux). Thus installing the build-essential package will install

the GCC compilers, libraries, and other utilities needed to compile in Ubuntu. **Note:** This is an x86 compiler that will be used for compiling code to run in the Ubuntu VM that is running on an x86 CPU. We will also need to compile code to run on the ARM CPUs on the DE10-Nano, which we will also do in the Ubuntu VM. Thus we will need a second compiler, a cross compiler, since we will be running on a x86 machine, but compiling for the ARM CPU. We will install this cross compiler later in Sect. 11.1.4.1 Configure and Set up the Linaro GCC ARM Tools in Ubuntu VM (page 238).

Step 6: Install **Guest Additions** that are needed for USB connectivity, shared folders, etc. Install by typing the command:

```
$ sudo apt install virtualbox-guest-utils
```

Listing 11.3: Linux Command: sudo apt install virtualbox-guest-utils

a. The **virtualbox-guest-utils** package is a collection of device drivers and system applications that allows closer integration between the Ubuntu VM and Windows (shared folders, cut and paste, etc.). This is useful for moving files between Windows and the Ubuntu VM. **Note 1:** If for some reason the window resizing function does not properly work after using the apt command above (and after the Ubuntu VM has been restarted), try installing the guest additions using this method:
   i. In the Ubuntu VM, select: Devices → Insert Guest Additions CD Image...
   ii. Run the installer.
   iii. When done, you can right click on the CD folder that is on the desktop and click eject to remove it.

   Note 2: On some machines, to get window resizing to work properly, it appears the apt command should be:

```
$ sudo apt-get install virtualbox-guest-dkms ↲
   ↪virtualbox-guest-utils virtualbox-guest-↲
   ↪x11
```

Listing 11.4: Linux Command: Another method for installing the guest additions. Note: The red arrows show that the command line is longer than what can be displayed in the listing. If you cut and paste the command these arrow characters need to be deleted.

b. Restart the Ubuntu VM after installing.

Step 7: Power off the Ubuntu VM.

11.1.2.8 Configure VirtualBox

This configuration is done in the VirtualBox Manager running in Windows 10. It is not done in a virtual machine that has been started up from the VirtualBox Manager.

Step 1: **Enable 3D Acceleration**

a. In VirtualBox Manager (not in the Ubuntu VM) select Ubuntu 20.04 LTS (but do not open).
b. Select: Settings → Display → Check Enable 3D Acceleration (window resizing can lock up if you do not do this, though it works best if you have a graphics card).

Step 2: **Allow Cut and Paste between Windows and Ubuntu VM**.

a. In VirtualBox Manager (not in the Ubuntu VM), select Ubuntu 20.04 LTS (but do not open).
b. Select: Settings → General → Advanced → Shared Clipboard → Bidirectional.
c. Click OK.
   The Bidirectional setting makes sure that both Ubuntu's and Window's clipboards contain the same data.
   Note: Ctrl-C does not work in Ubuntu's Terminal Window like it does in Windows. To copy/past in Ubuntu's Terminal Window, do either of the following:
   i. Highlight the desired text, right click, and select copy/paste.
   ii. Use Ctrl-Shift-C or Ctrl-Shift-V.

Step 3: **Allow Drag and Drop between Windows and Ubuntu VM**.

a. In VirtualBox Manager (not in the Ubuntu VM), select Ubuntu 20.04 LTS (but do not open).
b. Select: Settings → General → Advanced → Drag'n'Drop → Bidirectional.
c. Click OK.

Step 4: **Create a shared folder between Windows 10 and Ubuntu VM**.

a. In Windows 10, create a new folder, e.g.,
   `F:\Oracle\VMs\shared_folder`
   that will be used as the shared folder.
b. In VirtualBox Manager (not in the Ubuntu VM), select Ubuntu 20.04 LTS (but do not open).
c. Select: Settings → Shared Folders.
   i. Click on the icon at the right that has a plus sign on a folder and browse to the shared folder you created.
   ii. Check the Auto-mount option. The auto-mount option will automatically mount this folder in Ubuntu. Otherwise, you will need to mount this manually.

Note 1: To be able to access this folder, you will need to give yourself permission to access this folder (see instruction below) when you log into Ubuntu. You should see sf_shared_folder on the desktop in Ubuntu. The path location to this folder in Ubuntu is `/media/sf_shared_folder`.
Note 2: Additional shared folders will be created later on (e.g., quartus_project).

iii. Click OK.

Step 5: Click OK.

11.1.2.9 Give Yourself Permission to Access Shared Folders in Ubuntu VM

Step 1: Startup the Ubuntu VM in VirtualBox and log in.
Step 2: Open a Terminal Window.
Step 3: Set Linux permissions so that you can access the shared folders in Ubuntu (otherwise you will get the message: "You do not have the permissions necessary to view the contents of sf_shared_folder").
To view ownership, i.e., who owns the directory, change to the `/media` folder by typing the Linux command:

```
$ cd /media
```

Listing 11.5: Linux Command: cd /media

and then type the command:

```
$ ls -l
```

Listing 11.6: Linux Command: ls -l

to see the permissions on the shared folder. Note that root is the owner and vboxsf is the group for sf_shared_folder. To gain access to this directory, you need to add yourself to the vboxsf group. You do this by typing the following command:

```
$ sudo adduser <user_account_name> vboxsf
```

Listing 11.7: Linux Command: sudo adduser <user_account_name> vboxsf

where <user_account_name> means use the user name you created in Sect. 11.1.2.6 Setting up Ubuntu in VirtualBox (do not type the <> characters).

Note: In order for the adduser command to take effect, you need to log out and then back in. Otherwise, you will not have permission to access the shared folder yet. To log out, click the small down arrow symbol in the upper right corner and select Power Off/Log Out. Then log back in. You should now be able to see files contained in the shared Windows directory when you type the command:

```
$ cd /media/sf_shared_folder
```

Listing 11.8: Linux Command: cd /media/sf_shared_folder

11.1.2.10 Install More Software in Ubuntu VM

Step 1: Startup the Ubuntu VM in VirtualBox and log in.

Step 2: Open a Terminal window by right clicking on the desktop and selecting Open Terminal.

Step 3: It is always a good idea to check for updates before installing. In a terminal window, execute the following commands:

```
$ sudo apt update
```

Listing 11.9: Linux Command: sudo apt update

If you see the message that packages can be upgraded after running update, you can update them with the command:

```
$ sudo apt full-upgrade
```

Listing 11.10: Linux Command: sudo apt full-upgrade

The **full-upgrade** option for apt will remove installed packages if this is needed to upgrade the system as a whole.
To see if a package has already been installed, use the command:

```
$ dpkg -s <packagename>
```

Listing 11.11: Linux Command: dpkg -s <packagename>

although installing the package again using apt will not be a problem if it has already been installed.

Step 4: **Installing a Text Editor** You will need a text editor. The text editor you use comes down to personal preference and here are some options. You only need to install one of these, and you can substitute your favorite editor if it is not listed):

a. **Visual Studio Code.** In a terminal window, execute the following command:

```
$ sudo snap install --classic code
```

Listing 11.12: Linux Command: sudo snap install - -classic code

Note 1: snap is an app install utility that comes preinstalled in Ubuntu 16.04 or later, so there is no need to install it.
Note 2: You run Visual Studio Code by typing code on the command line.
Note 3: When you first open a .c file, the Visual Studio Code editor will ask if you want to install the recommended extensions for C. Install the extensions by clicking install.

b. **Atom.** In a terminal window, execute the following command:

```
$ sudo snap install atom --classic
```

Listing 11.13: Linux Command: sudo snap install atom –classic

c. **Vim and Nano.** In a terminal window, execute the following commands:

```
$ sudo apt install vim nano
```

Listing 11.14: Linux Command: sudo apt install vim nano

Step 5: **Install Git** (a version control system used to manage a distributed software repository). In a terminal window, execute the following command:

```
$ sudo apt install git
```

Listing 11.15: Linux Command: sudo apt install git

Step 6: **Install mkimage** (a U-Boot tool to create bootscript images). In a terminal window, execute the following command:

```
$ sudo apt install u-boot-tools
```

Listing 11.16: Linux Command: sudo apt install u-boot-tools

11.1.2.11 Testing the Ubuntu VM by Compiling "Hello World"

Step 1: Start up the Ubuntu VM in VirtualBox and log in.

Step 2: Open a Terminal Window by right clicking on the desktop and selecting Open Terminal.

Step 3: Open a text editor by typing the command (The Visual Studio Code editor is used in this example.):

```
$ code hello.c
```

Listing 11.17: Linux Command: code hello.c

Step 4: Enter the following C source code:

```
#include <stdio.h>

int main(){
    printf("Hello World\n");
    printf("My Name is <full_name>\n");
    return 0;
}
```

Listing 11.18: Hello World

Note: Replace <full_name> in the Hello World code with your first and last name.

Step 5: Save the file, open a terminal window if one is not already open, and make sure you are in the same directory as the Hello World .c file.

Step 6: Compile the code by typing the command:

```
$ gcc -o hello hello.c
```

Listing 11.19: Linux Command: gcc -o hello hello.c

This will use the GNU C compiler gcc to create the output (-o) file *hello*, which is an executable that uses the source file hello.c as input.

Step 7: Run the "Hello World" program by typing the command:

```
$ ./hello
```

Listing 11.20: Linux Command: ./hello

You should see "Hello World" and your name printed on the command line in the Terminal Window.

11.1.3 Developer's Boot Mode Setup

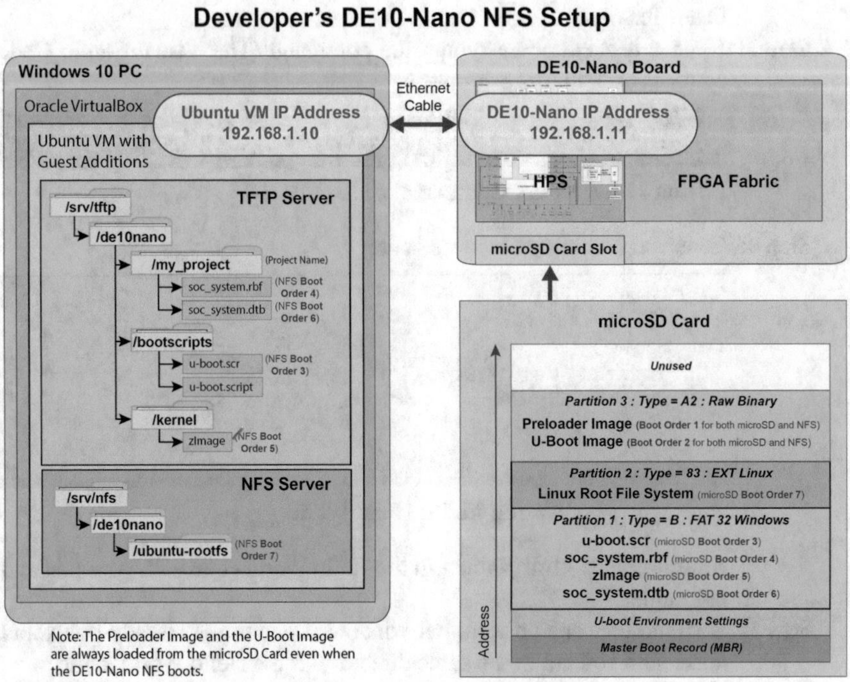

Fig. 11.13: In the developer's boot mode setup, files that are normally loaded from partitions 1 and 2 of the microSD card are served by the TFTP and NFS servers in the Ubuntu VM

The DE10-Nano board ships with a microSD card that contains all the information needed to boot the SoC FPGA system, which includes the bitstream to configure the FPGA fabric and Linux to run on the ARM CPUs. This setup is what you would expect when a hardware product is shipped. However, it is not the setup that you want as a developer (Fig. 11.13).

The developer setup uses your Ubuntu VM so that the files you change as a developer will be pulled from directories within the Ubuntu VM when then DE10-Nano board is powered up. This makes it much easier for development since changes to the SoC FPGA system can be made in the Ubuntu VM rather than having to constantly pop the microSD card out and re-image or copy files to the microSD card. Instead, you just power cycle the DE10-Nano board, and it reboots pulling in the new files from the Ubuntu VM with the changes you have made. **Note:** Reboot is only required for files served by TFTP since these files are associated with booting and configuring the SoC FPGA on power-up. Files served by the NFS server will be immediately available, which means that software being developed that can be run by Linux will be immediately available.

We will use two static IP addresses in this developer's boot mode setup where they will be set as private IP addresses, which fall in the range 192.168.0.0–192.168.255.255. We will pick two convenient addresses to remember, so we will choose the following static IP addresses for our setup where the NFS and TFTP servers will be housed in the Ubuntu VM. The even address will be for the Ubuntu VM, and the FPGA board will be assigned the next odd address.

Static IP for Ubuntu VM: 192.168.1.10
Static IP for DE10-Nano Board: 192.168.1.11

Note: We will be using a dedicated USB Ethernet adapter so there will not be any IP conflicts as determined in Step 4 of Sect. 11.1.3.3 Network Setup in the Ubuntu VM.

11.1.3.1 Install the USB Ethernet Adapter

If this is the first time installing the USB Ethernet adapter, the steps to follow are:

Step 1: Before installation of the USB Ethernet adapter, open the Windows 10 *Device Manager* and note what *Network adapters* currently exist (you will need to expand this section).
Step 2: Install the USB Ethernet adapter by following the manufactures installation instructions.
Step 3: In the Windows 10 Device Manager, see what new Ethernet adapter shows up. Make note of this name since you will need to select it later in the VirtualBox Manager when you set up Adapter 2 in item Step 2: Network Configuration in the VirtualBox Manager (page 215).

If the USB Ethernet adapter has already been installed, the steps to follow are:

Step 1: Open the Windows 10 *Device Manager* and expand the *Network adapters* section.
Step 2: Plug the USB Ethernet adapter into your computer's USB port.
Step 3: Note what Network adapter shows up. Make note of this name since you will need to select it later in the VirtualBox Manager when you set up Adapter 2 in item Step 2: Network Configuration in the VirtualBox Manager (page 215).
Step 4: Unplug the USB Ethernet adapter from your computer's USB port.
Step 5: The Network adapter should disappear.

Note: If the new Ethernet adapter does not show up in the Windows 10 Device Manager, unfortunately this USB Ethernet adapter will not work for the VM setup, which means that you will need to get a different USB Ethernet Adapter.

11.1.3.2 Network Configuration in the VirtualBox Manager

The Ubuntu VM will be set up to use two network adapters. This setup needs to be done in two different software locations. The first configuration is done in the VirtualBox Manager, and the second is down in a terminal window in the Ubuntu VM. This section covers what needs to be done in the VirtualBox Manager. The network setup that needs to be done in the Ubuntu VM is covered in Sect. 11.1.3.3 Network Setup in the Ubuntu VM.

Two network adapters will be created for the Ubuntu VM in the VirtualBox Manager. **Adapter 1** will be set up as a *Network Address Translation* (**NAT**) adapter so that the Ubuntu VM can access the Internet through Windows. It is used to map one IP address into another IP address. This allows the use of one Internet-routable IP address of the NAT gateway (i.e., the Internet IP address of your computer) to be used for our entire private network (Ubuntu VM and DE10-Nano board). We use it to allow the Ubuntu VM to connect to the Internet and route all traffic through Windows. Figure 11.14 illustrates this NAT mapping.

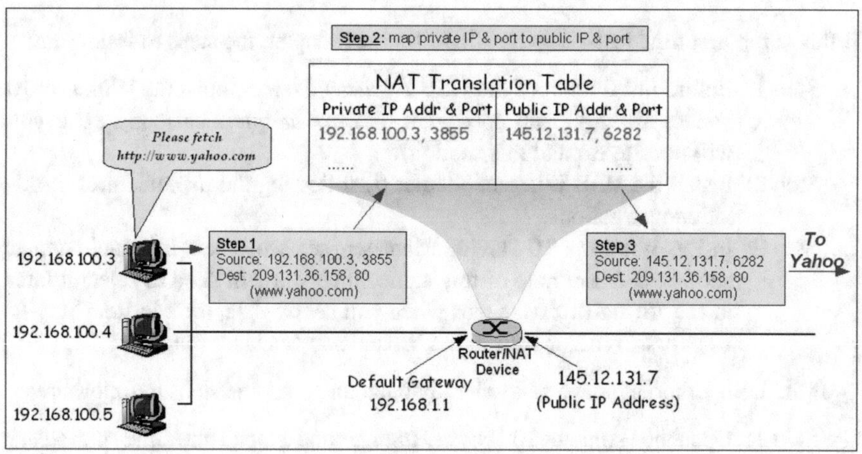

Fig. 11.14: Illustration of network address translation (NAT). [3]

Adapter 2 will be set up for bridged networking, which is used for more advanced networking needs such as running servers in the Ubuntu VM, which we will be doing. When enabled, VirtualBox connects to one of your installed network cards and exchanges network packets directly, circumventing Windows network stack. We will use bridged networking so that the TFTP and NSF servers in the Ubuntu VM can serve files to the DE10-Nano board.

The steps for setting up the two network adapters are:

Step 1: With Ubuntu 20.04 LTS selected (select but do not run the Ubuntu VM yet) in the **VirtualBox Manager**, select: Settings -> Network.
Click on the **Adapter 1** tab and select the following:

- Enable Network Adapter (check the box).
- In the *Attached to:* pull down menu, select **NAT**.
- Accept the default for *Name*. (There is no name for NAT.)
- Expand the *Advanced* menu. **Make a note of the MAC Address** (e.g., 080027A53AE7). This is how we will tell which Ethernet adapter the Ubuntu VM is using for the NAT.
- Click OK.

Step 2: With Ubuntu 20.04 LTS selected (select but do not run the Ubuntu VM yet) in the **VirtualBox Manager**, select: Settings -> Network.
Click on the **Adapter 2** tab and select the following:

- Enable Network Adapter (check the box).
- In the *Attached to:* pull down menu, select **Bridged Adapter**.
- In the pull down menu associated with *Name*, select the name of your USB Ethernet adapter that you determine in the Windows 10 Device Manager in Sect. 11.1.3.1 Install the USB Ethernet Adapter (page 213). Note: Your USB Ethernet Adapter needs to be connected to be seen.
- Expand the *Advanced* menu. **Make a note of the MAC Address** (e.g., 0800276638E7). This is how we will tell which Ethernet adapter the Ubuntu VM is using for the Bridged Adapter.
- Make sure that the **Cable Connected** checkbox is checked.
- Click OK.

11.1.3.3 Network Setup in the Ubuntu VM

Our next step is to configure networking in the Ubuntu VM so go ahead and start up the Ubuntu VM. The steps for configuring the network are:

Step 1: In VirtualBox, open the Ubuntu VM and open a terminal window.
Step 2: Install **iproute2** (tools to manage network configuration).

```
$ sudo apt install iproute2
```

Listing 11.21: Linux Command: sudo apt install iproute2

Note: In the past, net-tools (for networking utilities and support, e.g., ifconfig) have been used, and to install these tools, use the command:

```
$ sudo apt install net-tools
```

Listing 11.22: Linux Command: sudo apt install net-tools

Step 3: Determine the networking setup in Ubuntu VM by typing the command:

```
$ ip addr
```

Listing 11.23: Linux Command: ip addr

Note: In the past, the command has been:

```
$ ifconfig
```

Listing 11.24: Linux Command: ifconfig

In Ubuntu, **Ethernet interfaces are identified by the name enp0s<x>**, where <x> is the unique number assigned by Linux (e.g., see [4]). In my case, I can see two interfaces named **enp0s3** and **enp0s8**. Under the enp0s3 section, I see a line that says:

link/ether **08:00:27:a5:3a:e7** brd ff:ff:ff:ff:ff:ff

which matches the MAC address 080027A53AE7 that we saw in the VirtualBox Network manager for Adapter 1. Thus we know that the enp0s3 Ethernet interface is connected to the network through the NAT Adapter, and thus:

enp0s3 = NAT.

Under enp0s8, I see a line that says:

link/ether **08:00:27:66:38:e7** brd ff:ff:ff:ff:ff:ff

which matches the MAC address 0800276638E7. Thus we know that the enp0s8 Ethernet interface is connected to the Bridged Adapter, and thus:

enp0s8 = Bridged Adapter

Therefore, **enp0s8 is the bridged adapter that we will want to set up with a static IP address.**
Note: You can also find the names of your network interfaces by using the command:

```
$ ls /sys/class/net/
```

Listing 11.25: Linux Command: ls /sys/class/net/

or

```
$ ip link
```

Listing 11.26: Linux Command: ip link

Step 4: **Determine the static IP for the Bridged Adapter**

Note: You can skip this step. *We are assuming that you are using a dedicated USB Ethernet adapter, which means that there will not be any IP conflicts*. If you are not using a dedicated USB Ethernet, you are ultimately on your own to configure the local network, and this is one of the steps that you will have to perform. This first step of finding IP addresses that are being used is not strictly necessary when using just the Ubuntu VM on your laptop with the DE10-Nano board and where these are the only two devices on your private network. However, if you are setting this up in a scenario where there are many devices on the private network (e.g., at home where your PC is connected to a router), you will want to check what IP addresses currently exist. Since we are assigning static IP address 192.168.1.10, it is possible that the IP addresses we want to use have already been assigned. We need to check what addresses have already been assigned and are being used on the private network. We will do this by using the **nmap** utility, where **nmap** stands for **"Network Mapper."** It is a utility for network discovery and will find all the live hosts on the network.

a. Install **nmap** if it has not already been installed:

```
$ sudo apt install nmap
```

Listing 11.27: Linux Command: sudo apt install nmap

Type the nmap command:

```
$ nmap -sn 192.168.1.*
```

Listing 11.28: Linux Command: nmap -sn 192.168.1.*

- The **-sn** option (no port scan) tells Nmap not to do a port scan after host discovery, and only print out the available hosts that responded to the scan.
- The third option **192.168.1.*** tells nmap to scan all 256 IP addresses that start with 192.168.1.
- All the addresses that come back in the response have been already taken, so **make sure the IP address you plan on using is not in this list.** See [5] for more examples of what you can do with nmap.

Step 5: Determine the **netmask** (typically 255.255.255.0) and **gateway IP address** of the Bridged Adapter using the following command:

```
$  route -n
```

Listing 11.29: Linux Command: route -n

which gives the response shown in Fig. 11.15.

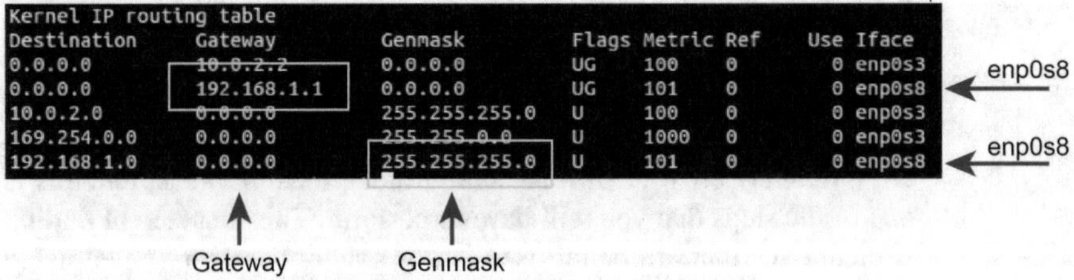

Fig. 11.15: Kernel routing table from the route command that determines the gateway IP address and netmask of the bridge adapter

First we note three columns with the headings **Gateway**, **Genmask**, and **Use Iface**. Under the **Use Iface** column, we make note of two rows that correspond to the Bridged Adapter (enp0s8). On these two rows, we then note the Gateway (192.168.1.1) and Genmask (255.255.255.0) entries. Therefore, we have:

Bridged Adapter Gateway IP Address = 192.168.1.1
Bridged Adapter Netmask = 255.255.255.0

Note: Your numbers may be a bit different depending on what the Linux route command returns.

Step 6: **Set the static IP for the Bridged Adapter in Ubuntu VM.**
Click on the **System Menu** icon in the upper right corner of Ubuntu and open the window as shown in Fig. 11.16. Then click on *Ethernet (enp0s8) Connected* and click on *Wired Settings*.

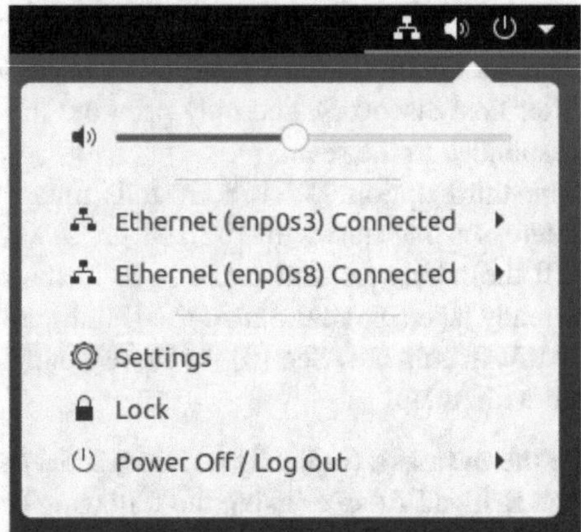

Fig. 11.16: Opening "Settings" in Ubuntu from the system menu

Then with **Network** selected on the left side, click the gear icon on the right side for the Bridged Adapter (enp0s8). In the window that pops up, click on the **IPv4 tab** and select:

- **Manual** as the IPv4 Method.
- Enter the IP address of **192.168.1.10** (assuming no conflicts found in step 4).
- Enter the Netmask and Gateway determined from step 5.
- For the DNS entry, add Google's name servers 8.8.8.8, 8.8.4.4.
- Click the **Apply** button when done.

Your configuration should look something like Fig. 11.17. To verify that the Bridged Adapter (enp0s8) has changed, type the command:

```
$ ip addr
```

Listing 11.30: Linux Command: ip addr

and look for the interface enp0s8. The IP address should now be 192.168.1.10.

Note: The Ubuntu VM may need to be shut down and restarted for the static IP address to take effect.

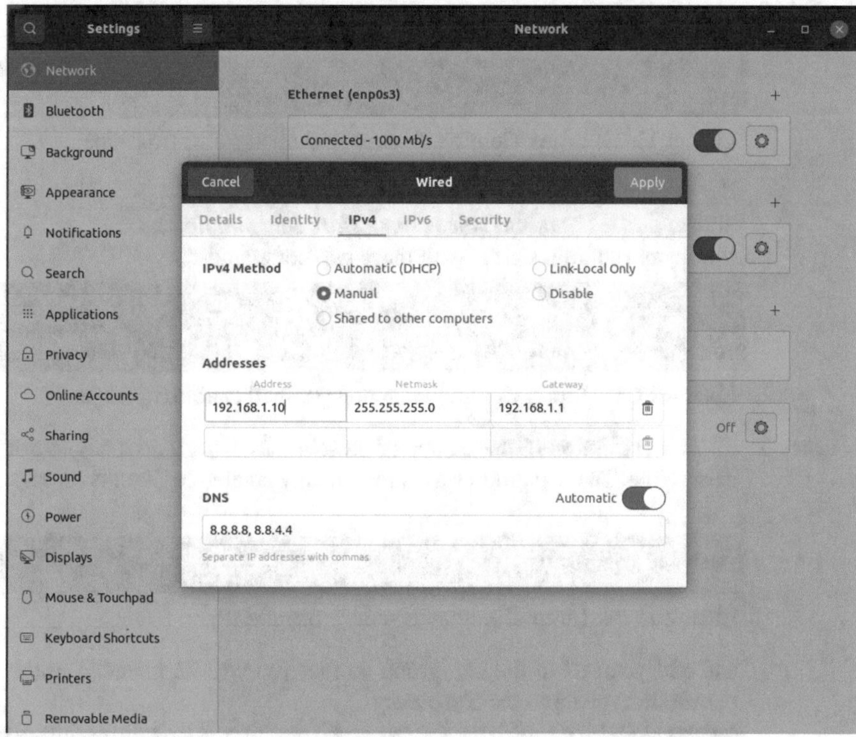

Fig. 11.17: Setting the static IP address for the bridged adapter in Ubuntu VM

11.1.3.4 TFTP Server Setup in the Ubuntu VM

We need a TFTP server that U-boot can access when it is configured to boot over the network. The TFTP server will serve the files that configure the DE10-Nano board during boot and are the files found in Partition 1 of the microSD card (see Fig. 11.13) when the DE10-Nano board normally boots from the microSD card. Changing these files will be much easier if they are served by the TFTP server rather than have to remove the microSD card from the DE10-Nano board, copy files, and reinsert the microSD card. This process quickly becomes annoying when it has to be done for each code change iteration. The acronym **TFTP** stands for Trivial File Transfer Protocol.

The steps to set up the TFTP server are:

Step 1: **Install TFTP** by typing the command in a Ubuntu terminal window:

```
$ sudo apt install tftpd-hpa
```

Listing 11.31: Linux Command: sudo apt install tftpd-hpa

Verify that TFTP is running by typing the command:

```
$ sudo systemctl status tftpd-hpa
```

Listing 11.32: Linux Command: sudo systemctl status tftpd-hpa

If it is running, you should see a line that says:
Active: active (running) since (some start date and time).
If it is not running, start it with these two commands:

```
$ sudo systemctl enable tftpd-hpa
$ sudo systemctl restart tftpd-hpa
```

Listing 11.33: Linux Command: sudo systemctl enable tftpd-hpa

Step 2: **TFTP Permissions Setup.** First, let us create the tftp group if it does not exist (Note: The tftp group was automatically created in Ubuntu 20.04.):

```
$ sudo groupadd tftp
```

Listing 11.34: Linux Command: sudo groupadd tftp

and add yourself to the tftp group so that you will be included in the permissions given to the tftp group:

```
$ sudo adduser <user_account_name> tftp
```

Listing 11.35: Linux Command: sudo adduser <user_account_name> tftp

where *<user_account_name>* is your user account name that you created when you installed Ubuntu.
Verify that you have been added to the tftp group by typing the command:

```
$ getent group tftp
```

Listing 11.36: Linux Command: getent group tftp

where you should see your user account name listed.
Now give the tftp group ownership of everything created in `/srv/tftp`, which will be the location of the files served by the TFTP server. First, change directories and go to the `/srv` directory:

```
$ cd /srv
```

Listing 11.37: Linux Command: cd /srv

and change the ownership of the `/tftp` directory by typing the command (in the `/srv` directory):

```
$ sudo chown :tftp tftp
```

Listing 11.38: Linux Command: sudo chown :tftp tftp

- chown is a command used to change the ownership of files and directories.
- :tftp signifies that the tftp group will be the new owner.
- tftp is the directory that group tftp is taking ownership of.

Now ensure that any new subfolder or file created in `/srv/tftp` will inherit the tftp group ownership. While in the `/srv` directory, type:

```
$ sudo chmod g+s tftp
```

Listing 11.39: Linux Command: sudo chmod g+s tftp

- The chmod command changes the file mode bits.
- g+s means that all new files and subdirectories created within the directory will inherit the group ID of the directory, rather than the primary group ID of the user who created the file.
- tftp is the directory that group tftp is taking ownership of.

Since it is possible that some subdirectories were already in `/tftp` before we changed ownership, let us change the group ownership of *all* the subdirectories and files in `/tftp` to tftp.

```
$ sudo chgrp -R tftp /srv/tftp
```

Listing 11.40: Linux Command: sudo chgrp -R tftp /srv/tftp

- The chgrp command changes the group ownership.
- The -R option means change all directories and files below in a recursive manner.
- The tftp option means the group to change to.
- The /srv/tftp means to change the group on this directory (and below because of the -R option).

Next, let us give rwx permissions (r=read, w=write, x=execute) to the tftp group to everything in `/tftp`.

```
$ sudo chmod -R g+rwx /srv/tftp
```

Listing 11.41: Linux Command: sudo chmod -R g+rwx /srv/tftp

- The chmod command changes the file mode bits.
- The -R option means change all directories and files below in a recursive manner.
- The g+rwx option means to add rwx access to the group settings.
- The /srv/tftp means to change the file mode bits in this directory (and below because of the -R option).

We can see what the file *access Control List* (**ACL**) looks like for `/srv/tftp` by typing the command:

```
$ getfacl /srv/tftp
```

Listing 11.42: Linux Command: getfacl /srv/tftp

Note: The group permissions could also be set using the command:

```
$ sudo setfacl -d --set u::rwx,g::rwx tftp
```

Listing 11.43: Linux Command: sudo setfacl

Step 3: **Create the following subdirectories in the `/srv/tftp` directory:**

a. `/srv/tftp/de10nano/AudioMini_Passthrough`
Create the `/AudioMini_Passthrough` directory. This is where the .rbf and .dtb files will reside associated with the Quartus project. We will learn about these files later. Note: Typically you will create a new directory like this for each new project (e.g., `/AudioMini_Passthrough`) associated with specific hardware (e.g., `/de10nano`).

Do this by typing the command:

```
$ sudo mkdir -p /srv/tftp/de10nano/↵
    ↪AudioMini_Passthrough
```

Listing 11.44: Linux Command: sudo mkdir -p /srv/tftp/de10nano/AudioMini_Passthrough

- The mkdir command creates directories.
- The -p in mkdir is short for −−parents. It will create parent directories if they do not exist. It is very useful for creating nested subdirectories without having to create each parent first (which you would have to do without the -p option).

b. `/srv/tftp/de10nano/kernel`
This is where the Linux kernel image zImage will reside.
Do this by typing the command:

```
$ sudo mkdir -p /srv/tftp/de10nano/kernel
```

Listing 11.45: Linux Command: sudo mkdir -p /srv/tftp/de10nano/kernel

c. `/srv/tftp/de10nano/bootscripts`
This is where the u-boot bootscripts will reside. U-Boot bootscripts allow customization of U-Boot without having to modify the U-Boot image on the microSD card. Typically, there will be a bootscript associated with each project that specifies where project files are found on the TFTP server. Do this by typing the command:

```
$ sudo mkdir -p /srv/tftp/de10nano/bootscripts
```

Listing 11.46: Linux Command: sudo mkdir -p /srv/tftp/de10nano/bootscripts

Step 4: **Change the default directory of TFTP to point to `/srv/tftp`**.
Change to the directory `/etc/default` by typing the command:

```
$ cd /etc/default
```

Listing 11.47: Linux Command: cd /etc/default

Edit the file **tftpd-hpa**. Visual studio code editor instructions are given here, but feel free to use your favorite editor. Type the command:

```
$ code tftpd-hpa &
```

Listing 11.48: Linux Command: code tftpd-hpa &

Edit the line (line 4) that says:

```
TFTP_DIRECTORY="/var/lib/tftpboot"
```

and change it to:

```
TFTP_DIRECTORY="/srv/tftp/"
```

Note: Changing the default directory to `/srv/tftp` only has to be for Ubuntu versions prior to 20.04.

When you save the file, you will see vscode gives the error:
Failed to save 'tftpd-hpa':Insufficient permissions. Select "Retry as Sudo" to retry as superuser.
Click the Retry as Sudo option and give the appropriate sudo password.

Note: If you use another editor, set the terminal window shell to have root access by typing:

```
$ sudo -i
```

Listing 11.49: Linux Command: sudo -i

Then change to the desired directory, open the file with your favorite editor, save the file (the editor will now have permission to write to the directory), close the editor, and exit the root access shell by typing:

```
#  exit
```

Listing 11.50: Linux Command: exit

Step 5: **Restart the TFTP server** so that it will now point to `/srv/tftp` . Restart the TFTP server by typing the following command:

```
$ sudo systemctl restart tftpd-hpa
```

Listing 11.51: Linux Command: sudo systemctl restart tftpd-hpa

Verify that TFTP is running by typing the following command:

```
$ sudo systemctl status tftpd-hpa
```

Listing 11.52: Linux Command: sudo systemctl status tftpd-hpa

11.1.3.5 NFS Server Setup in the Ubuntu VM

We need a NFS server so that when Linux boots on the DE10-Nano board it can mount a root file system contained in the Ubuntu VM rather than mounting the root file system from the microSD card. This allows us to change files easily in

the directories served by the NFS server rather than having to remove the microSD card from the DE10-Nano board, inserting the microSD into a Linux computer, and mounting the root file partition, just to add or change files.
The steps to set up the NFS server are:

Step 1: **Install the NFS server** by typing the command in a Ubuntu terminal window:

```
$ sudo apt install nfs-kernel-server
```

Listing 11.53: Linux Command: sudo apt install nfs-kernel-server

Create the following directory for the NFS server:

```
$ sudo mkdir -p /srv/nfs/de10nano/ubuntu-↲
    ↪rootfs
```

Listing 11.54: Linux Command: sudo mkdir -p /srv/nfs/de10nano/ubuntu-rootfs

This is the directory that will be served to the DE10-Nano board and will become the root directory when Linux boots on the DE10-Nano board. Files added to this directory (and in subdirectories) will show up under the root directory on the DE10-Nano board. It will be much easier to modify files in this directory than having to modify the rootfs image on the microSD card when you change a file during development.

Step 2: Edit **`/etc/exports`** as root, which is the access control list for filesystems that can be exported to NFS clients. First, go to the `/etc` directory:

```
$ cd /etc
```

Listing 11.55: Linux Command: cd /etc

Edit the file **exports**. Visual studio code editor instructions are given here, but feel free to use your favorite editor. Type the command:

```
$ code exports &
```

Listing 11.56: Linux Command: code exports &

The single line we will add to the exports file has the general format:

directory_to_share client(share_option1,. . . ,share_optionN)

where we will use:

directory_to_share = /srv/nfs/de10nano/ubuntu-rootfs/

client = 192.168.1.0/24

The client address specified here is using the *Classless Inter-Domain Routing* (**CIDR**) notation, which means clients can be in the address range of 192.168.1.0 to 192.168.1.255. The /24 means the subnet mask of 255.255.255.0 is being applied. We can set the DE10-Nano board to have an address of 192.168.1.11 since it is included in the allowable IP address range.

share_option1 = rw

Note: The rw option gives the client both read and write access to the directory.

share_option2 = no_subtree_check

The no_subtree_check option prevents subtree checking, which is the process of checking if the file is still available in the exported tree for every request. Subtree checking can cause problems if the file is renamed, while the client has it opened. We will not bother with subtree checking since subtree checking tends to cause more problems than it is worth and has only mild security implications.

share_option3 = sync

The sync option cause NFS to write changes to disk before replying to requests. This results in a more stable (but slower) environment.

share_option4 = no_root_squash

The no_root_squash option prevents NFS from mapping the root user on the client side to an anonymous user on the server side. This option is mainly useful for diskless clients such as our DE10-Nano board when booting using NFS.

Now add the following line to the /etc/exports file:

```
/srv/nfs/de10nano/ubuntu-rootfs/ 192.168.1.0/24(rw,↵
    ↪no_subtree_check,sync,no_root_squash)
```

Note: In the share options list, there should be no spaces after the commas.

Save the file and when you will see vscode give the error:

Failed to save 'tftpd-hpa':Insufficient permissions. Select "Retry as Sudo" to retry as superuser.

Click the Retry as Sudo option and give the appropriate sudo password.

Step 3: Verify that the changes to `/etc/exports` are correct by typing the command:

```
$ cat /etc/exports
```

Listing 11.57: Linux Command: cat /etc/exports

Step 4: **Restart the NFS server** by typing the command:

```
$ sudo service nfs-kernel-server restart
```

Listing 11.58: Linux Command: sudo service nfs-kernel-server restart

You can see all the exported file systems by typing the command:

```
$ sudo exportfs -v
```

Listing 11.59: Linux Command: sudo exportfs -v

Step 5: **Set NFS Permissions** by first creating the nfs group:

```
$ sudo groupadd nfs
```

Listing 11.60: Linux Command: sudo groupadd nfs

Then add yourself to the nfs group:

```
$ sudo adduser <user_account_name> nfs
```

Listing 11.61: Linux Command: sudo adduser <user_account_name> nfs

and then verify that you have been added to the nfs group:

```
$ getent group nfs
```

Listing 11.62: Linux Command: getent group nfs

11.1.3.6 Reimaging the microSD Card with the Developer's Image

We will next re-image the microSD card on the DE10-Nano board with an image that has been set up to work with the TFTP and NFS servers. The image is different from what ships with the DE10-Nano since the U-boot variables in the image have been modified to allow the DE10-Nano board to be easily connected to the TFTP and NFS in the Ubuntu VM.

To re-image the microSD card, you will need to download and install the following software on your Windows 10 machine, so go ahead and do that now:

1. **7-Zip**. You can download 7-Zip from https://www.7-zip.org.

2. **Win32DiskImager** or **balenaEtcher**. You can download Win32DiskImager from https://sourceforge.net/projects/win32diskimager, and you can download belenaEtcher from https://www.balena.io/etcher/.

First a couple of notes before we get started:

Note 1: ***Before proceeding, make sure that your computer files are backed up!*** If your computer hard drive (or SSD) got wiped or destroyed, could you restore the files you care about? If the answer is no, **stop right now and back up the files you care about before proceeding**.

Note 2: We are assuming that the microSD card you are going to image has a **16 GB** capacity. If your microSD card has a different size, you will need to translate the numbers accordingly.

The steps to re-image the microSD card are:

Step 1: Download the file **audiomini_nfs.img.xz** (76 MB), which is the new microSD card image. Download it from (here).

Step 2: **Uncompress the .xz file**. In Windows 10, go to the directory that contains the downloaded .xz file, right click on it, and select 7-Zip → Extract Here. This will uncompress the .xz file, which will end up as a ~16GB .img file.

Step 3: **Plug the microSD card into the USB card reader and plug the card reader into your Windows machine. Note:** Ideally, this is a new microSD card, or at least it is one that you do not care about erasing. If Windows pops up a couple of messages like the one in Fig. 11.18, this means the card has been previously used for the DE10-Nano, and there are two partitions (Partitions 2 and 3) that Windows does not recognize. Do not click the Format disk button, rather ignore the Windows pop-up messages (click Cancel for these pop-up windows). If this is the case, proceed to Step 1:. If your card is new and Windows recognizes it as a single FAT32 partition, then proceed to Step 2:.

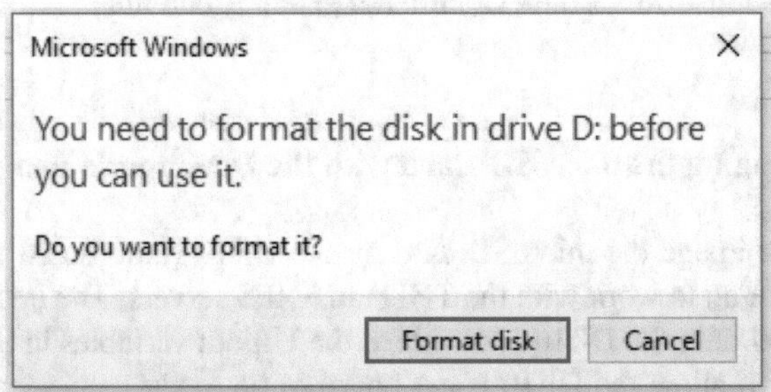

Fig. 11.18: Windows will ask if you want to format the disk if it does not recognize a partition. This will be the case if the microSD card already has a SoC FPGA image on it

Step 1: **Create a single partition on the microSD card.** In the Windows 10 search bar, type "disk management", and Windows 10 will suggest *Create and format hard disk partitions*. Click on that suggestion. The Disk Management utility will pop up, and there will be a disk listed (Disk 3 in this example) that is 16 GB (14.84 GB in this example). Here is what the disk utility shows for the microSD card that shows the three partitions (Fig. 11.19).

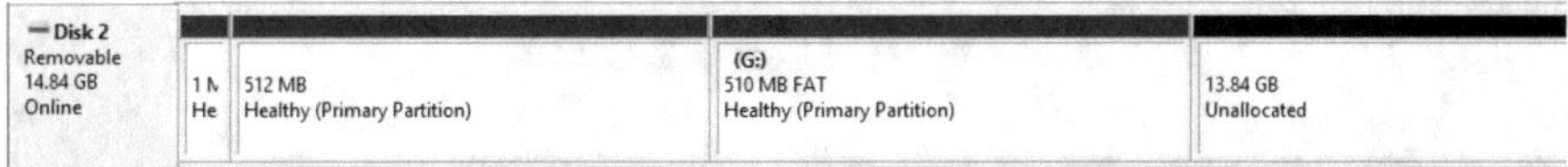

Fig. 11.19: The microSD card will have three partitions if it has been previously used for the DE10-Nano board

a. Verify that this is the 16 GB microSD card by unplugging it from Windows. This disk listing should disappear from the Disk Management utility, and when you plug it back into Windows, it should reappear (after a bit).
b. For each partition listed for the microSD card disk, right click and select "Delete Volume." **WARNING:** Triple check that this volume is associated with the microSD card. Windows will ask "Do you want to delete this partition." Select Yes (after making sure this is the microSD card disk). It should now look like (Fig. 11.20):

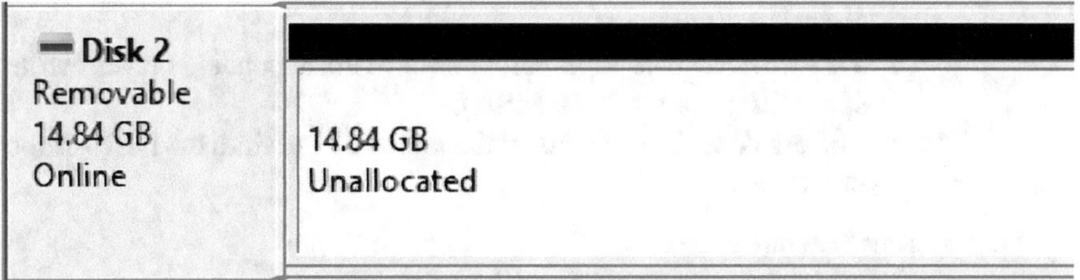

Fig. 11.20: The microSD card after the partitions have been deleted

c. Right Click on the area that says 14.84 GB Unallocated, and select "New Simple Volume" and create a volume that is the maximum disk space (15,192 MB) with FAT32 (accept default settings).
d. Close the Disk Management utility and restart it (otherwise, it may complain about some errors). It will say that the disk is a raw partition.

e. Right click on the partition and format with FAT32. The microSD card should now look like a new card with a single 14.84 GB FAT32 partition as in Fig. 11.21.

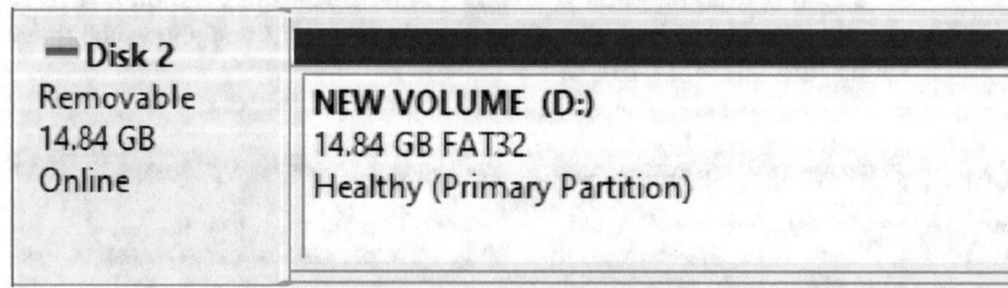

Fig. 11.21: The microSD card with a new single FAT32 partition

f. Proceed to Step 2:.

Step 2: Choose the imaging software that you will use. If it is **Win32DiskImager**, go to Step 3:. If it is **Etcher**, go to Step 4:.

Step 3: **Run Win32DiskImager**:

a. Browse to where **audiomini_nfs.img** is located and select it.
b. The device should be the drive that contains the microSD card. **WARNING:** Double check that this is the microSD card (14.8 GB capacity) since you will be overwriting this drive.
c. Click the Write button. Confirm again that you are targeting the drive with the microSD card. Do not worry about it complaining about extra space.
d. This will may take 15–25 minutes to write, depending on your write speed (16 minutes @ 15 MB/s).
e. You are done. You can insert the microSD card into the DE10-Nano board.

Step 4: **Run balenaEtcher**:

a. Click *Flash from file* and select audiomini_nfs.img from where you uncompressed and stored this image file.
b. Click *Select target* and select the microSD card.
c. Click *Flash!* This will take 15–25 minutes to write, depending on your write speed.
d. You are done. You can insert the microSD card into the DE10-Nano board.

11.1.3.7 DE10-Nano Board Switch Settings

The location of the configuration DIP Switch SW10 on the DE10-Nano board is shown in Fig. 11.22. The switch determines how the configuration bitstream for the FPGA fabric (.rbf file) gets interpreted.

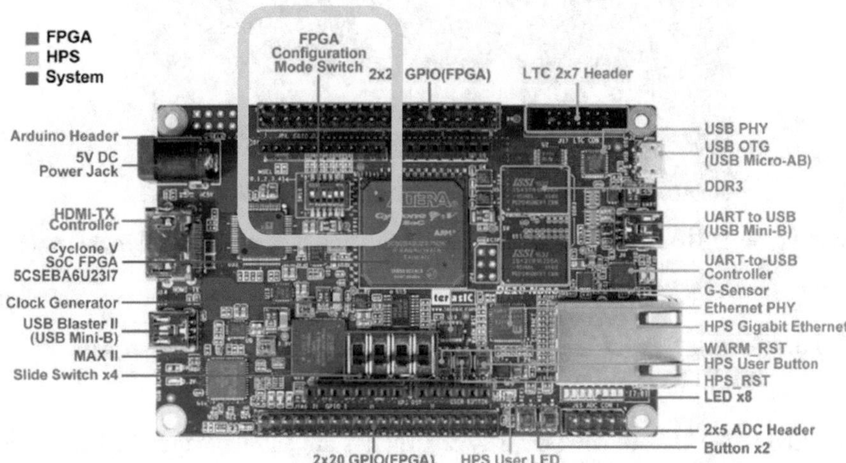

Fig. 11.22: Location of switch SW10 that determines how the .rbf file is interpreted

There are two switch settings to be familiar with.

Setting 1: **FPPx32** Fast Passive Parallel x32 mode, Compression Enabled, Fast POR. **This is the Default Setting on DE10-Nano board when it ships from Terasic.** This default switch setting is shown in Fig. 11.23.

Fig. 11.23: DE10-Nano Switch SW10 configuration that is shipped from Terasic

Setting 2: **FPPx16** Fast Passive Parallel x16 mode (no compression). ***This is the mode we will use for the labs in this book.*** All switches should be up in the zero (ON) state as shown in Fig. 11.24.

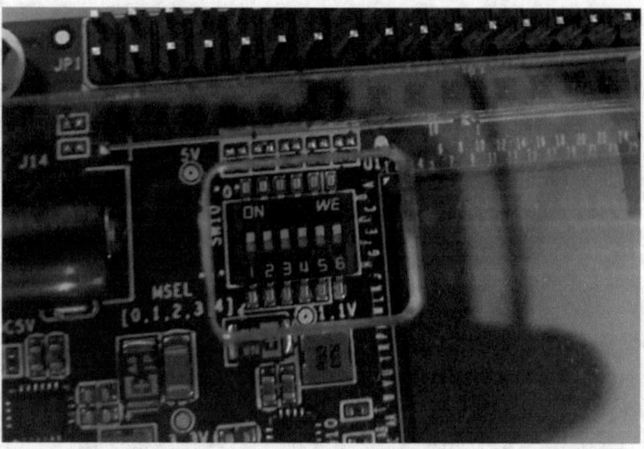

Fig. 11.24: DE10-Nano Switch SW10 configuration that will be used in this book

11.1.3.8 Setting U-boot Variables on the DE10-Nano Board to Boot via NFS/TFTP

We need to tell U-boot what the network settings are and the name the bootscript it should use since we can have multiple bootscripts in the bootscript folder. We do this by setting U-boot environment variables, which is done while U-boot is first booting and before U-boot loads the Linux image. The steps to modify the U-boot environment variables are:

Step 1: Insert the microSD card that contains the new image into your DE10-Nano board. (See Sect. 11.1.3.6 Reimaging the microSD Card with the Developer's Image (page 227) for instructions on flashing this new image.)

Step 2: Open a PuTTY Terminal Window in Windows and connect a USB cable to the DE10-Nano board to the UART port (right side above Ethernet port).

Step 3: Power up the DE10-Nano board. You should see the boot process start in the PuTTY Terminal Window.

Step 4: When it starts booting and when it says: **Hit any key to stop autoboot:**, **Hit a key to get to the U-boot prompt.**

Once the U-boot prompt appears, type the following U-boot commands to set up U-boot for the Developer's Boot Mode.

Setting 1: The U-boot environment variable **nfsboot** determines if the DE10-Nano boots from the microSD card (Ship Boot Mode when nfsboot=false) or if it boots from the Ubuntu VM (**Developer's Boot Mode when nfsboot=true**). To boot from the Ubuntu VM, set nfsboot=true by typing the following U-Boot command:

```
# setenv nfsboot true
```

Listing 11.63: setenv nfsboot true

Setting 2: **Set the IP Address for the DE10-Nano board** by typing the following U-Boot command:

```
# setenv ipaddr 192.168.1.11
```

Listing 11.64: setenv ipaddr 192.168.1.11

Setting 3: We need to tell U-Boot the IP Address of the server, which is the IP address of the Ubuntu VM. This is done by typing the following U-Boot command:

```
# setenv serverip 192.168.1.10
```

Listing 11.65: setenv serverip 192.168.1.10

Setting 4: We need to tell U-Boot the IP Address of the NFS server, which is also the IP address of the Ubuntu VM. This is done by typing the following U-Boot command:

```
# setenv nfsip 192.168.1.10
```

Listing 11.66: setenv nfsip 192.168.1.10

Setting 5: Next, we need to tell U-Boot the name of the bootscript to use by setting the environment variable **bootscript** to the name of the bootscript. This is done by typing the following U-Boot command:

```
# setenv bootscript lab3.scr
```

Listing 11.67: setenv bootscript lab3.scr

Note 1: In the U-Boot image that is in the new image that was flashed to the microSD card, the path that U-boot checks for bootscripts is `/srv/tftp/de10nano/bootscripts/`. If you want to change this path, you need to edit the U-boot environment variable **bootcmdnfs**.
Note 2: The bootscript tells U-boot the names of the files to load and where these files are located. You can edit (or copy and rename) lab3.script to change the directory locations and the names of the files. The .script file is the human readable text file for bootscripts. Typically, you will have a bootscript associated with each project since there is a project directory path specified in the bootscript.
Note 3: The human readable textfile .script needs to be converted into a machine readable image. This is done with the Linux command mkimage:

```
$ mkimage -A arm -O linux -T script -C none -a 0 -e 0 -n "My script" -d u-boot.txt u-boot.scr
```

Listing 11.68: Linux Command: mkimage

Note 4: If the **bootscript** U-boot environment variable is set to the name **u-boot.scr**, then this name does not need to change when you change bootscripts. Rather, you can just copy lab3.scr to u-boot.scr or my_project.scr to u-boot.scr in the Ubuntu VM and not bother having to stop and reset the U-boot environment variable to change the name.

Setting 6: Finally, we need to tell U-Boot to save the environment variables to flash so that it boots with these values the next time.

```
# saveenv
```

Listing 11.69: saveenv

To check the U-boot environment variables, we can print them out (in alphabetical order) with the U-boot command:

```
# printenv
```

Listing 11.70: printenv

To check that U-boot can communicate with the Ubuntu VM and assuming that the TFTP and NFS servers are set up and running, one can ping the server with the U-boot command:

```
# ping 192.168.1.10
```

Listing 11.71: ping 192.168.1.10

which should result in the response: *host 192.168.1.10 is alive*. Next, we need to put the files that U-boot will be looking for in their correct locations.

11.1.3.9 Setting Up the Audio Mini Passthrough Example

The naming convention for directories in the Ubuntu VM for the TFTP server is as follows:

`/srv/tftp/<board_name>/<project_name>`

In this example, we are setting up the Audio Mini Passthrough project on the DE10_Nano board, so the TFTP project path is:

`/srv/tftp/de10nano/AudioMini_Passthrough`

and will contain the .rbf and .dtb files for the project.

The Linux command to create this directory is:

```
$ sudo mkdir -p /srv/tftp/de10nano/↵
    ↪AudioMini_Passthrough
```

Listing 11.72: sudo mkdir -p /srv/tftp/de10nano/AudioMini_Passthrough

Copy the following files from (here):

File 1: soc_system.rbf
File 2: soc_system.dtb
File 3: lab3.scr
File 4: lab3.script
File 5: zImage
File 6: ubuntu-rootfs.tar

and put them in your *shared folder* that both Windows and the Ubuntu VM can access. Now in Ubuntu VM copy **soc_system.rbf** to the project folder:

```
$ sudo cp /media/sf_shared_folder/soc_system.rbf /srv/↵
    ↪tftp/de10nano/AudioMini_Passthrough
```

Listing 11.73: sudo cp /media/sf_shared_folder/soc_system.rbf /srv/tftp/de10nano/AudioMini_Passthrough

In a similar manner, copy **soc_system.dtb** to the project folder:

```
$ sudo cp /media/sf_shared_folder/soc_system.dtb /srv/↵
    ↪tftp/de10nano/AudioMini_Passthrough
```

Listing 11.74: sudo cp /media/sf_shared_folder/soc_system.dtb /srv/tftp/de10nano/AudioMini_Passthrough

Note: Make sure that the capitalization and spelling are correct for directory `/AudioMini_Passthrough` since it needs to be consistent with the line 3 entry in the bootscript lab3.script.

All the files and directories in `/srv/tftp` need to be owned by the tftp group. If you copy files and they are owned by root, you need to change them to have the tftp group ownership. Change the group ownership of everything contained in `/srv/tftp` to group **tftp** by using the **chgrp** command:

```
$ sudo chgrp -R tftp /srv/tftp
```

Listing 11.75: sudo chgrp -R tftp /srv/tftp

Note: The -R option for chgrp makes it recursive.

Now we need to change the permission so that everyone can read everything in `/srv/tftp`. We do this using the **chmod** command:

```
$ sudo chmod -R a+r /srv/tftp
```

Listing 11.76: sudo chmod -R a+r /srv/tftp

Note 1: The -R option for chmod makes it recursive. **Note 2:** The a+r option says to add read permission to all users.

The next two files, **lab3.scr** and **lab3.script**, need to be copied to the `/srv/tftp/de10nano/bootscripts` folder by using the commands:

```
$ sudo cp /media/sf_shared_folder/lab3.scr /srv/tftp/↲
  ↪de10nano/bootscripts
```

Listing 11.77: sudo cp /media/sf_shared_folder/lab3.scr /srv/tftp/de10nano/bootscripts

and

```
$ sudo cp /media/sf_shared_folder/lab3.script /srv/↲
  ↪tftp/de10nano/bootscripts
```

Listing 11.78: sudo cp /media/sf_shared_folder/lab3.script /srv/tftp/de10nano/bootscripts

Make sure that these files have group tftp ownership and that one can read them as was done previously in Listings 11.75 and 11.76.

Next, copy the Linux kernel image **zImage** into the kernel folder by using the command:

```
$ sudo cp /media/sf_shared_folder/zImage /srv/tftp/↲
  ↪de10nano/kernel
```

Listing 11.79: sudo cp /media/sf_shared_folder/zImage /srv/tftp/de10nano/kernel

and making sure that this image file has group tftp ownership and that one can read it as was done previously in Listings 11.75 and 11.76.

The next file needs to be put in the **nfs** folder at the location `/srv/nfs/de10nano` using the command:

```
$ sudo cp /media/sf_shared_folder/ubuntu-rootfs.tar.gz↲
  ↪ /srv/nfs/de10nano
```

Listing 11.80: sudo cp /media/sf_shared_folder/ubuntu-rootfs.tar.gz /srv/nfs/de10nano

When you untar (uncompress) the file, it will create the `/srv/nfs/de10nano/ubunut-rootfs` directory.

First, change to the `/de10nano` directory:

```
$ cd /srv/nfs/de10nano
```

Listing 11.81: cd /srv/nfs/de10nano

and then untar the file using the command:

```
$ sudo tar -zxpvf ubuntu-rootfs.tar.gz
```

Listing 11.82: sudo tar -zxpvf ubuntu-rootfs.tar.gz

Note: The tar options used are:

Option 1: The -z option is to unzip.
Option 2: The -x option is to extract (untar).
Option 3: The -p option is to preserve permissions.
Option 4: The -v option is to be verbose.
Option 5: The -f option is to specify the .tar file.

When done, you can remove the .tar file:

```
$ sudo rm ubuntu-rootfs.tar.gz
```

Listing 11.83: sudo rm ubuntu-rootfs.tar.gz

Note: When the rootfs is tarred up, it needs all the symbolic links preserved; otherwise, a kernel panic will occur since it will not be able to find the linked kernel `/sbin/init` file. Preserving the symbolic links when tarring the directory is done with the following command:

```
$ sudo tar -cpzf ubuntu-rootfs.tar.gz ubuntu-rootfs
```

Listing 11.84: sudo tar -cpzf ubuntu-rootfs.tar.gz ubuntu-rootfs

11.1.3.10 Running the Audio Mini Passthrough Example

Make sure that you have the following items set up before powering on the DE10-Nano board:

Setup 1: The Ubuntu VM is running with both the NSF and TFTP servers.
Setup 2: Connect the DE10-Nano board to your Laptop/computer using an Ethernet cable.
Setup 3: Connect the USB cable to the UART port (right side of DE10-Nano).
Setup 4: Open PuTTY in Windows on your laptop and connect to the UART port to watch the DE10-Nano board boot.

When your DE10-Nano board boots and you log into the HPS Linux via PuTTY as root (password root), you will see the following files:

```
root@localhost:~# ls
FE_AD1939.ko    hello_world           passthrough_scripts
FE_TPA613A2.ko  init_passthrough.sh
```

Fig. 11.25: Files in the root directory on the DE10-Nano board when using the Passthrough example

Note: These files have been served by the NFS server in the Ubuntu VM and exist in the Ubuntu VM in a directory located at:

`/srv/nfs/de10nano/ubuntu-rootfs/root`

Any files that you want to run on the DE10-Nano board when in the **Developer's Boot Mode** need to be put into (or below) this `/root` folder in the Ubuntu VM (Fig. 11.25).

In order to access this directory, you need root access, which you can get by issuing the following command:

```
$ sudo -i
```

Listing 11.85: sudo -i

which creates a terminal window with default root access. Then you can change to the `/root` that was not accessible without root access.

```
$ cd /srv/nfs/de10nano/ubuntu-rootfs/root
```

Listing 11.86: cd /srv/nfs/de10nano/ubuntu-rootfs/root

11.1.4 Cross Compiling "Hello World"

Our next step is to cross compile the "Hello World" example and run it on the DE10-Nano board on the ARM CPUs. We already compiled and ran it in the Ubuntu VM where it ran on the laptop x86 CPU in Sect. 11.1.2.11 Testing the Ubuntu VM by Compiling "Hello World" (page 211). We will now do this for the ARM CPUs on the DE10-Nano board and do this process in two different ways, but first we need to install the Linaro tools.

11.1.4.1 Configure and Set up the Linaro GCC ARM Tools in Ubuntu VM

The steps to set up the Linaro GCC tools for cross compiling from your x86 CPU and targeting the ARM CPU are:

Step 1: Install the Linaro tools in the Ubuntu VM by the command:

```
$ sudo apt install gcc-arm-linux-gnueabihf
```

Listing 11.87: sudo apt install gcc-arm-linux-gnueabihf

The ARM CPUs on the DE10-Nano board are the ARM Cortex-A9s, which implements the ARM v7 instruction set. The ARM cores also contain *Floating Point Units* (**FPU**). We note this since we want our cross compiler to take advantage of the FPUs (rather than using software libraries to do floating point). This is why we are installing the Linaro GCC tools with the **gnueabihf** option, which stands for ***GNU*** ***e**mbedded **a**pplication **b**inary **i**nterface **h**ard **f**loat*.

Step 2: Determine the version of Linaro that was installed by running the command:

```
$ /usr/bin/arm-linux-gnueabihf-gcc --version
```

Listing 11.88: /usr/bin/arm-linux-gnueabihf-gcc –version

11.1.4.2 Manual Cross Compilation of Hello World

The steps for manually cross compiling the Hello World program are:

Step 1: Create a **hello.c** file that will print the following three lines, replacing the text in the second printf function with your first and last name:

```c
#include <stdio.h>

int main(){
    printf("Hello ARM CPU World\n");
    printf(" My name is <first_name> <last_name>\n");
    printf("Manual Compilation\n");
    return 0;
}
```

Listing 11.89: Hello World

Step 2: First compile it so that it will run in the Ubuntu VM that is running on an x86 CPU on your Windows computer.

```
$ gcc -o hello hello.c
```

Listing 11.90: gcc -o hello hello.c

Step 3: Run the program from a Ubuntu VM terminal window:

```
$ ./hello
```

Listing 11.91: ./hello

Step 4: Now, determine the type of file that was created by using the *file* command:

```
$ file hello
```

Listing 11.92: file hello

Which will give the response:

```
hello: ELF 64-bit LSB shared object, x86-64, version 1 (SYSV),
dynamically linked, interpreter /lib64/ld-linux-x86-64.so.2, Bu
ildID[sha1]=13c23451a252ce49af91ee79c8bd4fa7821d86de, for GNU/L
inux 3.2.0, not stripped
```

Fig. 11.26: ELF x86-64

Notice that the file is a 64-bit executable (ELF) targeting the x86-64 architecture, which allows it to run in the Ubuntu VM on your Windows x86 laptop (Fig. 11.26).

Step 5: Now compile the hello.c file using the cross compiler:

```
$ /usr/bin/arm-linux-gnueabihf-gcc -o hello ↲
    ↪hello.c
```

Listing 11.93: /usr/bin/arm-linux-gnueabihf-gcc -o hello hello.c

If you try running the file, it will not run in the Ubuntu VM. Determine the type of this file using the *file* command:

```
$ file hello
```

Listing 11.94: file hello

which will now give the response:

```
hello: ELF 32-bit LSB shared object, ARM, EABI5 version 1 (SYSV
), dynamically linked, interpreter /lib/ld-linux-armhf.so.3, Bu
ildID[sha1]=2fe936d962883c0e9bc66241c09aa075a9b5aaf7, for GNU/L
inux 3.2.0, not stripped
```

Fig. 11.27: ELF ARM

Notice that the file is now a 32-bit executable (ELF) targeting the ARM architecture, which is why it will not run in the Ubuntu VM on your Windows x86 laptop (Fig. 11.27).

Step 6: To run the cross compiled program on the DE10-Nano board, you need to copy the file to the `/root` folder in the ubuntu-rootfs root file system that is being served by the NFS server. You will need root permission to copy into the `/root` folder. Assuming that you are in the directory containing the cross compiled hello world program, copy the file using the command:

```
$ sudo cp hello /srv/nfs/de10nano/ubuntu-↲
    ↪rootfs/root
```

Listing 11.95: sudo cp hello /srv/nfs/de10nano/ubuntu-rootfs/root

Remember: Any program that you want to run on the DE10-Nano board needs to be cross compiled and then placed in the root folder (or in a subdirectory below root) that is located in the Ubuntu VM at: `/srv/nfs/de10nano/ubuntu-rootfs/root`.

Step 7: To test that you can now run the hello world program on the DE10-Nano board, implement the following steps (a check list is provided for all the pieces that need to be in place):

Check 1: Make sure that the Ubuntu VM is set up and running the TFT and NFS servers.
Check 2: Connect your laptop to the USB UART port using a USB cable with a USB Mini-B connector.
Check 3: A PuTTY window is open and connected to the correct COM port in Windows.
Check 4: Boot (power cycle) the DE10-Nano board.
Check 5: Login as root on the DE10-Nano board using the PuTTY window.
Check 6: Run the "Hello World" program from the PuTTY terminal window by issuing the following command:

```
$ ./hello
```

Listing 11.96: ./hello

11.1.4.3 Makefile Cross Compilation of Hello World

In order to keep things organized, create a directory called `/software/lab3` under your home directory:

```
$ mkdir -p ~/software/lab3
```

Listing 11.97: mkdir -p /software/lab3

In this directory, create a new *hello.c* source file that will print the following three lines, replacing the text on line two with your first and last name:

```
Hello ARM CPU World
My name is <first_name> <last_name>
Makefile Compilation
```

Copy the following two files and put them in your `/software/lab3` folder:

File 1: arm_env.sh (Click here for file)
File 2: Makefile (Click here for file)

Open a terminal window in the Ubuntu VM and execute the commands in arm_env.sh by using the source command (the source command is used to load shell commands from a file):

```
$ source arm_env.sh
```

Listing 11.98: source arm_env.sh

or the shortcut for source using the dot command (**.**):

```
$ . arm_env.sh
```

Listing 11.99: . arm_env.sh

Notice that the prompt has been changed to:

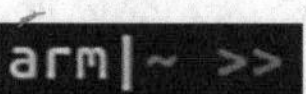

Fig. 11.28: ARM prompt that signifies that the terminal window is set up for ARM cross compilation

This new prompt tells you that this terminal window has been set up for cross compiling, which is useful when you have multiple terminal windows open. If you did not source the shell script, you would need to manually export the variable *CROSS_COMPILE* so that any child processes (any program that you run in the terminal window) would know where the Linaro toolchain is. Furthermore, you would need to execute the command every time you opened a new terminal window to compile targeting the ARM processors. This is why we are using the shell script so that we do not have to type the following lines each time we open a terminal window for cross compiling.

```
export CROSS_COMPILE=/usr/bin/arm-linux-gnueabihf-
export ARCH=arm
```

If you did not change the prompt to signify that the terminal windows have been set up for cross compiling, or if you wanted to check if the CROSS_COMPILE variable was set correctly, you could type:

```
$ export -p
```

Listing 11.100: export -p

which shows ALL the environmental variables are listed alphabetically. If you just want to see what the CROSS_COMPILE variable has been set to, use the following command:

```
$ env | grep CROSS_COMPILE
```

Listing 11.101: env | grep CROSS_COMPILE

To see what the architecture variable has been set to (i.e., the target CPU architecture), use the following command:

```
$ env | grep ARCH
```

Listing 11.102: env | grep ARCH

Now open the Makefile with an editor of your choice and make the following changes:

Change 1: Change line 26 from:
EXEC=
to:
EXEC=hello.

Change 2: Change line 29 from:
SRCS=
to:
SRCS=hello.c.

Save the Makefile and then compile using the Makefile by typing the command:

```
$ make
```

Listing 11.103: make

The Makefile will build both the ARM and x86 versions. To run the x86 version, type the command:

```
$ ./exec/x86/hello
```

Listing 11.104: ./exec/x86/hello

Copy the file `./exec/arm/hello` to the `/root` folder in the ubuntu-rootfs so that it can be run by the DE10-Nano board. You will need root permission to copy into this `/root` folder.

```
$ sudo cp ./exec/arm/hello /srv/nfs/de10nano/ubuntu-↲
    ↪rootfs/root
```

Listing 11.105: sudo cp ./exec/arm/hello /srv/nfs/de10nano/ubuntu-rootfs/root

Now boot the DE10-Nano from the Ubuntu VM and run the "Hello World" program from the DE10-Nano board using the PuTTY terminal window:

```
# ./hello
```

Listing 11.106: ./hello

11.2 List of U-boot Commands

Below is a list of alphabetized U-boot commands that are used in the book:

- **ping 192.168.1.10** used in 11.1.3.8
- **printenv** used in 11.1.3.8
- **saveenv** used in Setting 6:
- **setenv bootscript lab3.scr** used in Setting 5:
- **setenv ipaddr 192.168.1.11** used in Setting 2:
- **setenv nfsboot true** used in Setting 1:
- **setenv nfsip 192.168.1.10** used in Setting 4:
- **setenv serverip 192.168.1.10** used in Setting 3:

References

1. Canonical. *List of releases*. https://wiki.ubuntu.com/Releases. Accessed 23 Jun 2022.
2. Canonical. *The story of Ubuntu*. https://ubuntu.com/about. Accessed: 23 Jun 2022
3. Yangliy. *File:Network Address Translation (file2).jpg*. https://commons.wikimedia.org/w/index.php?curid=61795882. Transferred from en.wikibooks to Commons., Public Domain, Accessed 23 Jun 2022.
4. Canonical. *Network Configuration*. https://ubuntu.com/server/docs/network-configuration. Accessed 23 Jun 2022
5. T. Shrivastava. *29 Practical Examples of Nmap Commands for Linux System/Network Administrators*. https://www.tecmint.com/nmap-command-examples/. Accessed 23 Jun 2022.

Chapter 12
Creating a LED Pattern Generator System

We will create a LED Pattern Generation System, shown in Fig. 12.1, that will allow both hardware and software to generate LED patterns. We will start on the hardware side by creating the **LED_Patterns** component in the FPGA fabric, which will be connected to the switches, push buttons, and LEDs. We will take a hierarchical design approach where we will create larger hardware blocks out of smaller ones. We will then create registers for the control signals used to control LED_Patterns and connect them to the ARM CPUs (i.e., hard processor system or HPS) so that software running on Linux can control the LEDs. This will be done by creating the component **HPS_LED_Patterns**, which will connect the registers to the HPS Lightweight Bus. When finished, we will then have a component where LED patterns can be controlled by either hardware or software.

The LED Pattern Generation System is tackled following the steps listed below.

- Step 1: Creating the **LED_Patterns** hardware component. This is covered in Lab 4.
- Step 2: Debugging custom hardware with **Signal Tap**, an embedded logic analyzer tool. This is covered in Lab 5.
- Step 3: Creating the *HPS_LED_Patterns* hardware component using **Platform Designer**. This is covered in Lab 6.
- Step 4: Testing *HPS_LED_Patterns* using **System Console** and `/dev/mem` in Linux (which requires root access). This is covered in Lab 7.
- Step 5: Writing a **C program to create LED patterns**. This uses `/dev/mem` in Linux that requires root access. This is covered in Lab 8.

R. K. Snider, *Advanced Digital System Design using SoC FPGAs*,
https://doi.org/10.1007/978-3-031-15416-4_12

Step 6: **Creating a Linux kernel module**. This is covered in Lab 9.
Step 7: **Modifying the Linux device tree** so that Linux will know about your custom hardware. This is covered in Lab 10.
Step 8: Writing a **Linux Device Driver** for *HPS_LED_Patterns*. This allows user space to access your custom hardware. This is covered in Lab 11.

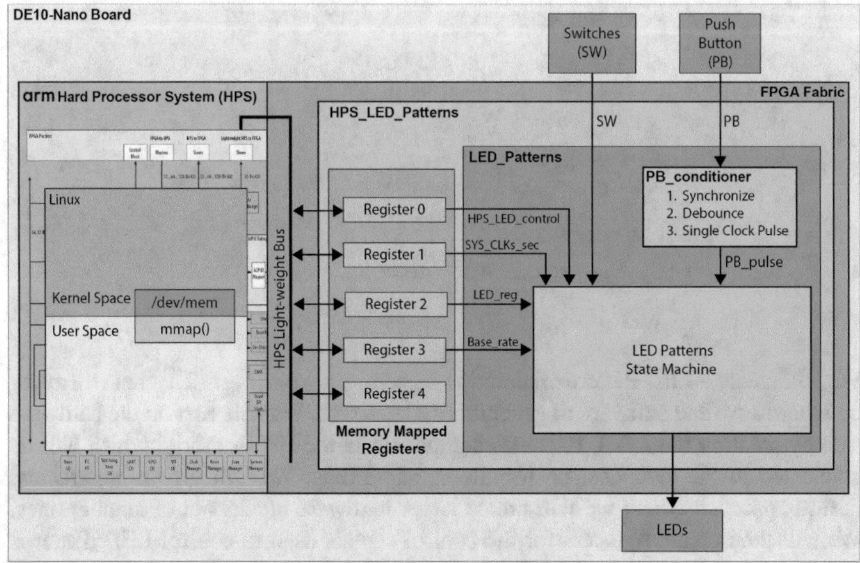

Fig. 12.1: LED Pattern Generation System

12.1 LED_Patterns Component

The LED_Patterns component is shown in Fig. 12.1 and can function in the FPGA fabric as a standalone hardware component that generates LED patterns. This component conditions the external switch and push-button signals, creates LED patterns using state machines, and drives the LEDs to display these patterns.

12.1.1 LED_Patterns Entity

The VHDL entity for LED_Patterns is shown in Listing 12.1. You can download this entity from here.

```
entity LED_patterns is
    port(
        clk             : in  std_logic;                        -- system clock
        reset           : in  std_logic;                        -- system reset (assume active high, change at top level if needed)
        PB              : in  std_logic;                        -- Pushbutton to change state (assume active high, change at top level if needed)
        SW              : in  std_logic_vector(3 downto 0);     -- Switches that determine the next state to be selected
        HPS_LED_control : in  std_logic;                        -- Software is in control when asserted (=1)
        SYS_CLKs_sec    : in  std_logic_vector(31 downto 0);    -- Number of system clock cycles in one second
        Base_rate       : in  std_logic_vector(7 downto 0);     -- base transition period in seconds, fixed-point data type (W=8, F=4).
        LED_reg         : in  std_logic_vector(7 downto 0);     -- LED register
        LED             : out std_logic_vector(7 downto 0)      -- LEDs on the DE10-Nano board
    );
end entity LED_patterns;
```

Listing 12.1: Entity of LED_patterns

The entity signals for the LED_Patterns component are described in Table 12.1.

Table 12.1: Entity Signals in the LED_patterns Component

Signal	Description
clk	System clock running at 50 MHz
reset	System reset (typically tied to KEY 0)
PB	Push-button signal that will be used to change state. Tied to KEY 1
SW	The four switches that will be used to determine the next state
HPS_LED_control	A signal that controls if the LEDs are controlled from the hardware state machine or from software via the LED_reg register. If set to "0," which it will be for this lab, the hardware state machine controls the LEDs. When set to "1," the LED output signal is connected directly to register LED_reg, which will be done in a future lab and controlled via software
SYS_CLKs_sec	Set to how many system clock periods are in one second
Base_rate	This signal is set to control the base rate of LED transitions in seconds and is how the system clock rate is converted into seconds. Note: This is an unsigned 8-bit fixed-point word with 4 fractional bits. For example, if Base_rate = 1.0, then transitions occur every 1 second
LED_reg	LED register signal for software control of LEDs
LED	Output signal to drive the LEDs

12.1.2 Functional Requirements for LED_Patterns

The first functional requirement of the LED_patterns component that you need to implement is dictated by the HPS_LED_control input signal. If this is set to one, the LEDs are controlled from software by the ARM HPS. If it is set to zero, then the LEDs are controlled by the LED_Patterns component in the FPGA fabric. In *pseudocode*, the requirement is:

```
if (HPS_LED_control == 1) then
    LED <= controlled by software that writes to register LED\_reg
else
    LED <= controlled by hardware state machines in LED\_Patterns.
```

12.1.2.1 State Machine

When the **LED_patterns** component is in the *hardware control mode* (HPS_LED_control=0), the requirements for the state machine are:

Requirement 1: **LED7 (LEDs[7]) always blinks at a 1 * Base_rate seconds** regardless of what the state machine is doing. This will allow you to verify that your base rate is set correctly.

Requirement 2: **The state machine needs to have 5 states** described below. When the state machine is in the noted state, you need to implement the described LED pattern. Note that these run at different rates.

- State 0: LEDs[6:0] show **one lit LED shifting right** at **1/2 * Base_rate seconds** (circular right shifting). **This is the default reset/power-up state.**
- State 1: LEDs[6:0] show **two lit LEDs, side-by-side, shifting left** at **1/4 * Base_rate seconds** (circular left shifting).
- State 2: LEDs[6:0] show the output of a **7-bit up counter** running at **2 * Base_rate seconds** (counter wraps).
- State 3: LEDs[6:0] show the output of a **7-bit down counter** running at **1/8 * Base_rate seconds** (counter wraps).
- State 4: **User defined pattern. Implement your own pattern**. It cannot be an up/down counter or a right/left shifter or any pattern that any of your classmates are implementing. Define your pattern and **your own pattern transition rate**, i.e., x * Base_rate seconds.

Requirement 3: State transitions. When the push button (PB) is pressed, the following sequence needs to happen each time the push button is pressed:

- Sequence 1: The binary code of the switches (SW) is displayed on LEDs[6:0] for 1 second. There are no LED patterns shown during this time.
- Sequence 2: The next_state is determined by the binary code of the switches (SW). If the switches specify a next_state of 5 or greater, the next_state is ignored and the current state is kept (the switch (SW) value is still displayed for 1 second even if it is 5 or greater).
- Sequence 3: The next_state implements the functionality from the state descriptions above and stays in

this state until the switches are changed and the push button (PB) is pressed again.

12.1.2.2 Conditioning the Push-Button Signal

The external push-button signal needs to be conditioned to operate synchronously and correctly with the state machine that it will control. We will take a general approach here that can be used with any state machine and external push button that needs debouncing (not necessarily the push button on the DE10-Nano).

You will need to create a component called **Conditioner** that does three things to the input signal:

Conditioning 1: Synchronizes the signal (two D flip-flops).
Conditioning 2: Debounces the signal.
Conditioning 3: Creates a single pulse with a period of 1 system clock no matter how long the push button is pressed or how many times it bounces (small state machine).

Creating these components is left as homework exercises.

12.1.3 Suggested Architecture

You should sketch out a block diagram of all the pieces you need in order to implement the LED_Patterns component. As you do so, you will want to think about how your design can be hierarchical. Build the component out of blocks (entities or processes) that you have tested (or will test) and know to be correct. What you do not want to do is to create the entire component and then test it to see if it is working correctly. If you take this approach, you will end up spending much more time debugging your design than if you took a systematic approach and implemented a hierarchical process in the first place. It is tempting to throw code together without first thinking about how all the pieces fit together (i.e., the architecture), and it will feel like the systematic hierarchical implementation process is slow going and may appear to be pointless. However, you will end up finishing faster by taking a systematic hierarchical approach since you will not be spending nearly as much time debugging your system. Jumping in and starting a design without thinking it through will cause the system to be fragile, and it will be hard to see where the logic is going wrong.

A suggested hierarchical design is shown in Fig. 12.2. You are free to implement your own design and encouraged to do so. The suggestions below are to get you started if you are unsure where to start.

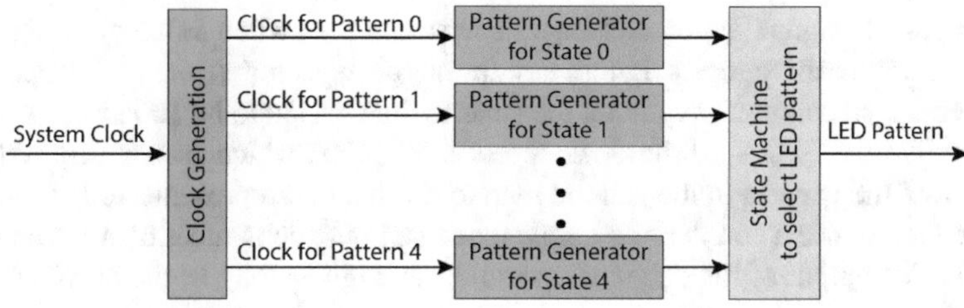

Fig. 12.2: Suggested architecture for the LED_patterns component

Suggestion 1: Start by sketching a block diagram of the system.

Suggestion 2: Create a separate pattern generator for each pattern described in Sect. Requirement 2:. Each pattern generator will have its own state machine with a clock input that will be set with the appropriate rate. Each pattern can then be tested individually and sent out to the LEDs.

Suggestion 3: Create a state machine that controls what is sent to the LEDs. To handle the switch display state (Requirement 3:), a suggestion is to:

1: Enter the switch display state from any state when the push button is pressed.
2: Start a One Second Counter when the state is entered.
3: Display the switch value on the LEDs during this time.
4: Leave state when one second is up and go to the state specified by the switches.

Suggestion 4: Create a clock component that generates all the appropriate clocks for the different pattern generators.

12.2 Homework Problems

Problem 12.1

Create a Synchronizer component and call this code **synchronizer.vhd**. When a person pushes a push button, the signal is asynchronous with respect to the FPGA fabric clock. Create a VHDL component that synchronizes the input signal, which is the first thing that needs to be done to an asynchronous signal coming into a synchronous design. You can do this with two D flip-flops (use behavioral VHDL for this, i.e., just make two signal assignments on the rising edge of the clock).

Problem 12.2

Create a Debouncer component and call this code **debouncer.vhd**. We will assume that the push button needs to be debounced, so create a VHDL component that

debounces the signal. When a change occurs, i.e., the switch value goes high (we are assuming positive logic), ignore any further changes for a time period specified by a **generic** in debouncer.vhd. Set the generic time period to be 100 msec. You can assume that the FPGA fabric clock speed is 50 MHz, which means that you will need to set the generic in the generic map to a value that represents 100 msec. You are required to use a counter and a state machine in the debouncer component.
Note 1: The output of the debouncer should stay high as long as the push button is pressed.
Note 2: Any component that uses a state machine needs to have a reset signal in the entity where the state machine can be reset to the starting state when the reset signal is asserted.

Problem 12.3
Create a **OnePulse** component and call this code **onepulse.vhd**. This component will create a single pulse with a period of 1 system clock no matter how long the push button has been pressed (use a state machine).

Problem 12.4
Create a **Conditioner** component and call this code **conditioner.vhd**. This component instantiates the components Synchronizer, Debouncer, and OnePulse. The signal flow should be Input → Synchronizer → Debouncer → OnePulse → Output.

Part III
SoC FPGA System Examples

"A good stock of examples, as large as possible, is indispensable for a thorough understanding of any concept, and when I want to learn something new, I make it my first job to build one." A quote by mathematician Paul Halmos.[1]

[1] P.Halmos, I Want to be a Mathematician: An Automathography (page 63, Springer, 1985).

Example 1
Audio FPGA Passthrough

1.1 Audio Passthrough System Overview

The Passthrough System example is the first system developed where the Audio Mini Board is interfaced to the DE10-Nano board. The overview of the system can be seen in Fig. 1.1. Stereo audio signals are converted into digital by the analog-to-digital converter of the AD1939. The digital signals in I2S format are sent to the FPGA fabric where the serial data words are converted into 24-bit words and made to conform to the Avalon Streaming Interface, which uses the data–channel–valid protocol (see Sect. 6.2.3). These signals are then converted back into I2S serial data, converted by the AD1939 back into analog, and then amplified by the headphone amplifier. These data conversions require a VHDL component to convert I2S into Avalon Streaming and back, which is covered in Sect. 1.2.1. This VHDL component is ported into Platform Designer (described in Sect. 1.4.1) so that it can be easily used when creating other audio processing systems. There is no audio processing done in this example since the design goal is to only get the data and control interfaces working.

You will not see any SPI or I2C control signals associated with the **AD1939_hps_audio_mini** data conversion component. To use these control signals, we take advantage of the Hard Processor System (HPS) that is part the Cyclone V SoC FPGA. The HPS has SPI and I2C interfaces already in place but must be exported, and this HPS setup is discussed in Sect. 1.4.3.2. The associated SPI and I2C Linux device drivers also need to be in place to be able to control the AD1939 audio codec and TPA6130A2 headphone amplifier, and these are covered in Sect. 1.5.

R. K. Snider, *Advanced Digital System Design using SoC FPGAs*,
https://doi.org/10.1007/978-3-031-15416-4_13

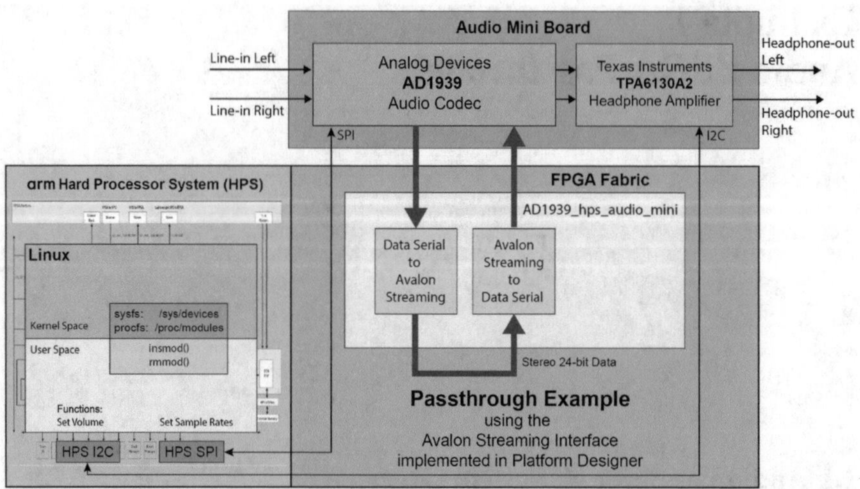

Fig. 1.1: Audio Passthrough System. Stereo audio signals are converted into digital by the analog-to-digital converter of the AD1939. The digital signals in I2S format are sent to the FPGA fabric where the serial data words are converted into 24-bit words and converted into the Avalon Streaming Interface that uses the data–channel–valid protocol (see Sect. 6.2.3). These signals are then converted back into I2S serial data and converted by the AD1939 back into analog where the signals are then amplified by the headphone amplifier. The SPI and I2C control interfaces are handled by the Hard Processor System (HPS) in the Cyclone V SoC FPGA and their associated Linux device drivers

1.2 Audio Data Streaming

Although the AD1939 audio codec can sample up to a f_s = 192 kHz sample rate, we will set the sample rate to 48 kHz since it is good enough for our audio processing designs that we will start with. Our goal is to route stereo audio data to the FPGA fabric and back out with minimal latency.

We will set the FPGA fabric clock to run at 98.304 MHz, which is eight times greater than the AD1939 master clock of 12.288 MHz. This clock will be created by using one of the FPGA *Phase Locked Loops* (**PLLs**). This is done so that we can implement a synchronous design where all the clock edges align. Having a system clock of 98.304 MHz processing audio with a sampling rate of 48 kHz means that there will be 2048 system clock cycles between each audio sample. Thus a lot of parallel processing can be done in the FPGA fabric in between audio samples.

1.2.1 I2S to Avalon Streaming Data Conversion

The first conversion that we must do is convert the I2S serial data coming out of the AD1939 into an Avalon streaming format so that we can easily make streaming data connections using Intel's Platform Designer in Quartus. Information on the Avalon Streaming Interface is covered in Sect. 6.2.3. The conversion to be performed is shown in Fig. 1.2 where the I2S signals are shown at the top and the Avalon data–channel–valid signals are shown at the bottom. The VHDL component that performs this conversion is named **AD1939_hps_audio_mini**, and the entity of this component can be seen in Listing 1.1 (click here for the source file).

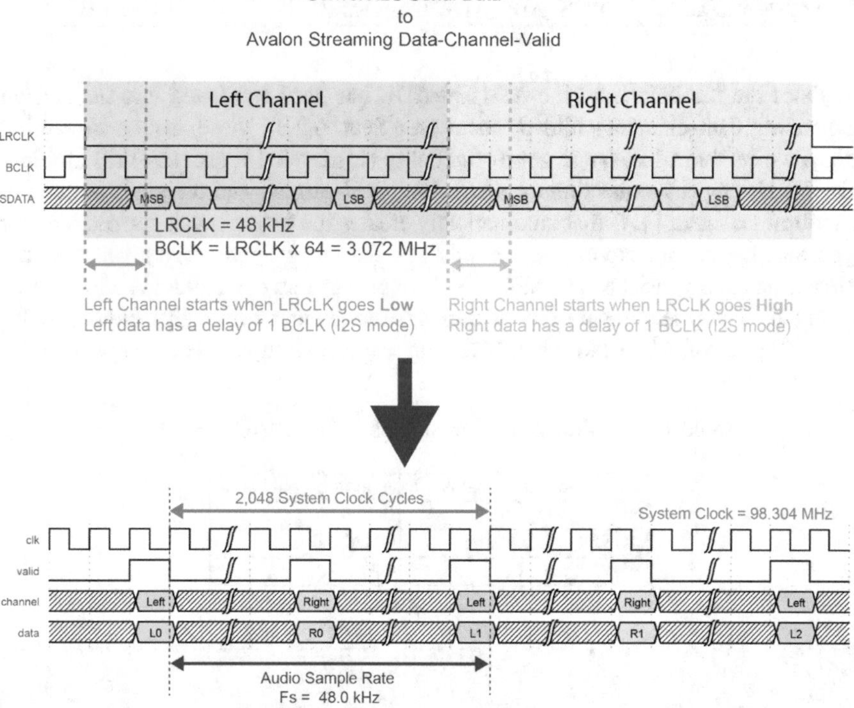

Fig. 1.2: Converting I2S data from the AD1939 audio codec into Avalon Streaming data using the data–channel–valid protocol. The Avalon Streaming Interface allows easy data connections to be made within Platform Designer

In the entity, the clock signal sys_clk (line 79) is assumed to be 98.304 MHz and created by an on-board PLL that is using the AD1939 12.288 MHz master clock as a reference. We do this so that we will have a synchronous design where we do not have to worry about any issues with data crossing boundaries between two different clock domains. The sys_reset signal (line 81) is present since we have a state machine in the component.

The I2S serial data signals coming from the AD1939 ADC (see Listing 1.1 lines 87–93) are physical signals that come into the FPGA. The associated entity, top level, I2S, and AD1939 names are shown in tabled 1.1. When creating a Platform Designer component, these signals must be exported (i.e., conduit signals) and connected to the associated top level signals (click here for the top level source file).

Table 1.1: Serial data signals from the AD1939 ADC

Entity Signal Name	Top Level Name	I2S Name	AD1939 Pin	AD1939 Name
ad1939_adc_asdata2	AD1939_ADC_ASDATA2	SDATA	26	asdata2
ad1939_adc_abclk	AD1939_ADC_ABCLK	BCLK	28	abclk
ad1939_adc_alrclk	AD1939_ADC_ALRCLK	LRCLK	29	alrclk

Once the I2S data from the AD1939 ADC has been converted into the Avalon Streaming data–channel–valid protocol (see Sect. 6.2.3), these Avalon Streaming signals (see Table 1.2) come out of the entity (Listing 1.1 lines 109–113) and into the FPGA fabric. Connecting up these Avalon Streaming audio signals is done in Platform Designer (1.4) that automatically creates the bus/network connections for you, and they are not exported to the top level. The clock signal is not given since the streaming clock is assumed to be the FPGA fabric system clock (98.304 MHz in this example). Thus the number of clock cycles between the left (or right) audio samples is 2048 system clock cycles when assuming a 48 kHz sample rate (see Fig. 1.2).

Table 1.2: Avalon streaming signals from the AD1939 ADC

Entity Output Signal Name	Avalon Signal Name
ad1939_adc_data	DATA
ad1939_adc_channel	CHANNEL
ad1939_adc_valid	VALID

```
74 entity ad1939_hps_audio_mini is
75   port (
76     -- fpga system fabric clock  (note: sys_clk is assumed to be faster and
77     -- synchronous to the ad1939 sample rate clock and bit clock, typically
78     -- one generates sys_clk using a pll that is n * ad1939_adc_alrclk)
79     sys_clk              : in    std_logic;
80     -- fpga system fabric reset
81     sys_reset            : in    std_logic;
82     ----------------------------------------
83     -- Physical signals from adc (serial data)
84     ----------------------------------------
```

```
    -- serial data from ad1939 pin 26 asdata2, adc2 24-bit
    --normal stereo serial mode
    ad1939_adc_asdata2 : in    std_logic;
    -- bit clock from adc (master mode) from pin 28 abclk on ad1939;
    -- note: bit clock = 64 * fs, fs = sample rate
    ad1939_adc_abclk   : in    std_logic;
    -- left/right framing clock from adc (master mode) from pin 29 alrclk
    -- on ad1939;  note: lr clock = fs, fs = sample rate
    ad1939_adc_alrclk  : in    std_logic;
    -------------------------------------------
    -- Physical signals to dac (serial data)
    -------------------------------------------
    -- serial data to ad1939 pin 20 dsdata1, dac1 24-bit
    -- normal stereo serial mode
    ad1939_dac_dsdata1 : out   std_logic;
    -- bit clock for dac (slave mode) to pin 21 dbclk on ad1939
    ad1939_dac_dbclk   : out   std_logic;
    -- left/right framing clock for dac (slave mode)
    -- to pin 22 dlrclk on ad1939
    ad1939_dac_dlrclk  : out   std_logic;
    ---------------------------------------------------
    -- Avalon streaming interface from ADC to fabric
    ---------------------------------------------------
    -- Data: w=24; f=23; signed 2's complement
    ad1939_adc_data    : out   std_logic_vector(23  downto 0);
    -- Channels: left <-> channel 0;  right <-> channel 1
    ad1939_adc_channel : out   std_logic;
    -- Valid: asserted when data is present
    ad1939_adc_valid   : out   std_logic;
    ---------------------------------------------------
    -- Avalon streaming interface to DAC from fabric
    ---------------------------------------------------
    -- Data: w=24; f=23; signed 2's complement
    ad1939_dac_data    : in    std_logic_vector(23 downto 0);
    -- Channels: left <-> channel 0; right <-> channel 1
    ad1939_dac_channel : in    std_logic;
    -- Valid: asserted when data is present
    ad1939_dac_valid   : in    std_logic
  );
end entity ad1939_hps_audio_mini;
```

Listing 1.1: VHDL Snippet: Entity of AD1939_hps_audio_mini.vhd

In this passthrough example, Avalon streaming audio data that would normally be processed in the FPGA fabric are instead sent directly back to the AD1939_hps_audio_mini component. These Avalon Streaming output signals (see Table 1.3) go into the component and are converted into I2S serial data before being sent to the AD1939 DAC.

Table 1.3: Avalon Streaming Signals going to the AD1939 DAC

Entity Input Signal Name	Avalon Signal Name
ad1939_dac_data	DATA
ad1939_dac_channel	CHANNEL
ad1939_dac_valid	VALID

The I2S serial data signals going to the AD1939 DAC (Listing 1.1 lines 99–104) are physical signals that leave the FPGA and connect to the AD1939 DAC. The associated entity, top level, I2S, and AD1939 names are shown in Table 1.4. When creating a Platform Designer component, these signals must be exported (i.e., conduit signals) and connected to the associated top level signals in the top level VHDL file DE10Nano_AudioMini_System.vhd.

Table 1.4: Serial Data Signals going to the AD1939 DAC

Entity Signal Name	Top Level Name	I2S Name	AD1939 Pin	AD1939 Name
ad1939_dac_dsdata1	AD1939_DAC_DSDATA1	SDATA	20	dsdata1
ad1939_dac_dbclk	AD1939_DAC_DBCLK	BCLK	21	dbclk
ad1939_dac_dlrclk	AD1939_DAC_DLRCLK	LRCLK	22	dlrclk

1.2.1.1 Serial to Parallel Conversion

The first step in processing the I2S serial data coming into the FPGA (see signals in Table 1.1), which is shown in Fig. 1.2, is to perform a serial to parallel conversion of the 32-bit framing window that contains the 24-bit sample word. This is done using a shift register as illustrated in Fig. 1.3. The shift register is configured so that the bits enter from the right side and shift left. Thus, when we capture the 24-bit sample word (by capturing all 32 bits in the left/right framing window), the MSB will be on the left side, i.e., bit 23 of the extracted signal will hold bit 23 of the ADC sample. The shift register is 32 bits since there are 32 BCLK cycles in each of the left and right framing windows that are controlled by the LRCLK framing clock.

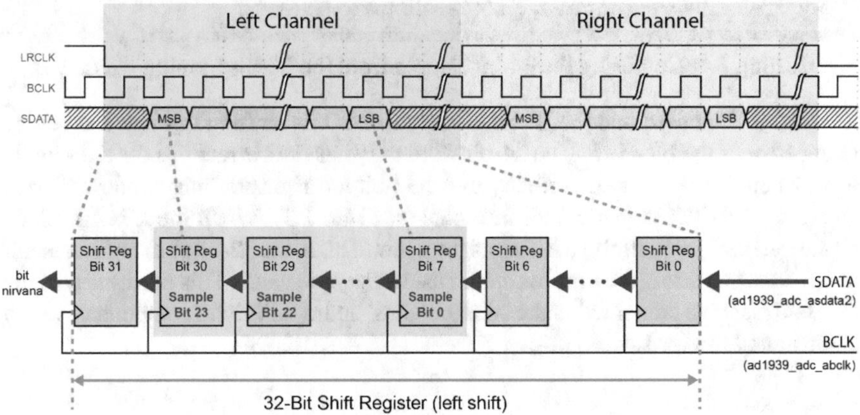

Fig. 1.3: Converting the serial I2S data into parallel 24-bit sample words

We utilized Quartus' IP library to create the 32-bit shift register (click here for the generated VHDL file). We can see that the shift register is 32 bits, and the shift direction is left from the generic map as shown in Listing 1.2 (lines 75–79).

```
75  GENERIC MAP (
76      lpm_direction => "LEFT",
77      lpm_type => "LPM_SHIFTREG",
78      lpm_width => 32
79  )
```

Listing 1.2: The generic map configures the shift register to be 32 bits and to shift left

The instantiation of the shift register can be seen in lines 200–205 of Listing 1.3. The clock into the serial to parallel converter is the I2S bit clock (BCLK) that is named ad1939_adc_abclk. The serial data (SDATA) signal, named ad1939_adc_asdata2, is connected to the shiftin port.

```
200 s2p_adc2 : component serial2parallel_32bits
201   port map (
202     clock   => ad1939_adc_abclk,
203     shiftin => ad1939_adc_asdata2,
204     q       => sregout_adc2
205   );
```

Listing 1.3: Instantiating the serial to parallel converter in component AD1939_hps_audio_mini

The output q is 32 bits and is connected to the signal named sregout_adc2. We capture the 24-bit sample word by taking the appropriate 24-bit signal slice as shown in Listing 1.4.

```
adc2_data <= sregout_adc2(29 downto 6);
```

Listing 1.4: Capturing the 24-bit sample from the 32-bit framing window

It should be noted that the 24-bit signal still has bits shifting through it on every rising edge of the bit clock. The 24-bit word only has the correct bits for the sample word when it is positioned correctly in the 32-bit left/right framing window. This is where the LRCLK signal comes into play (see Fig. 1.3). When the LRCLK clock signal (ad1939_adc_alrclk) transitions from low to high, the 32-bit shift register is full of left data, and this is the time to capture the left sample value. This timing is done by the state machine contained in the AD1939_hps_audio_mini.vhd component, which is illustrated in Fig. 1.4.

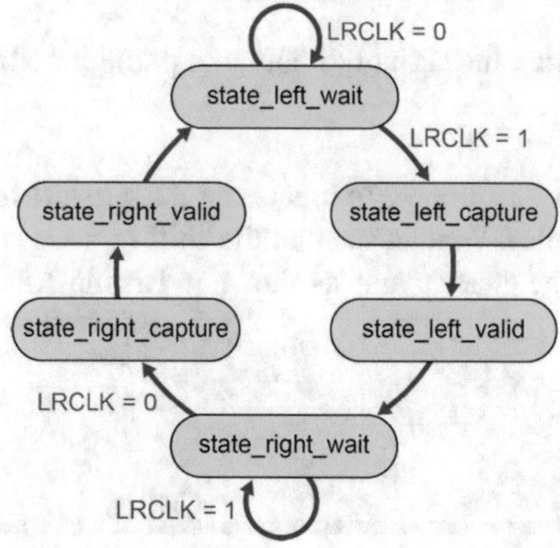

Fig. 1.4: State machine used to capture the 24-bit I2S audio words using the LRCLK signal as shown in Fig. 1.3

The state machine is running at the FPGA fabric system clock speed (98.304 MHz), and for this discussion, we assume that it is currently in the state *state_left_wait* and that the LRCLK clock (ad1939_adc_alrclk) is *low*. We will only discuss the left channel since the right channel is similar.

As soon as the LRCLK goes high (see Fig. 1.3 and associated VHDL code in Listing 1.5), the state machine transitions immediately through states *state_left_capture* and *state_left_valid*, taking only one system clock cycle for each state. It is in these states that we create the Avalon Streaming signals, which is described in the next section.

```
      when state_left_wait =>
        -- The 32-bit shift register is full of left data
        -- when alrck goes high
        if (ad1939_adc_alrclk = '1') then
          state <= state_left_capture;
        else
```

```
            state <= state_left_wait;
          end if;

        when state_left_capture => -- state to capture data
          state <= state_left_valid;

        when state_left_valid => -- state to generate valid signal
          state <= state_right_wait;
```

Listing 1.5: Left channel states

1.2.1.2 Avalon Streaming Output Interface

When the state machine is in the state *state_left_capture*, it saves the 24-bit sample word into the output signal ad1939_adc_data since the 24 bits are correctly aligned in the 32-bit left/right framing window. The output signal ad1939_adc_data is the data channel of the Avalon Streaming interface as shown at the bottom of Fig. 1.3, and it keeps this value until the state machine enters the *state_left_capture* again and captures a new value. In the next state *state_left_valid*, the Streaming Interface signals *valid* (1) and *channel* (0 for left) are sent out. The associated VHDL code is shown in Listing 1.6. The state machine then deals with the right channel in a similar manner before repeating the left channel states again.

```
        when state_left_capture =>
          ad1939_adc_data <= adc2_data; -- capture left 24-bit word

        when state_left_valid =>
          ad1939_adc_valid   <= '1'; -- sample is now valid
          ad1939_adc_channel <= '0'; -- data is from the left channel
```

Listing 1.6: State machine signals that implement the Avalon Streaming Interface (left channel)

1.2.1.3 Avalon Streaming Input Interface

When streaming data is sent into the AD1939_hps_audio_mini component's Avalon Streaming interface, the valid signal *ad1939_dac_valid* is asserted when data is present. The valid signal is continually checked (every system clock cycle) to see if it has been asserted as shown in Listing 1.7 (line 332). If the valid is asserted, we then need to determine which channel the data belongs to, which is done by the case statement block on lines 334–349. However, before the 24-bit data is saved to either the left or right DAC signal, it needs to be converted to a 32-bit I2S signal. The data is converted into I2S format by adding a leading zero bit and adding seven trailing zeros to make the data 32 bits.

```
      if (ad1939_dac_valid  = '1') then -- data has arrived

         case ad1939_dac_channel is

            -- data is in i2s-justified mode, which has one ↲
                                  ↪empty bit
            -- before the MSB. See Fig. 23 on page 21 of the ↲
                                  ↪AD1939 datasheet
            when '0' => -- left data
               -- pack into 32-bit word for L/R framing slot
               dac1_data_left <= '0' & ad1939_dac_data & "↲
                                  ↪0000000";

            when '1' => -- right data
               -- pack into 32-bit word for L/R framing slot
               dac1_data_right <= '0' & ad1939_dac_data & "↲
                                  ↪0000000";

            when others =>
               null;

         end case;

      end if;
```

Listing 1.7: Capturing Avalon Streaming data

Once the left input data has been captured and converted into the 32-bit I2S signal dac1_data_left, it needs to be converted into serial since I2S is the data serial format that the AD1939 DAC expects.

1.2.1.4 Parallel to Serial Conversion

The parallel to serial conversion is performed by a 32-bit shift register that performs the inverse of the serial to parallel shift register discussed in Sect. 1.2.1.1. The parallel to serial shift register also shifts left since the MSB needs to come out first. However, there is a control signal *load* that when asserted (1) loads the 32-bit word into the shift register. When *load* is deasserted (0), the shift register shifts left each bit clock cycle. It turns out that LRCLK signal (see Fig. 1.3) is exactly what we need for the load signal since we want the load signal to be low during the left frame to shift out the left channel bits. The right channel is done similarly except we need the inverse of the LRCLK signal for the load signal for the right parallel to serial converter.

```
  p2s_dac1_left : component parallel2serial_32bits
    port map (
      clock      => ad1939_adc_abclk,
```

```
        data      => dac1_data_left,
        load      => ad1939_adc_alrclk,
        shiftout => ad1939_dac_dsdata1_left
    );
```

Listing 1.8: Creating the I2S serial data stream for the left channel

1.2.1.5 Serial Data Multiplexing

Since we have serial data coming from both the left and right channel parallel to serial shift registers, we need to select the serial data that goes out with the LRCLK signal, which is low for left data and high for right data. This is easily done by creating a signal multiplexer as shown in Listing 1.9. The serial data has been delayed by one BCLK to align appropriately in the left/right framing window.

```
  interleave : process (sys_clk) is
  begin

    if rising_edge(sys_clk) then
      if (ad1939_adc_alrclk = '0') then
        ad1939_dac_dsdata1 <= ad1939_dac_dsdata1_left_delayed;
      else
        ad1939_dac_dsdata1 <= ad1939_dac_dsdata1_right_delayed;
      end if;
    end if;

  end process interleave;
```

Listing 1.9: Multiplexing the left and right data serial channels

1.2.1.6 Clocking

The AD1939 is configured in the *ADC Control 2* register so that both the ADC bit clock BCLK and the left/right framing clock LRCLK are set as clock masters (i.e., clock sources, see Table 5.3). We keep the BCLK and LRCLK for the DAC as clock slaves, which is the default configuration (see DAC Control 1 register in Table 5.2). This means that we need to connect the DAC BCLK and LRCLK to the ADC BCLK and LRCLK so that the sample rate for the DAC is controlled by the ADC (f_s=48 kHz). A timing wrinkle crops up since we have added latency to the DSDATA data line by virtue of capturing and registering the Avalon data channel. Furthermore, we would like to add some setup time for the I2S serial data going to the DAC. We can do this by delaying both the DAC BCLK and LRCLK clocks using the delay_signal component. A delay of eight system clock cycles (81.4 nanoseconds) is added for the BCLK as shown in Listing 1.10. The LRCLK is done similarly.

```
delay_dac_bclk : component delay_signal
  generic map (
    signal_width => 1,
    signal_delay => 8
  )
  port map (
    clk                 => sys_clk,
    signal_input(0)     => ad1939_adc_abclk,
    signal_delayed(0)   => ad1939_dac_dbclk
  );
```

Listing 1.10: Delaying the DAC BCLK

If a delay is not added to the bit clock *ad1939_dac_dbclk* going out to the DAC, the serial data does not have the required setup time as can be seen in Fig. 1.5. The rising edge of the bit clock is shown in blue, and transitions of the serial data are shown in green. The serial data is transitioning on the clock edge and sometimes afterward rather than the requirement of being set up *before* the clock edge.

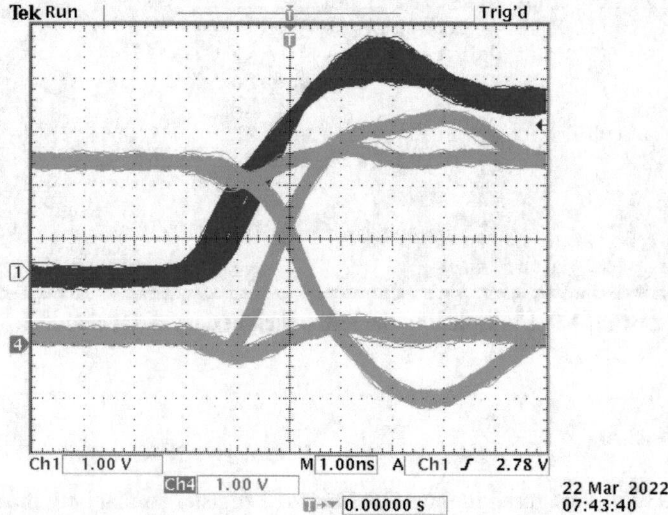

Fig. 1.5: Setup time being violated for the DAC serial data with respect to the bit clock. The rising clock edges are shown in blue. The data transitions are shown in green. Figure courtesy of Trevor Vannoy

By delaying the bit clock, we shift the serial data transitions well before the rising edge of the bit clock as shown in Fig. 1.6.

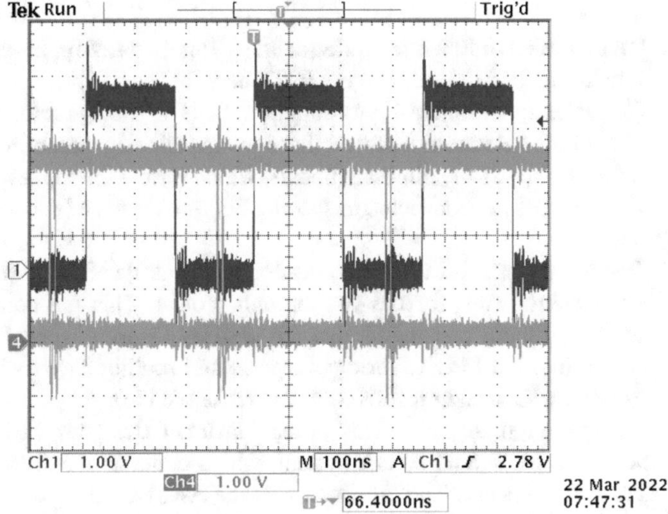

Fig. 1.6: Setup time being met for the DAC serial data with respect to the bit clock. The rising clock edges are shown in blue. The data transitions are shown in green. Figure courtesy of Trevor Vannoy

1.3 Quartus Passthrough Project

The Passthrough Quartus project is given as a *Golden Hardware Reference Design* (**GHRD**) for the Audio Mini board working with the DE10_Nano board. The example passthrough system is a minimal working system (no audio processing) with all the data and control interfaces working. To create this GHRD, create a Quartus project folder called `\passthrough` and copy the following six files into the `\passthrough` folder.

Note: The easiest way to download all the passthrough project files for Windows is to go to the book's GitHub Code repository (click here for the Code Repository), click on the green Code dropdown menu/button, and select *Download ZIP* from the dropdown menu. Unzip the Code-main.zip file that gets downloaded, and go to `\Code-main\examples\passthrough` where the Passthrough Quartus project files are located (with the exception of *delay_signal.vhd*, which is located in `\Code-main\lib\vhdl`).

Copy the following files into the `\passthrough` folder:

File 1: **DE10Nano_AudioMini_System.qpf** This is the **Quartus Project File** that you open in Quartus to open the Passthrough project (click here for the file).

File 2: **DE10Nano_AudioMini_System.vhd** This is the **Top Level VHDL File** for the Passthrough project (click here for the file).

File 3: **DE10Nano_AudioMini_System.qsf** This is the **Quartus Settings File** for the Passthrough project that contains all assignments and project settings, such as the pin assignments, which is how the top level VHDL signal names are connected to specific FPGA I/O pins (click here for the file).

File 4: **DE10Nano_AudioMini_System.sdc** This is the **Synopsys Design Constraints File** for the Passthrough project. This file contains the project's timing constraints that are used by Quartus when fitting the design into the FPGA fabric (i.e., place and routing have a significant impact on the timing results) (click here for the file).

File 5: **soc_system_hps.qsys** This is the **Platform Designer File** that describes a basic SoC FPGA system, which contains the HPS or Hard Processor System. The HPS contains the ARM CPUs and related peripherals. Platform Designer was previously called Qsys, thus the file extension (click here for the file). **Note:** We will create a hierarchical design where *soc_system_hps.qsys* will be modified by adding two Platform Designer subsystems *clk_reset_subsystem.qsys* (File 6:) and *ad1939_subsystem.qsys* (to be created shortly). The resulting hierarchical design will be called *soc_system_passthrough.qsys*.

File 6: **clk_reset_subsystem.qsys** This is the **Platform Designer File** that implements and encapsulates the clocks and resets for the passthrough system (click here for the file).

File 7: **soc_system_passthrough.ipx** This is the Platform Designer **IP Index File** that describes where IP associated with Platform Designer can be found. This is included so that Platform Designer will see the AD1939 IP located in `\passthrough\ip\ad1939` (click here for the file).

Create the project subdirectories: `\passthrough\ip\ad1939`, and copy the following files into `\ad1939`:

File 9: **ad1939_hps_audio_mini.vhd** This is the VHDL file that provides the Avalon Streaming Interface for the AD1939 audio codec. The VHDL file name has the string *_hps_* in the name to signify that it requires the SoC FPGA Hard Processor System (HPS) to fully work. This is because the SPI control interface for the AD1939 is not part of ad1939_hps_audio_mini.vhd code but is implemented using the SPI interface through the HPS (click here for the file).

File 10: **serial2parallel_32bits.vhd** This is the 32-bit serial to parallel shift register used to convert I2S data into 24-bit words. It was generated by Quartus' IP wizard (click here for the file).

File 11: **parallel2serial_32bits.vhd** This is the 32-bit parallel to serial shift register used to convert 24-bit audio words into I2S data. It was generated by Quartus' IP wizard (click here for the file).

File 12: **delay_signal.vhd** This is a component that can delay signals by the specified number of clock cycles (click here for the file).

1.4 Platform Designer

We will use Intel's Platform Designer system integration tool to connect up audio streaming interfaces using the Avalon Streaming Interface. To do so, we need to import the *AD1939_hps_audio_mini.vhd* file into Platform Designer so that it will show up as a library component where we can easily drop it into our passthrough and future systems. We also need to set up the clocks associated with the AD1939 component so that all audio related processing will be done synchronously with respect to the AD1939 master clock. This will be done by creating a Platform Designer subsystem for the AD1939. We then need to configure the HPS so that the I2C and SPI signals are exported so that we can control the Audio Mini board. Finally, we need to generate the Platform Designer system and connected it to the top level VHDL that has all the associated Audio Mini signals.

1.4.1 Creating the AD1939 Platform Designer Component

Once the Passthrough project files have been copied to their respective directories (see Sect. 1.3), open Quartus, and open the Quartus project file *DE10Nano_AudioMini_System.qpf* that is located in the project folder `\passthrough`.

Open Platform Designer in Quartus, which is located under *Tools* → *Platform Designer*. After some initializations, it will ask to open a .qsys file. Select *soc_system_hps.qsys* and click open. After some checking, the Platform Designer system will show up as shown in Fig. 1.7. This is the basic SoC FPGA system to which we will add our AD1939 subsystem once we create the AD1939 component.

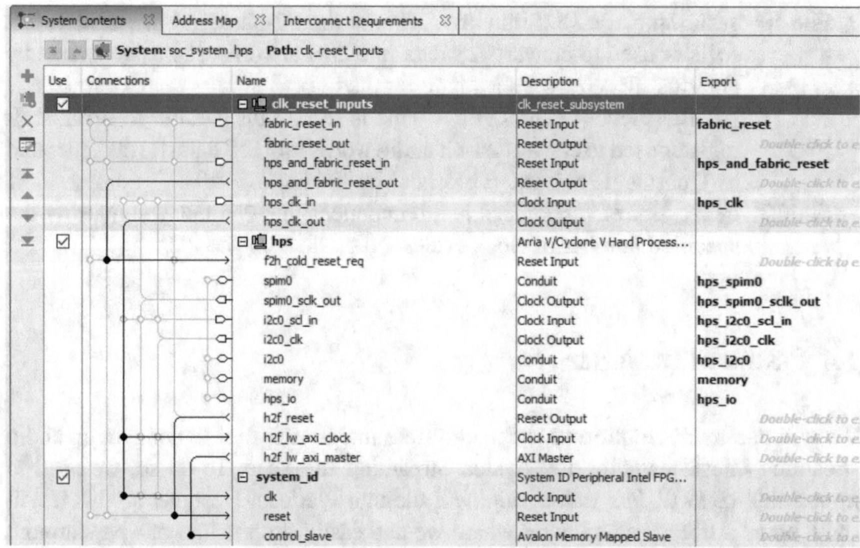

Fig. 1.7: The initial Passthrough Platform Designer System

On the left hand top side, you should see the *IP Catalog* Window. If not, select *View* → *IP Catalog*. We will be creating a new component in Platform Designer so double click *New Component* under `\Project` as shown in Fig. 1.8 (or click the *New...* button at the bottom of the IP Catalog window). The Component Editor will pop up.

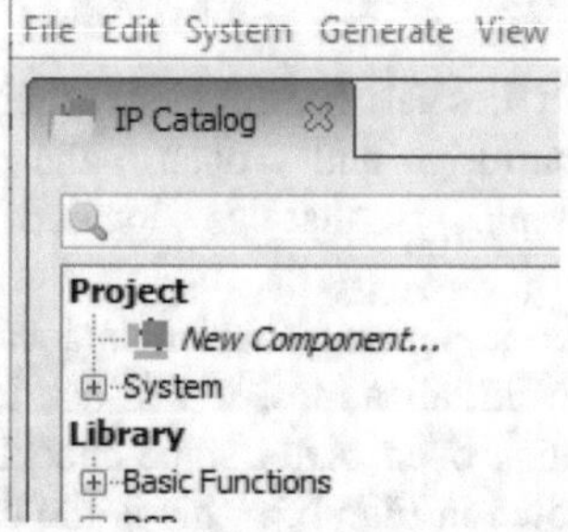

Fig. 1.8: Creating a new Platform Designer Component for the AD1939

In the **Component Type** tab of the Component Editor, enter **AD1939_Audio_Mini** for both the *Name:* and *Display name:* as shown in Fig. 1.9.

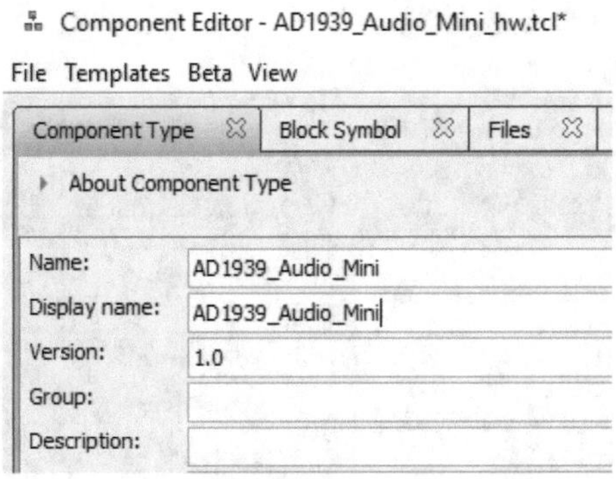

Fig. 1.9: Component Type tab of the Component Editor

In the **Files** tab of the Component Editor and in the ***Synthesis Files*** section at the top, click the *Add File...* button. Browse to `\passthrough\ip\ad193` and open *AD1939_hps_audio_mini.vhd*. Then click the *Analyze Synthesis Files* button. After getting the green message: "Analyzing Synthesis Files: completed successfully," click close. **Note:** When importing code into Platform Designer, make sure that the VHDL is correct before importing since the error messages from Platform Designer are not as helpful as the error messages from a Quartus compilation. You can check a VHDL file in Quartus by going to the menu and selecting *Processing* → *Analyze Current File*. It is suggested that you do this first and eliminate any VHDL errors before importing a file into Platform Designer (you do not need to do it for *AD1939_hps_audio_mini.vhd* since it is ready to go).

When first importing *AD1939_hps_audio_mini.vhd*, you will see a bunch of errors in the Messages window at the bottom of the Component Editor after it has been analyzed. We will fix these errors next since Platform Designer has assumed wrong interfaces for ***all*** the signals associated with the entity in *AD1939_hps_audio_mini.vhd*. First, click on the **Signals and Interfaces** tab in the Component Editor. The wrong signal associations are shown in Fig. 1.10 since it has assumed that they are all Avalon Memory-Mapped Slave signals, of which none of them are. Notice in Sect. 1.10 at the bottom of the signal list the command *«add interface»* that has been marked with an arrow. We now need to add seven interfaces. To add an interface, click on *«add interface»* and select from the drop down menu the interface type.

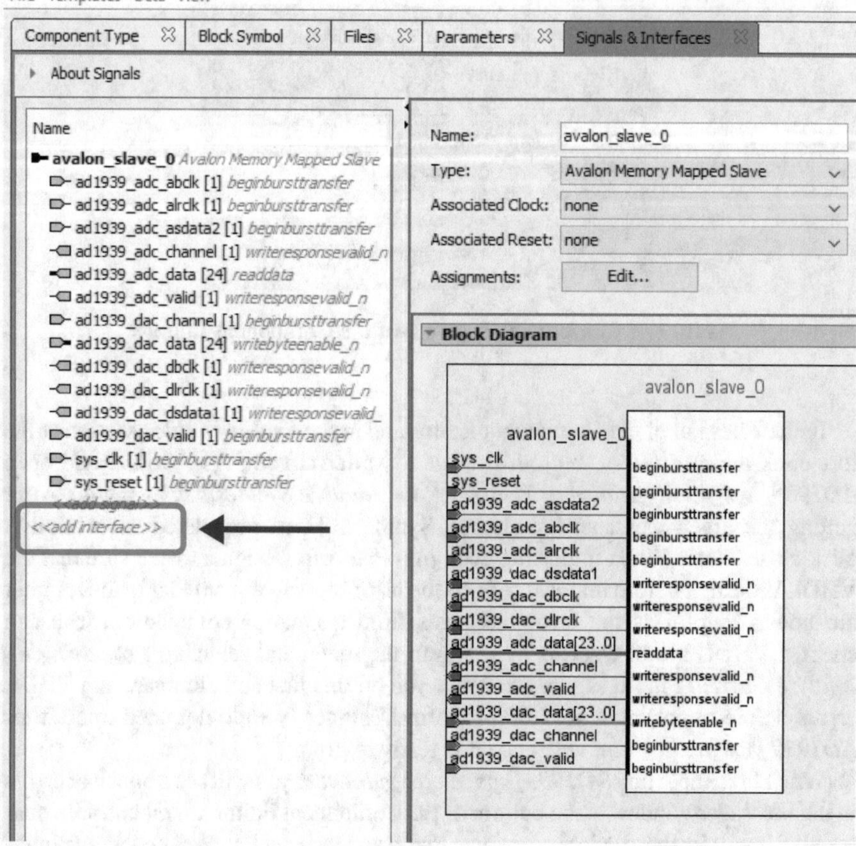

Fig. 1.10: Component Editor making the wrong associations regarding the AD1939 signals

Let us start by adding three clock interfaces since we have three clocks associated with the component. These are the 98.304 MHz FPGA fabric system clock, the bit clock BCLK, and the left/right framing clock LRCLK. Click *«add interface»*, select *Clock Input* from the drop down menu, and name the clock interface **sys_clk**. Set the clock rate to be 98304000. Now drag the signal *sys_clk* that is under the avalon_slave_0 interface to be under the **sys_clk** interface. The signal type will be wrong (probably beginbursttransfer) so click in the signal to highlight it and select **clk** from the *Signal Type:* pull down menu. The result should be like the figure in Fig. 1.11, and the summary of the interface is:

Interface 1: *Name:* **sys_clk**, *Interface Type:* **Clock input**

Parameter 1: *Clock rate:* **98304000**

Grouped Signal:

Signal 1: *Name:* **sys_clk[1]** *Signal Type:* **clk**

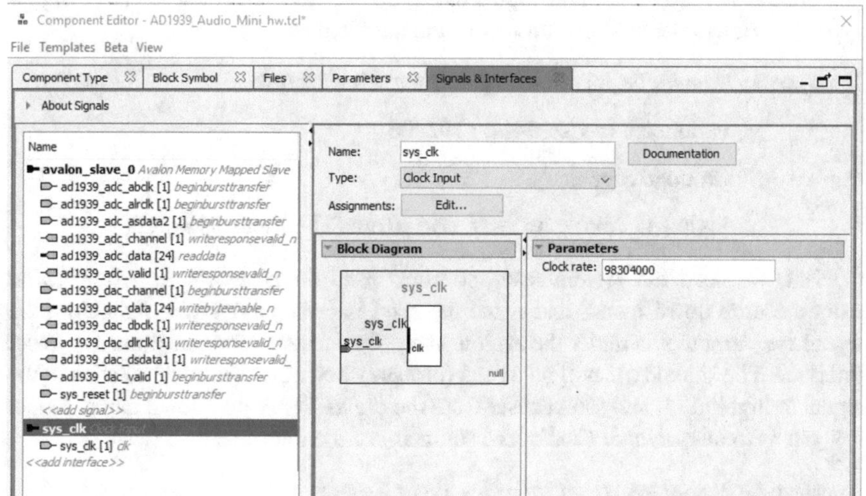

Fig. 1.11: Creating the FPGA fabric system clock interface for the AD1939 Platform Designer component

In a similar manner for the bit clock, click *«add interface»*, select *Clock Input* from the drop down menu, and name the clock interface **clk_abclk**. Set the clock rate to be 12288000 since this is the fastest the bit clock can be if the sample rate is 192 kHz (192000 * 64). We will be running the AD1939 sample rate at 48 kHz, but we want the Platform Designer clock constraint to allow us the possibility of running at 192 kHz so we will pick the fastest clock speed that it can be. Now drag the signal *ad1939_adc_abclk* that is under the avalon_slave_0 interface to be under the **clk_abclk** interface. The signal type will be wrong (probably beginbursttransfer) so click in the signal to highlight it and select **clk** from the *Signal Type:* pull down menu. The summary of the interface is:

Interface 2: *Name:* **clk_abclk** *Interface Type:* **Clock input**

Parameter 1: *Clock rate:* **12288000**

Grouped Signal:

Signal 1: *Name:* **ad1939_adc_abclk[1]** *Signal Type:* **clk**

Again, in a similar manner for the left/right framing clock, click *«add interface»*, select *Clock Input* from the drop down menu, and name the clock interface **clk_alrclk**. Set the clock rate to be 192000 since this is the fastest the framing clock can be if the sample rate is 192 kHz. We will be running at 48 kHz, but we want the Platform Designer clock constraint to allow us the possibility of running at 192 kHz. Now drag the signal *ad1939_adc_alrclk* that is under the avalon_slave_0 interface to be under the **clk_alrclk** interface. The signal type will be wrong (probably beginbursttransfer) so click in the signal to highlight it and select **clk** from the *Signal Type:* pull down menu. The summary of the interface is:

Interface 3: *Name:* **clk_alrclk** *Interface Type:* **Clock input**

Parameter 1: *Clock rate:* **192000**

Grouped Signal:

Signal 1: *Name:* **ad1939_adc_alrclk[1]** *Signal Type:* **clk**

Next, we need the system reset, so click *«add interface»*, select *Reset Input* from the drop down menu, and name the reset interface **sys_reset**. Now drag the signal *sys_reset* that is under the avalon_slave_0 interface to be under the **sys_reset** interface. The signal type will be wrong (probably beginbursttransfer) so click in the signal to highlight it and select **reset** from the *Signal Type:* pull down menu. Select **sys_clk** as the *Associated Clock:* The summary of the interface is:

Interface 4: *Name:* **sys_reset** *Interface Type:* **Reset input**
Associated Clock: **sys_clk**
Grouped Signal:

Signal 1: *Name:* **sys_reset[1]** *Signal Type:* **reset**

We will now create the Avalon Streaming sink interface, so click *«add interface»*, select *Avalon Streaming Sink* from the drop down menu, and name the Avalon Streaming sink interface **to_headphone_out**. Set the *Associated Clock:* to *sys_clk* and the *Associated Reset:* to *sys_reset*. Now drag the 24-bit signal *ad1939_dac_data* that is under the avalon_slave_0 interface to be under the **to_headphone_out** streaming interface. The signal type will be wrong (probably writebyteenable_n) so click in the signal to highlight it and select **data** from the *Signal Type:* pull down menu. Drag the channel signal *ad1939_dac_channel* that is under the avalon_slave_0 interface to be under the **to_headphone_out** streaming interface. The signal type will be wrong (probably beginbursttransfer) so click in the signal to highlight it and select **channel** from the *Signal Type:* pull down menu. Drag the valid signal *ad1939_dac_valid* that is under the avalon_slave_0 interface to be under the **to_headphone_out** streaming interface. The signal type will be wrong (probably beginbursttransfer) so click in the signal to highlight it and select **valid** from the *Signal Type:* pull down menu. Click on the *to_headphone_out* streaming interface, and under the *Parameters* section, set

Data bits per symbol to **24** since the data signal is 24 bits and set *Maximum channel* to **1**. If you keep the max channel setting at zero, the right channel (channel 1) will not be implemented in the system. The result should be like the figure in Fig. 1.12, and the summary of the interface is:

Interface 5: *Name:* **to_headphone_out** *Interface Type:* **Avalon Streaming Sink**
Associated Clock: **sys_clk**
Associated Reset: **sys_reset**

Parameter 1: *Data bits per symbol:* **24**
Parameter 2: *Maximum channel:* **1**

Grouped Signals:

Signal 1: *Name:* **ad1939_dac_channel[1]** *Signal Type:* **channel**
Signal 2: *Name:* **ad1939_dac_data[24]** *Signal Type:* **data**
Signal 3: *Name:* **ad1939_dac_valid[1]** *Signal Type:* **valid**

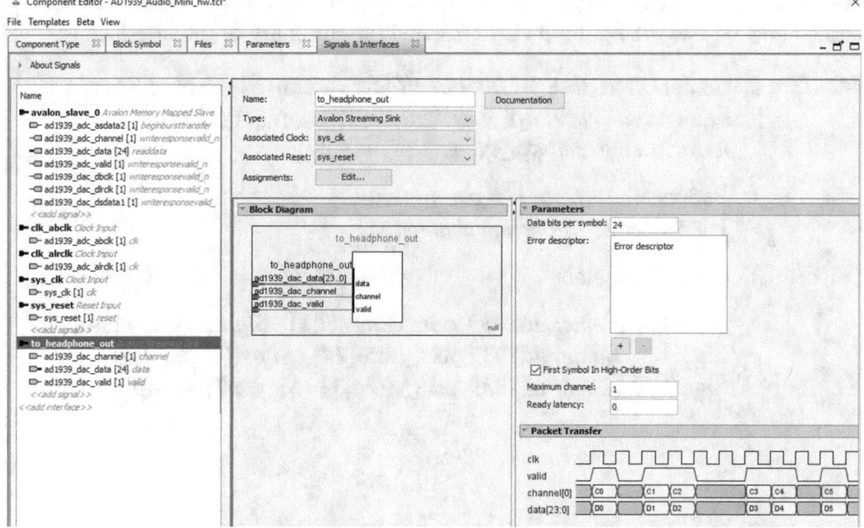

Fig. 1.12: Creating the Platform Designer Avalon Streaming Sink interface that goes to the Headphone amplifier and Headphone out on the Audio Mini board

Create the Avalon Streaming source interface in a similar manner. Click *«add interface»*, select *Avalon Streaming Source* from the drop down menu, and name the Avalon Streaming source interface **from_line_in**. Set the *Associated Clock:* to *sys_clk* and the *Associated Reset:* to *sys_reset*. Now drag the 24-bit signal *ad1939_adc_data* that is under the avalon_slave_0 interface to be under the **from_line_in** streaming interface. The signal type will be wrong (probably readdata) so click in the signal to highlight it and select **data** from the *Signal Type:* pull down menu. Drag the channel signal *ad1939_adc_channel* that is under the avalon_slave_0 interface to be under the **from_line_in** streaming interface. The signal type will be wrong (probably writeresponsevalid_n) so click in the signal to highlight it and select **channel** from the *Signal Type:* pull down menu. Drag the valid signal *ad1939_adc_valid* that is under the avalon_slave_0 interface to be under the **to_headphone_out** streaming interface. The signal type will be wrong (probably writeresponsevalid_n) so click in the signal to highlight it and select **valid** from the *Signal Type:* pull down menu. Click on the *from_line_in* streaming interface, and under the *Parameters* section, set *Data bits per symbol* to **24** since the data signal is 24 bits and set *Maximum channel* to **1**. If you keep the max channel setting at zero, the right channel (channel 1) will not be implemented in the system. The result should now be like the figure in Fig. 1.13, and the summary of the interface is:

Interface 6: *Name:* **from_line_in** *Interface Type:* **Avalon Streaming Source**
Associated Clock: **sys_clk**
Associated Reset: **sys_reset**

Parameter 1: *Data bits per symbol:* **24**
Parameter 2: *Maximum channel:* **1**

Grouped Signals:

Signal 1: *Name:* **ad1939_adc_channel[1]** *Signal Type:* **channel**
Signal 2: *Name:* **ad1939_adc_data[24]** *Signal Type:* **data**
Signal 3: *Name:* **ad1939_adc_valid[1]** *Signal Type:* **valid**

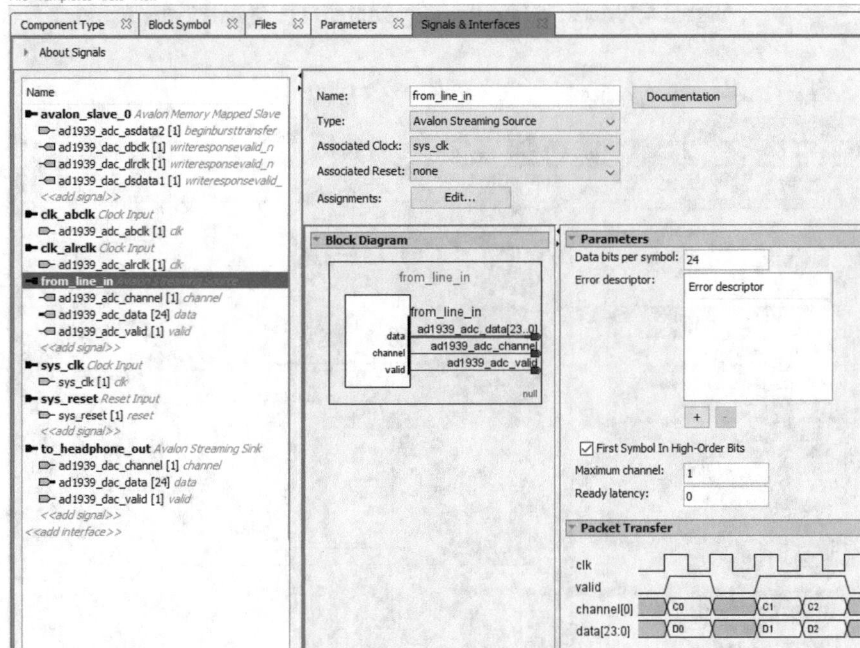

Fig. 1.13: Creating the Platform Designer Avalon Streaming Source interface that comes from the line-in input on the Audio Mini board

The signals that are left are related signals but are under the wrong interface. These signals need to be exported since they need to physically connect with the AD1939 audio codec on the Audio Mini board. Click on the *avalon_slave_0* interface to change the interface type and change the *Type:* to **Conduit** and the *Name:* to **connect_to_AD1939**. The conduit interface will export these signals out of the soc_system component so that they can be connected at the top level in the passthrough project. Set the *Associated Clock:* to *clk_abclk* and the *Associated Reset:* to *sys_reset*. Click on the signal *ad1939_adc_asdata2* to highlight it, and type in **asdata2** as the *Signal Type:*. Click on the signal *ad1939_dac_dbclk* to highlight it, and type in **dbclk** as the *Signal Type:*. Click on the signal *ad1939_dac_dlrclk* to highlight it, and type in **dlrclk** as the *Signal Type:*. Click on the signal *ad1939_dac_dsdata1* to highlight it, and type in **dsdata1** as the *Signal Type:*. The result should be like the figure in Fig. 1.14, and the summary of the interface is:

Interface 7: *Name:* **connect_to_AD1939** *Interface Type:* **Conduit**
Associated Clock: **clk_abclk**
Associated Reset: **sys_reset**
Grouped Signals:

Signal 1: *Name:* **ad1939_adc_asdata2[1]** *Signal Type:* **asdata2**
Signal 2: *Name:* **ad1939_dac_dbclk[1]** *Signal Type:* **dbclk**

Signal 3: *Name:* **ad1939_dac_dlrclk[1]** *Signal Type:* **dlrclk**
Signal 4: *Name:* **ad1939_dac_dsdata1[1]** *Signal Type:* **dsdata1**

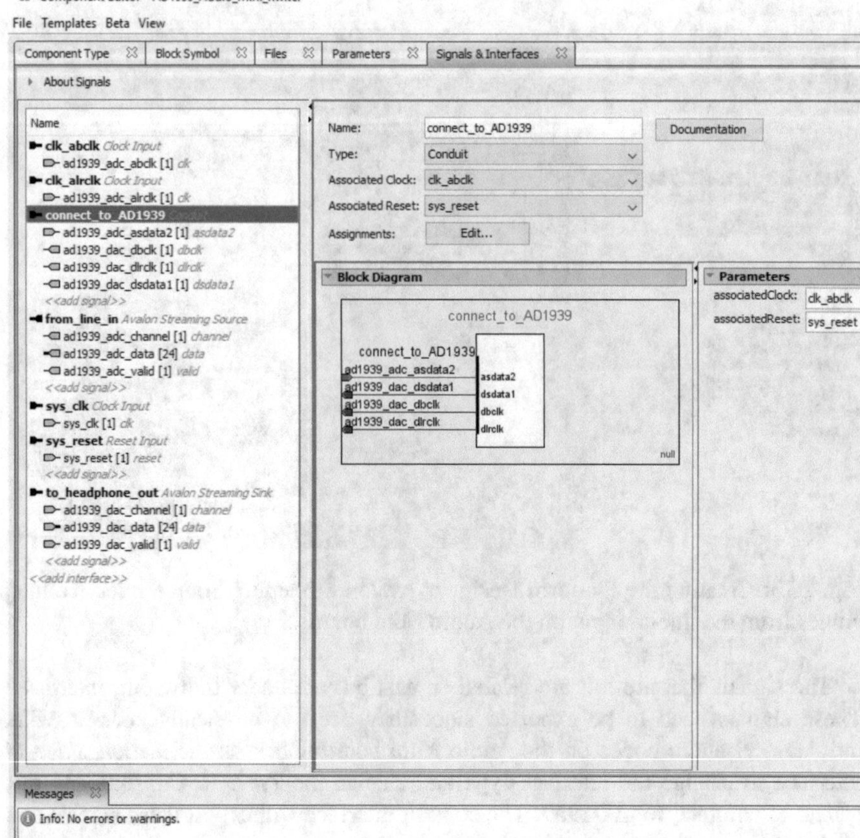

Fig. 1.14: Creating the Conduit interface that exports the AD1939 data serial signals that need to connect to the AD1939

At this point, there should be no more errors or warnings, which means that we are done with the Component Editor, so click *Finish*. . . and click Yes, Save. Notice that there is now the new Component *AD1939_Audio_Mini* under Project in the IP Catalog window. Click on this new component, and click the +Add. . . button at the bottom right of the IP Catalog window. The component will be added to the Platform Designer system, but with errors that we will fix shortly. So for now, just delete the AD1939 component since we first need to create a Platform Designer subsystem for it, which is covered in the next Sect. 1.4.2.

Now for a bit of housekeeping. When you clicked the *Finish*. . . and saved the component, the *AD1939_Audio_Mini_hw.tcl* file was created and put in the

\passthrough project folder. However, this file is associated with the VHDL file *AD1939_hps_audio_mini.vhd*. So move *AD1939_Audio_Mini_hw.tcl* into the IP folder \passthrough\ip\ad1939. Also, make a backup copy of the .tcl file since if this gets modified by Platform Designer, you can restore the file without having to import and making all the signal assignments again (yes, it does happen).

Note: One complicating factor of moving the .tcl file to \passthrough\ip\ad1939 is that the .tcl script seems to add a relative reference to the VHDL source code that will cause an error when adding the AD1939 component in Platform Designer (if the file was imported when in the project directory). If you get this error, a fix to this is to comment out the *add_fileset_file* command on line of *AD1939_Audio_Mini_hw.tcl* (in the *file sets* section ~line 42). Since we include this directory in the soc_system_passthrough.ipx file, Quartus will find the VHDL source file even though the path is commented out.

1.4.2 Creating a Platform Designer Subsystem for the AD1939

Open Platform Designer and select *File* → *New System...*, which will be named **ad1939_subsystem.qsys**. It will open up with a *Clock Source* named **clk_0**. However, we will not be using a clock source so delete it.

Now add the **ad1939_audio_mini** component that you just created by selecting it in the IP Catalog window under Project and clicking the +Add... button, and then Finish to close the pop-up window. Do not worry about errors in the Messages window since we will be fixing them. Rename the component from *ad1939_audio_mini_0* to *ad1939_audio_mini*.

We are creating a Platform Designer *subsystem*, which means that any signal that we want coming out of the subsystem we need to export. To export a signal, double click in the *Export* column and on the line that contains the signal that you want to export, and type in the name of the signal that will show up as a port in the subsystem. The result should be like Fig. 1.15.

Component *ad1939_audio_mini* Exports

Export 1: Signal *to_headphone_out* exported with name ***to_headphone_out***
Export 2: Signal *from_line_in* exported with name ***from_line_in***
Export 3: Signal *connect_to_AD1939* exported with name ***ad1939_physical***

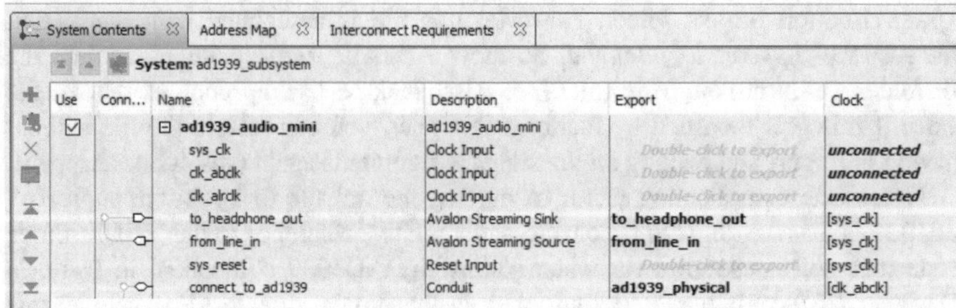

Fig. 1.15: Adding the ad1939_audio_mini component to the subsystem and exporting signals. The exported signals will show up as ports in the Platform Designer subsystem

The first clock we need to create and add to the subsystem is the 98.304 MHz system clock that will be derived from the AD1939 12.288 MHz MCLK (master clock of the AD1939). To create this clock, we need to use one of the *Phase-Locked Loops* (**PLLs**) that Cyclone V FPGA contains (there are six Cyclone V PLLs in the FPGA on the DE10-Nano board).

With Platform Designer opened, expand the Library in the IP Catalog window and expand the path to *Library → Basic Functions → Clocks; PLLs and Resets → PLL*. Select **PLL Intel FPGA IP** and click the +Add... button.

Right click on the *PLL_0* component that was added and select *Rename*. Change the PLL name to *sys_clk_from_ad1939_mclk_pll*, and move the PLL component to the top by selecting it and then using the up arrow at the left.

In the configuration window that pops up for the PLL component, and in the *General* tab, select the following parameters:

Component *sys_clk_from_ad1939_mclk_pll* Parameters

Parameter 1:	Set *Reference Clock Frequency:* to **12.288**.
Parameter 2:	**Disable** (uncheck) the *Enable locked output port* since we will not bother with monitoring whether the PLL is locked or not.
Parameter 3:	In the *outclk0* section, set the *Desired Frequency:* to **98.304**.

The result should be like the figure in Fig. 1.16. Leave all the other options with their default settings and click Finish.

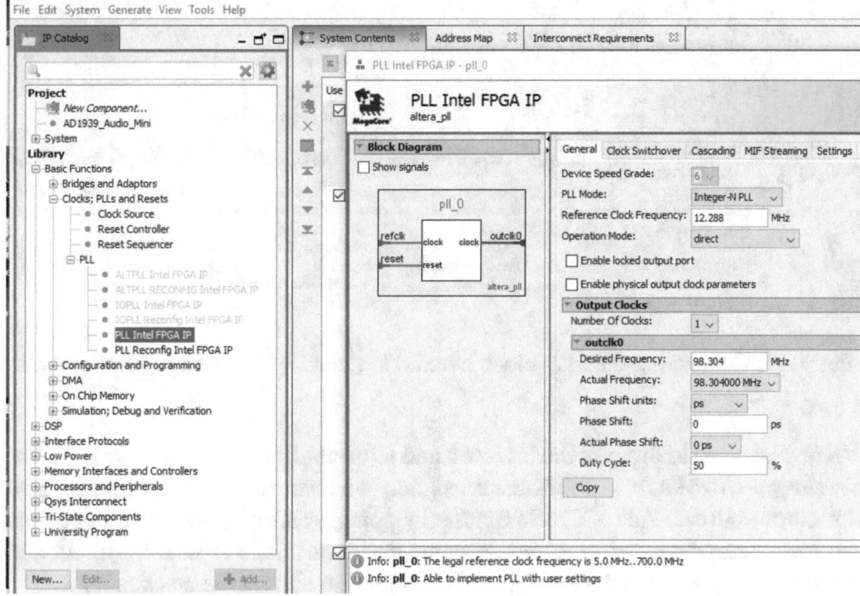

Fig. 1.16: Setting the PLL parameters for the system clock

Next, export the PLL reference clock:

Component ***sys_clk_from_ad1939_mclk_pll*** **Exports**
Export 4: Signal *refclk* exported with name ***ad1939_physical_mclk***

Now connect the clock output of the PLL (*outclk0*) to the *sys_clk* input of the ad1939_audio_mini component. The connection is made by clicking on the small circle that intersects these two signals in the *Connect...* column so that the circle becomes a black solid circle. An open circle signifies a possible connection but is not connected. If you click on the signal *outclk0*, it will make the signal bold in the connect column, which can make it easier to see what ports can connect to it, especially if there are a lot of signals in the connect column:

Component ***sys_clk_from_ad1939_mclk_pll*** **Connections**
Connection 1: Source: **outclk0** in *sys_clk_from_ad1939_mclk_pll* Target: **sys_clk** in *ad1939_audio_mini*

The result should be like the figure in Fig. 1.17.

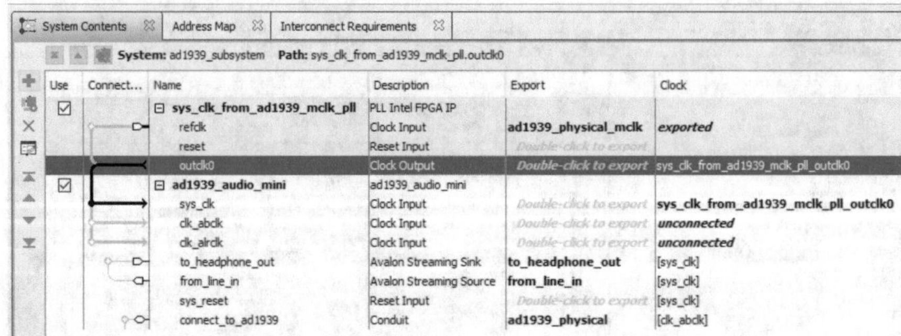

Fig. 1.17: Connecting the PLL clock outclk0 to the AD1939_Audio_Mini sys_clk

Now we need to bring in the bit clock and left/right framing clock. **Since we are creating a subsystem in Platform Designer, we need to use *clock bridges* and not clock sources.** Add a **Clock Bridge** by going to *Library* → *Basic Functions* → *Bridges and Adaptors* → *Clock*. Rename the clock bridge as **ad1939_abclk** and move the component to the top. Open the component and set the clock parameter:

Component *ad1939_abclk* Parameters

Parameter 4: *Explicit clock rate:* **12288000**

Make the following export:

Component *ad1939_abclk* Exports

Export 5: Signal *in_clk* exported with name **ad1939_physical_abclk**

Make the following clock connection:

Component *ad1939_abclk* Connections

Connection 2: Source: **out_clk** in *ad1939_abclk*
Target: **clk_abclk** in *ad1939_audio_mini*

In a similar fashion, add a clock bridge named **ad1939_alrclk**, and move the component to the top. Open the component and set the clock parameter:

Component *ad1939_alrclk* Parameters

Parameter 5: *Explicit clock rate:* **192000**

Make the following export:

Component *ad1939_alrclk* Exports

Export 6: Signal *in_clk* exported with name **ad1939_physical_alrclk**

Make the following clock connection:

Component *ad1939_alrclk* Connections	
Connection 3:	Source: **out_clk** in *ad1939_alrclk* Target: **clk_alrclk** in *ad1939_audio_mini*

We want to bring the system clock out of the subsystem, so add another clock bridge named **system_clock**, move the component to just below the PLL, and make the following export:

Component *system_clock* Exports	
Export 7:	Signal *out_clk* exported with name **audio_fabric_system_clk**

Make the following clock connection:

Component *system_clock* Connections	
Connection 4:	Source: **outclk0** in *sys_clk_from_ad1939_mclk_pll* Target: **in_clk** in *system_clock*

Finally, we need to add a reset bridge, which can be found at *Library* → *Basic Functions* → *Bridges and Adaptors* → *Reset*. Add a reset bridge named **reset**, move the component to the top, and set the following parameter:

Component *reset* Parameters	
Parameter 6:	*Synchronous edges:* **None**. **Note:** If you leave it as the default setting (deassert), it will end up holding the subsystem in reset and the passthrough system will not work.

Make the following export:

Component *reset* Exports	
Export 8:	Signal *in_reset* exported with name **subsystem_reset**

Make the following reset and clock connections:

Component *reset* Connections	
Connection 5:	Source: **out_reset** in *reset* Target: **reset** in *sys_clk_from_ad1939_mclk_pll*
Connection 6:	Source: **out_reset** in *reset* Target: **sys_reset** in *ad1939_audio_mini*
Connection 7:	Source: **outclk0** in *sys_clk_from_ad1939_mclk_pll* Target: **clk** in *reset*

The result should be like Fig. 1.18. Save this subsystem as **ad1939_subsystem.qsys**, and place it in the project folder `\passthrough`.

Fig. 1.18: The AD1939 Platform Designer subsystem *ad1939_subsystem.qsys*

1.4.3 Creating the Platform Designer Passthrough System

The starting point for the Passthrough system is the given system **soc_system_hps.qsys**. Open this system in Platform Designer.

1.4.3.1 Adding the AD1939 Subsystem

In IP Catalog window under `Project`, expand *System*, and you will see the *ad1939_subsystem*. Add this subsystem like adding any library component by selecting it and clicking the +Add. . . button.

Name the component **ad1939_subsystem**, move it above the *hps* component, and make the following exports:

Component *ad1939_subsystem* Exports

Export 1: Signal *ad1939_physical* exported with name **ad1939_physical**
Export 2: Signal *ad1939_physical_abclk* exported with name **ad1939_physical_abclk**
Export 3: Signal *ad1939_physical_alrclk* exported with name **ad1939_physical_alrclk**
Export 4: Signal *ad1939_physical_mclk* exported with name **ad1939_physical_mclk**

Then make the following connections:

Component *ad1939_subsystem* Connections	
Connection 1:	Source: **from_line_in** in *ad1939_subsystem*
	Target: **to_headphone_out** in *ad1939_subsystem*
	Note: This implements the audio passthrough.
Connection 2:	Source: **fabric_reset_out** in *clk_reset_inputs*
	Target: **subsystem_reset** in *ad1939_subsystem*

Save this Platform Designer Passthrough system with the name **soc_system_passthrough.qsys**.

1.4.3.2 Adding the SPI and I2C Interfaces to the hps Component

Open Platform Designer in Quartus and load *soc_system_passthrough.qsys* if it is not already opened. Double click on the **hps** component in the System Contents window to open the *Parameters* window for the Cyclone V Hard Processor System. Then click the *Peripheral Pins* tab.

Scroll down to the *SPI Controllers* section. Make the following configuration change:

Setting 1: *SPIM0 pin:* **FPGA**

Scroll further down to the *I2C Controllers* section. Make the following configuration change:

Setting 2: *I2C0 pin:* **FPGA**

Save the **soc_system_passthrough.qsys** Platform Designer Passthrough system.

The result is shown in Fig. 1.19 where both the SPI (SPI master 0) and I2C (I2C 0) interfaces were configured with the *FPGA* option. This means that the SPI and I2C signals show up in the hps component. The SPI signals (shaded blue) are then exported with the *hps_* prefix where they show up in the soc_system_passthrough entity that can be connected at the top level. Similarly, the I2C signals (shaded green) are exported with the *hps_* prefix where they show up in the soc_system_passthrough entity that can be connected at the top level.

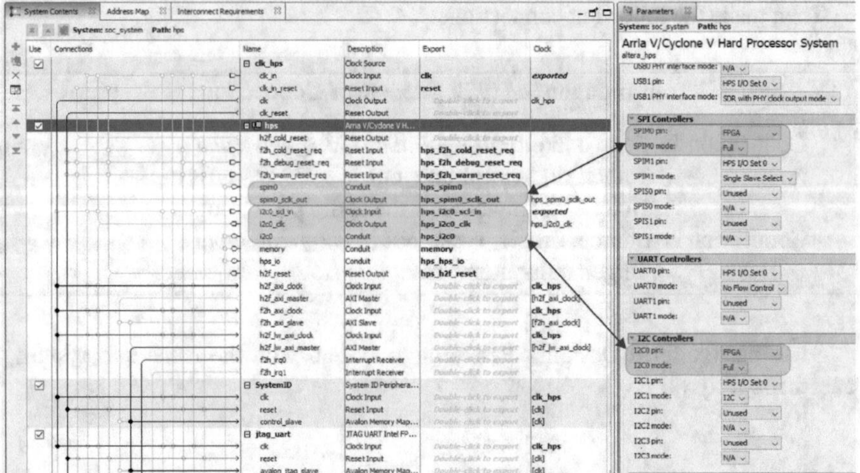

Fig. 1.19: Configuring the SPI and I2C interfaces in the HPS

The final Platform Designer Passthrough system can be seen in Fig. 1.20.

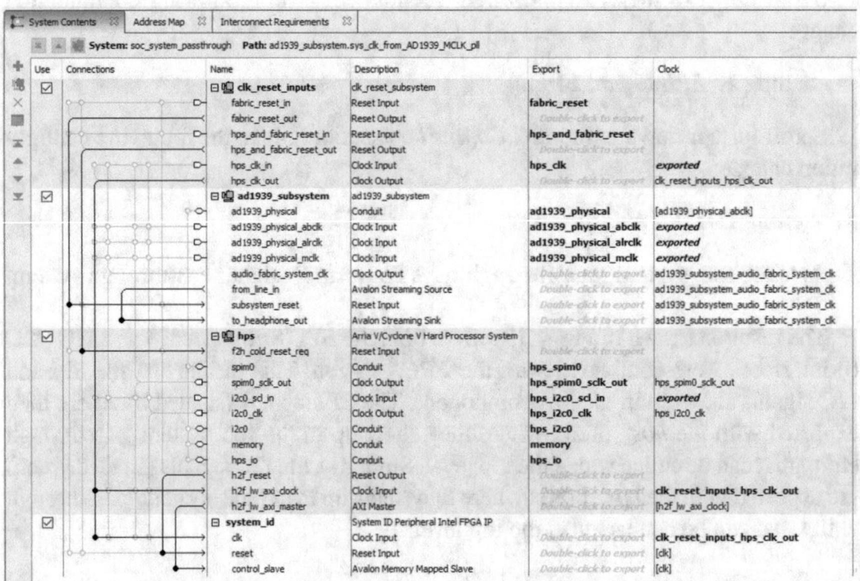

Fig. 1.20: The Platform Designer Passthrough SoC System

It should be noted that there are more signals inside the HPS than there are pins on the FPGA device packaging. When designing a PCB with the Cyclone V

SoC FPGA on it, one of the design choices is which HPS signals to bring out to which FPGA pins. If you scroll down to the bottom of the *Peripheral Pins* tab in the platform Designer HPS *Parameters* window, you will see the *Peripherals Mux Table*. This allows the designer to choose which HPS signals get routed to which physical pins, which are named in the left column. The complicated design choices can be seen in the pin information table. These choices have already been made for the DE10-Nano board, but one advantage of using FPGAs is that we can bring signals out to the FPGA fabric (exported in Platform Designer) and then route these signals to available pins in the I/O headers of the DE10-Nano board. This is what we did for the Audio Mini board for the SPI and I2C control.

1.4.4 Generating the Platform Designer System

Once the Platform Designer system has been finished like in Fig. 1.20, it is time to generate the HDL for the system. This is done by going to *Generate* → *Generate HDL...* where the Generate window will pop up. In the *Synthesis* section, select **VHDL** as the language. You can also uncheck the option *Create block symbol file* since we will not be using it. The output directory will be `.../passthrough/soc_system_passthrough` which is what we want. Then click the *Generate* button (and close). The generation process will take several minutes depending on how complicated the system is. Our system is not too bad. When done, click Close.

In Platform Designer, select *Generate* → *Show Instantiation Template...* and select VHDL as the language. This is how you get the component instantiation and port map for the Platform Designer system that must be connected at the top level. Fortunately, this has already been done for you.

The generated system is placed in the `/soc_system_passthrough` directory under the Quartus project directory. The system is added to the Quartus project by means of a .qip file that is located in the project folder under `/soc_system_passthrough/synthesis/soc_system_passthrough.qip`. In the Project Navigator in Quartus, select the *Files* view. You will see a *soc_system.qip* file listed. Delete this file and add the *soc_system_passthrough.qip* file. If you open this file, you will see thousands of lines of configurations and HDL files.

Note: A common developer's mistake that is made when working with Platform Designer systems is the following. You are making changes to the VHDL code that was imported by the Platform Designer Component Editor. When Platform Designer generates the soc_system_passthrough system, it **COPIES** the imported VHDL files to `<project_name>\soc_system_passthrough\synthesis\submodules`. The mistake that is made is making changes to your original VHDL file, not realizing that the file getting compiled resides in `<project_name>\soc_system_passthrough\synthesis\submodules`. The solutions to this problem are:

Solution 1: Regenerate the Platform Designer soc_system. This can take a bit of time and is cumbersome when performing code iterations, but the imported file(s) will get copied.

Solution 2: Copy over the file(s) into `<project_name>\soc_system_passthrough\synthesis\submodules` after you make edits. This is generally the easiest and fastest way since you do not need to re-generate the Platform Designer soc_system_passthrough system.

Solution 3: Create a VHDL wrapper that instantiates an entity whose VHDL code is placed in a directory that is in Quartus' search path and make edits in this location.

1.4.4.1 Platform Designer soc_system Entity and Port Map

Once the system has been generated in Platform Designer, go to *Generate* → *Show Instantiation Template…* and select the language to be VHDL. Click *Copy* and paste it into an editor. We will reference this shortly. Notice that there are many signals in this entity. Fortunately, you have been given this soc_system_passthrough component declaration and port map in the top level VHDL file *DE10Nano_AudioMini_System.vhd* so go ahead and open this file up in Quartus. You can see the file if you select *Files* in the Quartus *Project Navigator*. You can close down Platform Designer at this point.

1.4.4.2 Hooking Up the soc_system_passthrough System in the Top Level

Compare the soc_component_passthrough declaration in the top level file DE10Nano_AudioMini_System.vhd (lines 231–319) with the instantiation template you copied in Platform Designer. These should be identical. Now compare the port maps (lines 387–497 in the top level VHDL file). Notice that all these signals have been connected for you in the top level VHDL file, so do not copy over the port map in the instantiation template. If you do, there will be a lot of work connecting up signals. For this passthrough project, you will not need to do anything except compile the project in Quartus.

However, this will not always be the case. Any time you export (add) or remove a signal in Platform Designer, you will need to modify both the component declaration and port map in *DE10Nano_AudioMini_System.vhd*. Thus the two modifications to keep in mind for future developments are:

Modification 1: Add or remove any signals in the **soc_system component declaration** that were exported or removed in Platform Designer. Just compare the soc_system component in the top level VHDL file with the instantiation template and make the required few modifications.

Modification 2: Add or remove any signals in the **soc_system port map** and connect them with the appropriate signals in the top level VHDL file.

Go ahead and compile the project in Quartus.

1.5 Linux Device Drivers

In Fig. 1.1, you will see that there are no direct control lines to the *AD1939_hps_audio_mini* component. Rather, the AD1939 audio codec is controlled through a HPS SPI interface, and the TPA6130A2 headphone amplifier is controlled through a HPS I2C interface (see Sect. 1.4.3.2 for exporting the HPS SPI and I2C interfaces). Thus we need Linux device drivers that will use these interfaces to control both devices, which are on the Audio Mini board (see Chap. 5 for a discussion of the Audio Mini board). The AD1939 audio codec driver that uses the HPS SPI interface is discussed in Sect. 1.5.1, and the TPA6130A2 headphone amplifier driver that uses the HPS I2C interface is discussed in Sect. 1.5.2. These AD1939 register settings are found in Sect. 5.3.2.1, and the TPA6130A2 register settings are found in Sect. 5.4.1.1.

1.5.1 Linux SPI Device Driver for the AD1939 Audio Codec

The AD1939 audio codec is controlled through a HPS SPI interface where the register values that can be set are listed in Tables 5.1, 5.2, 5.3. There are several registers that need to be set that are different from the default power-up values. The registers that need to be explicitly set are colored blue in the tables.

The Linux device driver code for the AD1939 can be seen at the following GitHub link (click here for the source file). The steps for compiling the AD1939 Linux device driver are as follows:

Step 1: **Cross Compile the Linux kernel** so that the compiled AD1939 kernel module will have the same version as the Linux kernel. The instructions for compiling the Linux kernel are found in Sect. 9.2 Cross Compiling the Linux Kernel (page 129).
Step 2: **Download the files *ad1939.c*, *Kbuild*, and *Makefile*** from here.
Step 3: **Modify the KDIR variable in the *Makefile*** to point to where you installed `/linux-socfpga` (see Sect. 9.2.2) and run *make* (see Sect. 9.3.2).
Step 4: Copy the kernel module ad1939.ko to the DE10-Nano root file system, and put it in the folder `/lib/modules/` in the Ubuntu VM.

1.5.2 Linux I2C Device Driver for the TPA6130A2 Headphone Amplifier

The TPA6130A2 headphone amplifier is controlled through a HPS I2C interface where the register values that can be set are listed in Table 5.4. There are several registers that need to be set that are different from the default power-up values. The registers that need to be explicitly set are colored blue in the tables.

The Linux device driver code for the TPA6130A2 can be seen at the following GitHub link (click here for the source file). The steps for compiling the TPA6130A2 Linux device driver are as follows:

Step 1: **Cross Compile the Linux kernel** so that the compiled AD1939 kernel module will have the same version as the Linux kernel. Note: This should have already been done for the SPI driver.
Step 2: **Download the files *tpa613a2.c*, *Kbuild*, and *Makefile*** from here.
Step 3: **Modify the KDIR variable in the *Makefile*** to point to where you installed `/linux-socfpga` (see Sect. 9.2.2) and run *make* (see Sect. 9.3.2).
Step 4: Copy the kernel module tpa613a2.ko to the DE10-Nano root file system and put it in the folder `/lib/modules/` in the Ubuntu VM.

1.5.3 Linux Device Tree

The devices that are used to control the Audio Mini board are the SPI and I2C controllers that are contained in the HPS. There is already a device tree created for these SoC FPGA devices in the *socfpga.dtsi* file as shown in Fig. 9.10, and the device tree hierarchy is described in Sect. 9.4.2 Device Tree Hierarchy (page 145).

The device tree for the passthrough project is shown in Listing 1.11 (click here for the source file).

```
1 // SPDX-License-Identifier: GPL-2.0+
2 #include "socfpga_cyclone5_de10_nano.dtsi"
3
4 /{
5     model = "Audio Logic Audio Mini";
6
7     ad1939 {
8         compatible = "dev,al-ad1939";
9     };
10
11    tpa613a2 {
12        compatible = "dev,al-tpa613a2";
13    };
14 };
15
16 &spi0{
17     status ="okay";
18 };
```

Listing 1.11: The device tree for the passthrough project

The passthrough device tree includes all the previous device tree include files (line 2) and defines the *compatible* strings needed so that the Linux kernel will bind the associated device drivers to these devices. The spi0 device is defined in *socfpga.dtsi* file (lines 823–834) in the Linux repository, but on line 833, it says *status = "disabled."* We enable it in the device tree by referencing the spi0 node using &spi0 and changing its status to *status = "okay."* To compile the device tree, follow the steps in Sect. 9.4.3 Creating a Device Tree for Our DE10-Nano System (page 146).

1.5.4 Loading the Linux Device Drivers on Boot

The *ad1939.ko* and *tpa613a2.ko* kernel modules need to be automatically loaded when the DE10-Nano first boots up. To do this, we make use of **systemd**, which is a service manager in Linux. The steps for setting this service up are:

Step 1: **Download the service file *audio-mini-drivers.service*** from here.
Step 2: Copy *audio-mini-drivers.service* to the DE10-Nano root file system, and put it in the folder `/etc/systemd/system` in the Ubuntu VM.
Step 3: **Download the shell script *load-audio-mini-drivers.sh*** from here.
Step 4: Copy *load-audio-mini-drivers.sh* to the DE10-Nano root file system, and put it in the folder `/usr/local/bin` in the Ubuntu VM.
Step 5: **Enable the systemd service** by issuing the command:

```
$ systemctl enable audio-mini-drivers.service
```

Listing 1.12: systemctl enable audio-mini-drivers.service

In the *audio-mini-drivers.service* file, the *oneshot* service type tells Linux to execute the service once on boot and then exit. The line, *WantedBy=multi-user.target*, tells Linux to start the service when Linux reaches the login screen.

You can read the log of the *audio-mini-drivers* systemd service of issuing the command:

```
$ journalctl -u audio-mini-drivers.service
```

Listing 1.13: journalctl -u audio-mini-drivers.service

Example 2
Feedforward Comb Filter System

2.1 Overview

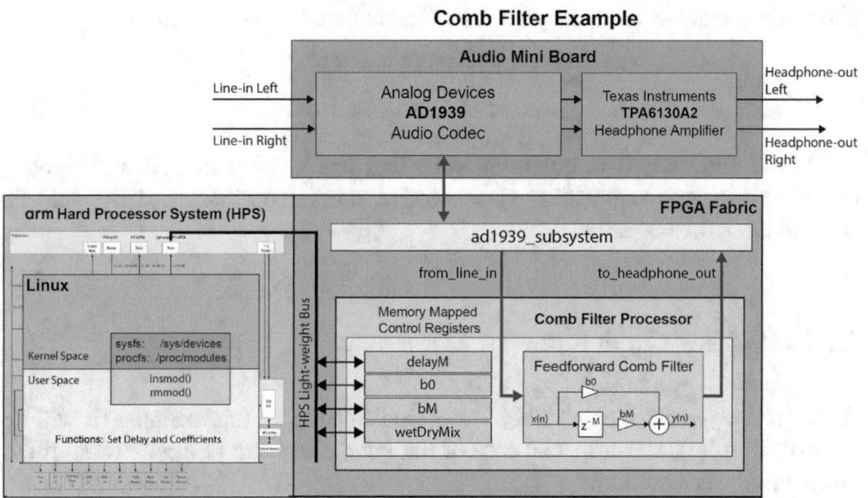

Fig. 2.1: The Feedforward Comb Filter System. Stereo audio signals are run through the comb filter processor where the left and right channels have the filter applied to them. This can simulate an echo with an appropriate value of the delay M

The feedforward comb filter [1] can simulate an echo and is one of the basic building blocks used in digital audio effects. It is a special case of the *Finite Impulse Response* (**FIR**) filter, which has the general form:

R. K. Snider, *Advanced Digital System Design using SoC FPGAs*,
https://doi.org/10.1007/978-3-031-15416-4_14

$$y[n] = \sum_{k=0}^{M} b_k x[n-k] \tag{2.1}$$

The feedforward comb filter has only two nonzero b_k coefficients:

$$b_k = \begin{cases} b_0 & k = 0 \\ 0 & 1 \le k \le M-1 \\ b_M & k = M \end{cases} \tag{2.2}$$

allowing it to be written as

$$y[n] = b_0 x[n] + b_M x[n-M] \tag{2.3}$$

This is the feedforward comb filter that is shown in Fig. 2.1. The delay of the echo is determined by M and the loudness of the echo is determined by b_M. The delay M is in samples, so if you want an echo with a delay of t_{delay} seconds when using an audio signal that has been sampled at F_s Hz, you would need a delay M of

$$M = F_s \times t_{\text{delay}} \tag{2.4}$$

Thus if you wanted to create an echo that had a delay of 100 milliseconds and the signal was sampled at $F_s = 48\,\text{kHz}$, the value of M needed would be $M = 48000 \times 0.1 = 4800$.

2.1.1 Feedback Comb Filter

Although the *feedback* comb filter is not implemented in this example (it will be a lab assignment), it is a special case of the *Infinite Impulse Response* (**IIR**) filter, which has the general form:

$$y[n] = \frac{1}{a_0}\left(\sum_{k=0}^{M} b_k x[n-k] - \sum_{i=1}^{N} a_i y[n-i]\right) \tag{2.5}$$

The feedback comb filter has only one nonzero b_k coefficient and only one nonzero a_i coefficient:

$$b_k = \begin{cases} b_0 & k = 0 \\ 0 & \text{otherwise} \end{cases} \tag{2.6}$$

$$a_i = \begin{cases} a_N & i = N \\ 0 & \text{otherwise} \end{cases} \tag{2.7}$$

allowing it to be written as

$$y[n] = b_0 x[n] - a_N y[n - N] \tag{2.8}$$

This feedback comb filter can simulate a series of echoes that decay over time. Care must be taken with the a_N coefficient that controls the echo decay rate since it must be less than 1 ($|a_N| < 1$) for the filter to be stable.

2.1.2 Comb Filter Uses

By time varying parameters of the comb filters, they can be used to create the following audio effects [1]:

- Reverberation
- Flanging
- Chorus
- Phasing

2.2 Simulink Model

The Simulink model of the comb filter is comprised of multiple parts and files. The common view that most people have when thinking about the Simulink model is shown in Fig. 2.8, which is a correct view. However, this view is the end state of the model. How did we get there? Having just this end view in mind can lead to some fairly disorganized models if targeted directly without a development framework in place, which makes it hard to reuse model code when developing new models. In this chapter we present a model organization and development methodology that keeps the model organized, allows for easy modifications, and can be used as a framework for new models. The guiding principles that helped create this framework are listed below:

Principle 1: A model should be hierarchical where each level is easy to view and understand.

Principle 2: Simulink blocks should be HDL compatible. Not all Simulink blocks can be converted to VHDL, so only use blocks that can be converted (see Sect. 2.4).

Principle 3: Parameters in Simulink blocks should not be hard coded. Rather, variables should be defined in an initialization file and these variable names used as entries for parameters in Simulink blocks. The reason for this is for ease of model modification, which happens all the time during model development. If you want to make a change to a parameter, you know right where it is in the initialization file. In contrast, it is practically impossible, and at best, very time consuming, to click through all the blocks in a large hierarchical trying to find where a certain value was entered.

Principle 4: Model callbacks should have their own associated files. This allows for easier access and modification using the Matlab editor. These are common enough that they can be easily reused in new models (see Sect. 2.2.2). For example, the file *initCallback.m* is called by the *InitFcn* callback (see Fig. 2.2 to see where this callback is accessed in the model).

Principle 5: Organize code based on functional groupings. The three functional groupings used in the model are (see Sect. 2.2.3):

Group 1: Model specific parameters. These are created in *createModelParams.m*. This includes the variables that are used as Simulink block parameter entries.

Group 2: Simulation specific parameters. These are created in *createSimParams.m*. This includes the signal data types that are sent as inputs into the model.

Group 3: HDL conversion specific parameters. These are created in *createHdlParams.m*. This includes parameters such as the targeted FPGA clock speed.

2.2.1 Overview of the Simulink Model Files

Before we get into describing how the model was developed, we first need to describe the files that will be associated with the Simulink models we develop. We first make the distinction between files that any model can and will use, which will be placed in a library folder called *simulink-common*, and model specific files that will be placed in the model's own folder.

2.2.2 Simulink Common Files

First, get the *simulink-common* folder by going to the book's code location and clone the entire repository or click on the green Code button and download the repository .zip file, so you can just copy the files/folders you want. The folder in the .zip file will be located at `\Code-main\examples\simulink-common` (GitHub link). Copy the *simulink-common* folder to your computer.

We now need to tell Matlab where this folder is located by adding the folder's path to Matlab's search path. There are multiple ways of doing this, but a convenient way is to add a command to Matlab's Startup File that is contained in Matlab's Startup Folder. Open (or create) this startup file:

```
>> edit startup.m
```

Listing 2.1: Modifying Matlab's startup file

and add the following command to the file:

```
>> addpath('\<path_to>\simulink-common')
```

Listing 2.2: Adding the Matlab search path to the folder *simulink-common*

The files in the *simulink-common* folder are associated with Simulink's Model Callbacks. These callback functions are functions that get called at specific times during the model's simulation. These callbacks are specified in the Simulink *Model Explorer*, which can be found in Simulink at *MODELING tab* → *Model Explorer*. Clicking on the Model Explorer button will pop up a window such as the one below:

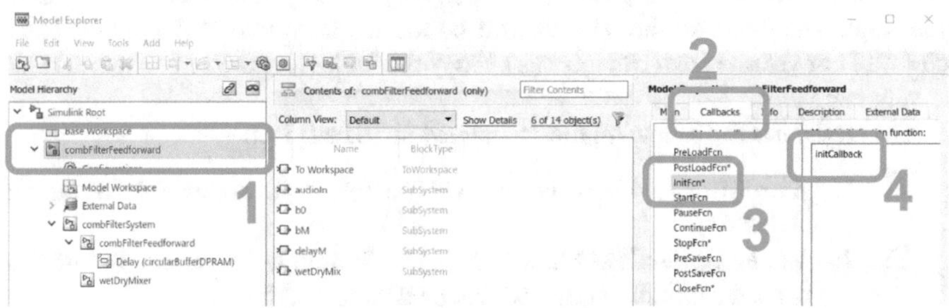

Fig. 2.2: Specifying the Simulink model callback functions

In order to see the callbacks, follow these steps that are outlined in Fig. 2.2:

Step 1: Select the *top level* of the Simulink model.
Step 2: Select the *Callback* tab.
Step 3: Select the *Callback Type*.
Step 4: Enter the *Callback Function Name* and click *Apply*.

Note: In Model Explorer, if a callback type ends with a *, it means that there is an entry for this callback.

In our models we use the following callback types (Table 2.1):

Table 2.1: Simulink Callback Functions in *simulink-common*

Callback type	Function called	Purpose
InitFcn	**initCallback.m**	Initialize the model parameters, the simulation parameters, and the HDL Coder parameters. It calls the following *model specific files*: **createModelParams.m** **createSimParams** **createHdlParams.m**
StopFcn	**stopCallback.m**	Verify the simulation and play the audio effect. It calls the following *model specific files*: **verifySimulation.m** **playOutput.m**
PostLoadFcn	**postLoadCallback.m**	Add the simulink model's directory to MATLAB's search path (for convenience).
CloseFcn	**closeCallback.m**	Remove the model's directory from MATLAB's search path (cleanup).

2.2.3 Model Specific Files

To get the Simulink model files for the Feedforward Comb Filter System, go to the book's code location and clone the entire repository or click on the green Code button and download the repository .zip file, so you can just copy the files/folders you want. The folder in the .zip file will be located at `\Code-main\examples\combFilter\simulink` (GitHub link). Copy the folder `\combFilter\simulink` to your computer.

The model has the following files contained in `\simulink`:

File 1: **combFilterFeedforward.slx**: This is the HDL compatible Simulink model file.

File 2: **circularBufferDPRAM.slx**: This is a Simulink Subsystem that implements the circular buffer, which is HDL compatible.

File 3: **createModelParams.m**: This file creates the model parameters that are stored in the data structure *modelParams*. These are the variable names that are used in the Simulink blocks, which allows parameter changes to be made without having to search through all the blocks in the model hierarchy. It is good programming practice to have a single place to change parameter values rather than searching everywhere for hard coded values.

File 4: **createSimParams.m**: This file creates the simulation specific parameters that are stored in the data structure *simParams*. These are parameters such as how long the simulation will run and it defines the data types of the signals that are fed as input into the model.

File 5: **createHdlParams.m**: This file creates the HDL Coder parameters that are stored in the data structure *hdlParams*. This sets parameters such as the targeted FPGA clock speed.

File 6: **playOutput.m**: This file plays the created sound effect.

File 7: **verifySimulation.m**: This file verifies that the simulation output is correct.

2.3 Creating the Simulink Model

When developing Simulink models that will be implemented in the FPGA fabric, a different mental model of the computation needs to be formed as compared to the mental model that is typically formed when using serial programming languages. When programming in a serial language, you are just concerned with the step-by-step logic that needs to be implemented. Programming gets harder when you need to deal with parallel programming and harder still when you need to think about how it is being implemented in hardware. This is true even when programming for desktop computers since programming for optimal performance requires one to be aware of cache effects and how one deals with chunks of data that fit in cache lines.

Fortunately, parallel data flows for DSP applications can be modeled easily in Simulink. However, when developing for FPGAs, we need to keep in mind how computation and memory will be implemented in the FPGA fabric. If you treat FPGA development like an abstracted computer system, which is typically done when targeting CPUs, you are likely to implement a model that cannot be placed in the FPGA fabric.

We will now go through the process of implementing the feedforward comb filter where we restate Eq. 2.3 here:

$$y[n] = b_0 x[n] + b_M x[n - M]$$

The addition and multiplications are straightforward, since adders can be constructed out of the logic elements (LEs) and multipliers exist has hardened resources in the FPGA fabric. However, where are we going to store the audio samples so that we can get a specific delayed sample? We are not dealing with a CPU with plenty of DRAM. Furthermore, how are we going to do this in Simulink?

To get to the solution, the first concept to understand is that of a circular buffer, which is shown in Fig. 2.3. The write pointer (i.e., write address) is where the current sample will be written, and after the write, the write address will get updated to the next memory location. When the write address falls beyond the buffer location in memory, it will be reset to the beginning of the buffer, so conceptually it constantly wraps around in a circular fashion.

The reason for using a circular buffer, which is true for applications using both FPGAs and CPUs, is that once a sample has been stored, it would be terribly inefficient to move a bunch of samples every time a new sample showed up. Rather, the sample is written once and never moved. It will, however, be overwritten at some point. When it gets overwritten is determined by the size of the circular buffer, which determines the maximum sample delay that can be achieved before the oldest sample gets overwritten with a new sample.

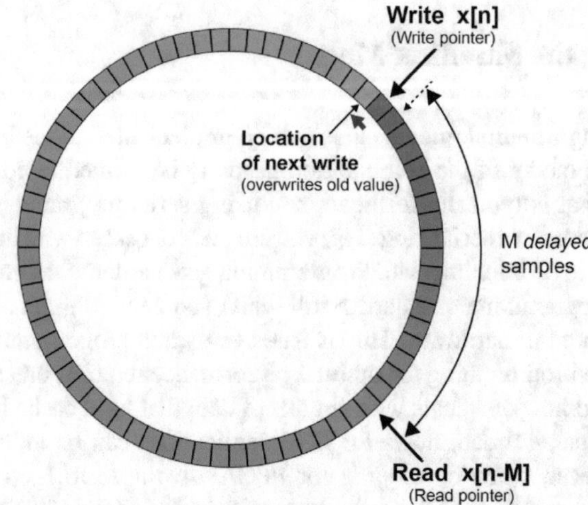

Fig. 2.3: The circular buffer (also called a ring buffer) eliminates having to move samples once they are written

Although the FPGA fabric *logic elements* (LEs) contain memory in their *Look-Up Tables* (**LUTs**) and *flip-flops* where the flip-flops store the outputs of the LUTs, a memory type that can store significantly more data is the *Block RAM* (**BRAM**). The Cyclone V FPGA on the DE10-Nano board (5CSEA6) has 557 of these BRAMs, which are M10K memory blocks (10.24 Kb) in the fabric. If all this memory (5.57 Mb) is used to store stereo samples (2 × 24-bit words @ 48 kHz), we could store 116,041 samples for each left and right channel. This means that we could create an echo with a maximum delay of 2.4 s on each channel. Of course if we did this, we would not have BRAMs for anything else.

The BRAMs in the FPGA fabric are dual port memories as shown at the top of Fig. 2.4 where the two ports are completely independent. This allows data to be written using clock A and then accessed with a different clock B that is asynchronous to clock A. Using BRAMs to transfer data is a technique for moving data between two independent clock domains that avoids metastability issues.

BRAMs can be configured to work together, which allows memories to be created with arbitrary data word sizes (width) and where the number of data words (depth) needs to be a power of 2, in order to reflect the memory locations accessed by the address bus.

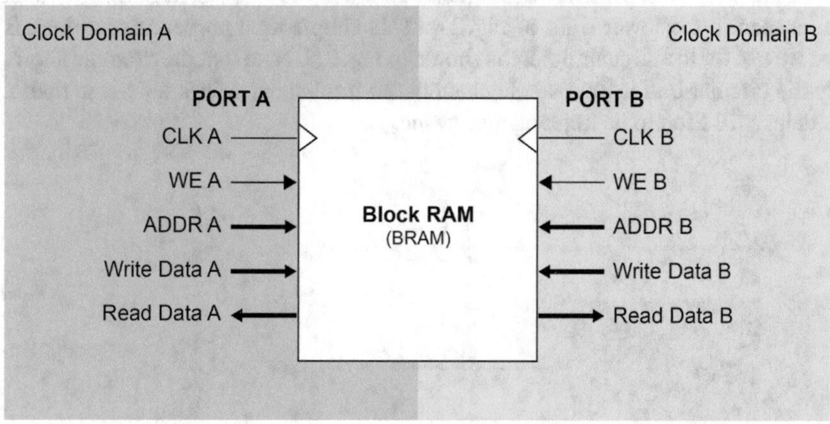

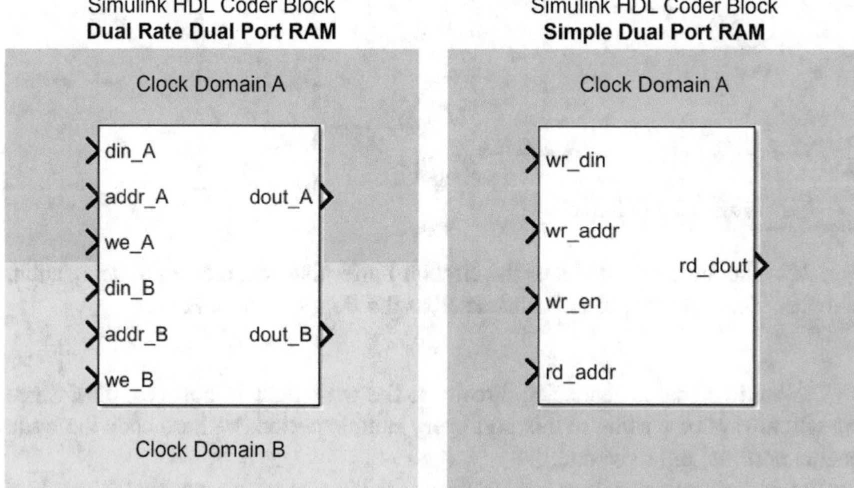

Fig. 2.4: Dual port memory available in the FPGA fabric

The general form of dual port memory can be modeled in Simulink by using the *Dual Rate Dual Port RAM* block found in the HDL Coder library. This block can be seen in the lower left of Fig. 2.4. This block has only one configuration parameter, the size of the address bus, and this determines the size of the memory. The clock is inferred from the sample rate of the signal going into the memory, so it is not explicitly set, and the word size of the memory is inherited from the width of the input signal.

For our purposes where we need to delay audio samples, we do not need the full complexity of the dual port memory. Rather, we have a single clock for both ports, we only want to write to port A, and we only want to read from port B. For this, Simulink has the *Simple Dual Port RAM* block that is found in the HDL Coder library, which can be seen in the lower right of Fig. 2.4. This simple dual port memory block is what we use for the circular buffer as shown in Fig. 2.5. Note that the example shows only the circular buffer for a single channel. To implement delays for stereo audio, this delay will need to be implemented twice.

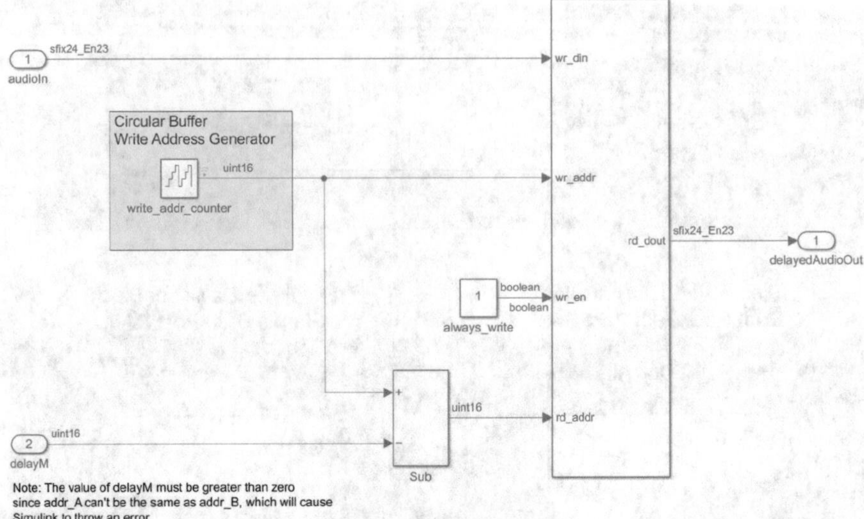

Fig. 2.5: The Simulink model of the circular buffer that was created to delay audio samples. This circular buffer is abstracted as the *Delay* block in Fig. 2.6

The audio signal is connected directly to the write data_in port (wr_din). Since we will always be writing to this port every sample period, we hard code the write enable port (wr_en) to be one.

The address bus signal *wr_addr* determines the size of the memory (Nwords = $2^{\text{wr\_addr\_size}}$). Here you need to know what device you are targeting (Cyclone V) and how much memory the device contains (557 M10K memory blocks). If you choose a memory size that is too big, the design will not synthesize and the fitter in Quartus will throw an error saying the design will not fit in the fabric.

In order to generate the write address, we next take advantage of the behavior of counters running in hardware. When the counter overflows because the carry bit cannot be stored, the counter gets reset back to zero. This is exactly the behavior that what we want for the circular buffer. We get this wrapping operation for free as compared to circular buffers written for CPUs where they have to always check to see if the DRAM write pointer has gone past the buffer in order to reset it back to the beginning of the buffer. The counter used in Fig. 2.5 is the *Counter Free-Running* block found in the HDL Coder library. The number of bits in the counter needs to match the address bus size (wr_addr) that it is connected to.

The *delayM* value is simply subtracted from the counter value to generate the read address (rd_addr) that is used to read out the delayed audio sample that has been delayed by *delayM* samples. When you run a simulation, you will get a warning that overflow has occurred in the subtraction block. This is OK since we want the same type of wrapping overflow that is occurring in the counter block (just in the opposite direction).

Note: Since both the constant one block and the counter block are *sources*, both the data type and sample period need to be explicitly set using variables found in the *modelParams* data structure. This is because Simulink cannot infer what these values should be. These values are set using modelParams variables defined in the initialization script *createModelParams.m* in order to allow easy modifications.

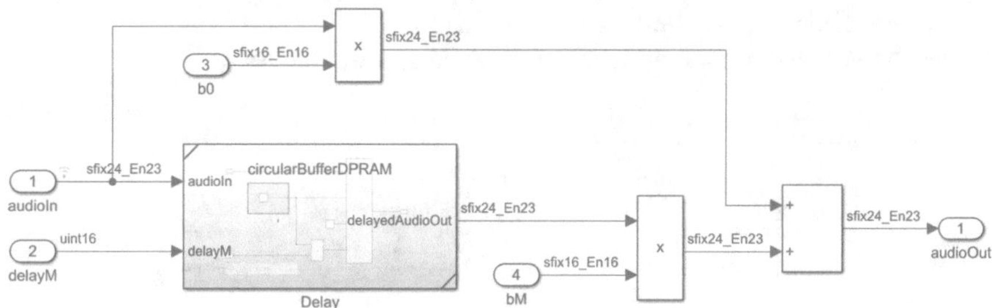

Fig. 2.6: The comb filter Simulink model

The comb filter Simulink model is shown in Fig. 2.6 where the circular buffer has been abstracted as the block named *Delay*. Abstraction in Simulink is simply done by selecting the blocks and signal paths that you want grouped together in a new block, then right clicking on the selection, and selecting *Create Subsystem from Selection* (Ctrl+G).

The default behavior for the product blocks in Simulink is to propagate a full precision data type, which is the sum of the bit widths of the signals. This is done so that there will not be any loss of precision. This can lead to unnecessary growth in the size of the signal widths. To keep this from happening, a common practice in audio DSP is to represent the audio signals and coefficients as fractional data types. Then when signals and coefficients are multiplied together, the lowest significant bits can simply be discarded because you cannot hear them anyway. This is why the output of the product blocks in Fig. 2.6 has been reset to the (W=24, F=23) data types using variables found in the data structure *modelParams*. W stands for the signal width in bits, and F stands for how many fractional bits are in the word. The adder has similarly been reset to (W=24, F=23). However, one additional option for the adder has been set in the Signal Attributes tab. The option *Saturate on integer overflow* has been selected. This is because there is the slight possibility of overflow if both signals are at their maximum amplitude, which will cause the resulting signal value to turn negative. This will be heard as a "pop" in the audio. The saturate option will clip the signal at its maximum value, which although can introduce a harmonic distortion is much more preferable than having noisy "pops" introduced into the audio.

The feedforward comb filter in Fig. 2.6 has been abstracted into the block named *combFilterFeedforward* in Fig. 2.7. The wet/dry mixer block takes the *wetDryMix* control value and proportionally mixes a dry signal (unprocessed/raw signal) with the wet signal (processed signal). The mixing computes the following:

$$\text{audio}_{\text{out}} = \text{audio}_{\text{dry}} \times (1 - \text{wetDryMix}) \; + \; \text{audio}_{\text{wet}} \times \text{wetDryMix} \tag{2.9}$$

where $0 \leq \text{wetDryMix} \leq 1$. Thus

$$\begin{aligned} \text{wetDryMix} = 0 \quad &\Longrightarrow \text{audio}_{\text{out}} = \text{audio}_{\text{dry}} \\ \text{wetDryMix} = 1 \quad &\Longrightarrow \text{audio}_{\text{out}} = \text{audio}_{\text{wet}} \\ \text{wetDryMix} = 0.5 &\Longrightarrow \text{audio}_{\text{out}} = 0.5 \; \text{audio}_{\text{dry}} + 0.5 \; \text{audio}_{\text{wet}} \end{aligned}$$

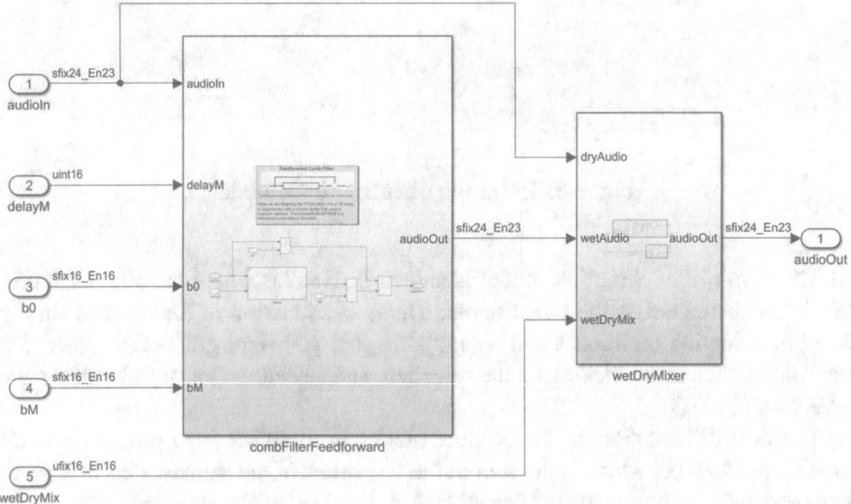

Fig. 2.7: The abstracted comb filter followed by wet/dry mixing

The comb filter is finally abstracted into the block named *combFilterSystem* as shown in Fig. 2.8. This is the block that we will convert to VHDL in Sect. 2.4. The blocks feeding the inputs to *combFilterSystem* get their values from the Matlab workspace. Since they are source blocks, they have to have both their data types and sample period defined, which will then be inferred through the system blocks. Since this is a single rate system, all the blocks are set with the same sample period, which is the sample rate of the audio signal ($F_s = 48\,\text{kHz}$).

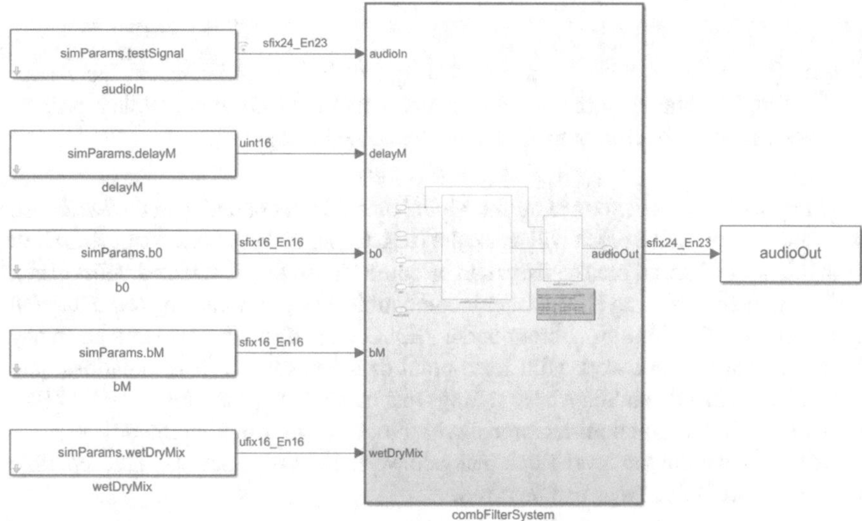

Fig. 2.8: The system block that implements the comb filter

2.4 VHDL Code Generation Using Simulink's HDL Coder

Now that we have built up our Simulink model that implements the feedforward comb filter as shown in Fig. 2.8, we now turn our attention to using Simulink's *HDL Coder* to convert this model to VHDL. This will provide us with the VHDL code needed for the feedforward comb filter block shown in Fig. 2.1. Note that this generated VHDL code is not the complete comb filter processor since we still need to interface it with the ARM CPUs in order to control the filter parameters from Linux user space. This interfacing with the HPS is covered in Sect. 2.5.

In order to generate VHDL code, everything that is to be converted to VHDL needs to be contained within a single block. Thus the block that we will convert is the block named *combFilterSystem* in Fig. 2.8. The blocks that provide the inputs and take the output signal will not be converted. Thus we do not care if these blocks are HDL compatible or not. This is analogous to creating a VHDL testbench where we do not care if the VHDL testbench code is synthesizable or not and we just care that the VHDL contained in the component being tested is synthesizable.

The first thing to be aware of when you are done creating a Simulink model and it performs simulations correctly is that the block you aim to convert to VHDL is most likely not fully HDL compatible. Hopefully, the block to be converted was developed using *only* the blocks that are HDL Coder compatible. The HDL compatible Simulink blocks can be seen in the Simulink Library by typing the Matlab command:

```
>> hdllib
```

Listing 2.3: Matlab command that displays only the HDL compatible blocks. Note: It does not eliminate all the incompatible blocks

If there are blocks in your Simulink Model that are not contained in the *hdllib* list, it is quite likely that the block will cause the HDL Coder to throw an error at you. Even with this set of filtered blocks, there can be some blocks that if selected with certain options will end up being HDL Coder incompatible. For example, the *Math Function* block found in the Simulink library under `/HDLCoder/MathOperations` has many functions that will not work with fixed-point data types, even for simulations, and even the ones that do simulate have settings that make them problematic for the HDL Coder (e.g., Math Function: reciprocal with Floor Integer rounding mode).

To check that the top level block being converted contains only HDL compatible blocks, the following steps can be taken:

Step 1: Make sure that you can first **Run** a simulation without any errors. Run a simulation and make sure that any paths that need to be added are added and you are in the correct directory. You probably want to reduce the simulation time by setting the parameter *simParams.stopTime* contained in the file *createSimParams.m* to a short time value. This is because the HDL Coder will run the simulation multiple times while it checks compatibility. Thus during development where multiple coding iterations are being done to get everything correct, it will save considerable time by making the simulation short since at this point we are more concerned with getting the model HDL compatible than with simulation results. Thus, to speed up simulations during HDL Coder operations, you can set in *createSimParams.m* :

- simParams.stopTime = 0.1
- simParams.verifySimulation = false
- simParams.playOutput = false

Step 2: To check that the top level block is HDL compatible, click on the top level block to select it and type the following Matlab command:

```
>> checkhdl(gcb)
```

The Matlab *get current block* command *gcb* returns the path of the currently selected block in Simulink. This is why you need to **Click** on the block *combFilterSystem* before running the command *checkhdl(gcb)*. This will result in a HDL Code Generation Check Report that will present Errors, Warnings, and Messages. At a minimum, you will need to correct any errors before proceeding with VHDL code generation.

2.4.1 Common Model Errors

As an example of some of the common errors that you can get when first creating a Simulink model targeting the FPGA fabric, the initial version of *combFilterFeedforward* model is provided, which is called *combFilterFeedforward_attempt1.slx*. If you run *checkhdl* on the block *combFilterSystem* in this model, you will get the following errors (Fig. 2.9):

HDL check for 'combFilterFeedforward_attempt1' complete with 4 errors, 0 warnings, and 0 messages.

The following table describes blocks for which errors, warnings or messages were reported.

Simulink Blocks and resources	Level	Description
combFilterFeedforward_attempt1/combFilterSystem/wetDryMixer	Error	Signals of type 'Double' will not generate synthesizable HDL. For synthesizable HDL code, set the "Library" option to "Native Floating Point". For non-synthesizable and simulation-only HDL code, set the "Check for presence of reals in generated HDL code" diagnostic to "Warning" or "None".
combFilterFeedforward_attempt1/combFilterSystem/wetDryMixer/Add1	Error	Mixed double, single, half, and non-real datatypes at ports of block is not allowed.
combFilterFeedforward_attempt1/combFilterSystem/wetDryMixer/Product1	Error	Mixed double, single, half, and non-real datatypes at ports of block is not allowed.
combFilterFeedforward_attempt1/combFilterSystem/wetDryMixer/Subtract	Error	Mixed double, single, half, and non-real datatypes at ports of block is not allowed.

Fig. 2.9: hdlcheck errors for model *combFilterFeedforward_attempt1*

These four errors are caused by just one block. If you click on the first error, it will take you to the *wetDryMixer* block where it is complaining about the *Double* data type that cannot be converted to HDL. This is the result of a block causing the same error to propagating through multiple blocks inside this *wetDryMixer* block.

To see the errors being propagated, make sure that you can see all of the data types associated with the signal paths. If you cannot see the signal types, turn them on by going to *Simulink* → *DEBUG Tab* → Click on *Information Overlays* in the DIAGNOSTICS section → Select *Base Data Types* in the PORTS section.

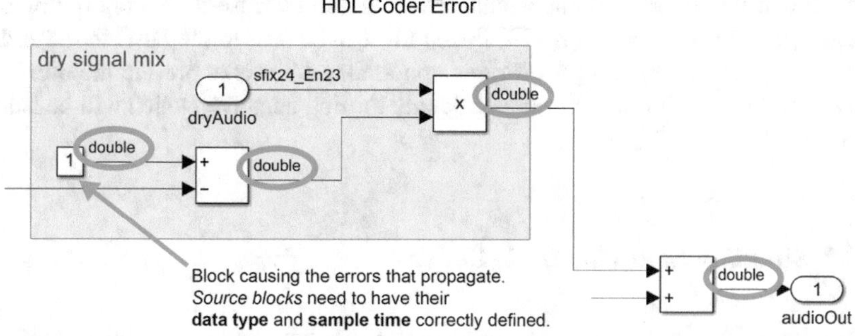

Fig. 2.10: HDL error caused by a constant *source block* not defining the data type correctly. The HDL Coder cannot convert double precision data types

These errors can crop up when a Simulink model is first being created. This is because you are initially just interested in how the model performs and you are exploring what the model can do and you are using the default double precision data types. In the case of developing an audio effect, you would primarily be interested to see if the model creates the sound effect you are after and you do not want to concern yourself with data types at this point. Once the model performs as expected, it is then time to think about hardware, which means converting to fixed-point data types (and hardware architectures).

Most of the fixed-point conversion work is simply done by defining the data type that goes into the model and Simulink propagates the data types throughout the rest of the model. However, when you create a source block (e.g., a constant value) as shown in Fig. 2.10, Simulink does not know what this data type should be, so it just assumes a double precision data type. It also does not know what the sample rate should be either. Thus anytime you create a source block, you need to define both the **data type** and **sample time**. This is best done using a variable name so that if the data type or sample rate needs to be changed, you can do it in the initialization script *createModelParams.m* rather than hunting through multiple levels of hierarchy in your model looking for the one source block to check and modify.

To fix the errors, go to the constant 1 block shown in Fig. 2.10 and double click on the block to open the *Block Parameters* window. Then, make the following two changes:

Fix 1: Enter **modelParams.audio.samplePeriod** in the *Sample Time* field.
Fix 2: Click on the *Signal Attributes* tab, and in the *Output data type* field, enter the string: **fixdt(0,1,0)**.

Running *checkhdl(gcb)* will show that these errors have been fixed.

The additional fixed-point conversion work is to reset the outputs of multipliers and adders back to (W=24, F=23) and also make sure that the adders have the option *Saturate on integer overflow* selected. This is to avoid unnecessary signal width growth that uses fabric resources unnecessarily and because ultimately the output audio signal must be set to (W=24, F=23) for it to be sent to the DAC in the audio codec. The *Saturate on integer overflow* option for adders is to prevent the overflow effect of a large audio sample value suddenly turning negative, which will be heard as a "pop."

2.4.2 Simulink Setup for HDL Coder

When the HDL Coder is run, errors can arise that have to do with how Simulink has or has not been set up for the HDL code conversion process. For example, the Simulink solver needs to be set to *Fixed-step* and *Discrete (no continuous states)*. If these parameters have not been set, they will be identified as issues when going through the HDL Coder process. Fortunately, MathWorks now has a function that will set the Simulink model to the common default values needed for HDL code generation. Assuming that only the model that you are converting is currently open, you can simply enter the command:

```
>> hdlsetup(gcs)
```

Listing 2.4: Matlab command that sets the default parameters for HDL code generation

This command will set the default values for HDL code generation. The parameter values that are affected can be seen here. The command *gcs* will get the current name of the Simulink model needed by *hdlsetup*, which is why you should have only this model open. If you have multiple models open, you will need to supply the specific name of the model to *hdlsetup*.

2.4.3 HDL Workflow Advisor

Before we can run the HDL Workflow Advisor, we need to tell Simulink what synthesis tool we are using and where it is located. We do this with the *hdlsetuptoolpath* command:

```
>> hdlsetuptoolpath('ToolName', 'Altera Quartus II', '↲
  ↪ToolPath', 'C:\<path_to_quartus>\quartus\bin64\↲
  ↪quartus.exe');
```

Listing 2.5: Matlab Command: hdlsetuptoolpath()

Note: If you need to set this frequently, add this command to your Matlab startup.m file.

In the Simulink model, right click on the top level block that is to be converted (*combFilterSystem*) and select *HDL Code* → ***HDL Workflow Advisor***. This will open the following window (Fig. 2.11):

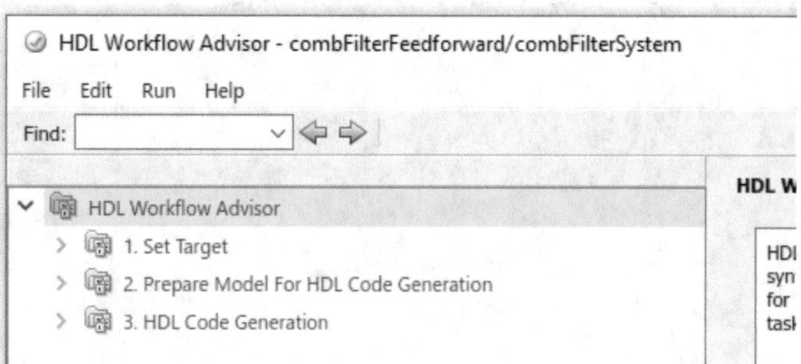

Fig. 2.11: HDL Workflow Advisor

On the left side, expand *1. Set Target* and select *1.1 Set Target Device and Synthesis Tool*. In the associated parameter window, select the following input parameters:

Target Device on DE10-Nano Board

Setting 1: Set *Target workflow* as **Generic ASIC/FPGA**
Setting 2: Set *Synthesis tool* as **Altera QUARTUS II**. (This should already be set.)
Setting 3: Set *Family* as **Cyclone V**
Setting 4: Set *Device* as **5CSEBA6U23I7** (and yes, it is annoying that they do not provide a filter that responds to text input because of all the options that are surprisingly not listed in alphabetical order).
Setting 5: Set *Project folder* as **hdl_prj**. (This should already be set as default.)

When done, click the *Run This Task* button. Next, select *1.2 Set Target Frequency* and enter 98.304 MHz and click the *Run This Task* button.

Target Frequency

Setting 6: Set *Target Frequency (MHz)* as **98.304**

Next, select *2. Prepare Model For HDL Code Generation* and click the *Run All* button.

Now, expand *3. HDL Code Generation*, select *Set HDL Options*, and click the *HDL Code Generation Settings...* button. The Simulink *Model Settings* window will pop up (you can also get the window by pressing Ctrl+E). Set the language to be **VHDL** as shown in Fig. 2.12.

Basic Options

Setting 7: Set *Language* as **VHDL**

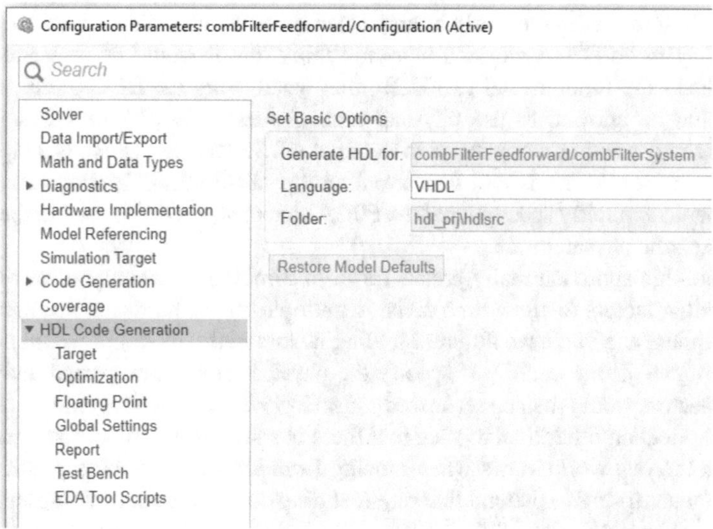

Fig. 2.12: Setting VHDL as the language for the HDL Coder

In the Simulink *Configuration Parameters* window, navigate to *HDL Code Generation* → *Global Settings* and click on the *Coding style* tab under *Additional settings*. In the *RTL Style* section, uncheck *Inline VHDL configuration* and click the Apply button. If you leave this checked, Quartus will complain about the included configuration statements, which are unnecessary since each entity has only one possible architecture anyway.

Coding Style

Setting 8: **Uncheck** *Inline VHDL configuration*

2.4.3.1 Clocking

We now come to the point where there is a collision of two world views. We have the hardware world of synchronous logic that marches in lockstep to a clock edge. Clocks are fundamental to digital design and obey the laws of physics in regard to timing. Time is a physical quantity. We also have computer scientists who have worked their entire careers to abstract away the low level details of computer systems. Simulink is such an abstraction because signal processing can be represented at a high abstraction level. When we run a Simulink simulation, we do not need to concern ourselves with the clock speed that the CPU is running at. Sure, faster CPUs will run simulations faster, but the correctness of the simulation does not depend on the CPU clock speed. Time is an abstraction in a Simulink simulation.

We now have to make the connection between Simulink's abstracted time and the physical time of clocks running in the FPGA fabric. The developers of Simulink try their best to stay in the world of abstracted time, which makes it annoying when connecting to an FPGA clock. You would think that it would be understood that since clocks are fundamental to VHDL, that when using the HDL Coder toolbox, you would be allowed to use physical time. An example of where this is done correctly in the author's opinion is in the Intel's DSP Builder, which is a high-level (and expensive) synthesis tool that layers on top of Simulink. It allows the FPGA developer to explicitly specify what the FPGA clock speed will be used in the FPGA fabric, i.e., the physical time.

Where this approach really breaks down in Simulink is when you start dealing with folding factors to allow time division multiplexing of hardware resources such as multipliers and you have different folding factors in the model. It would be much much simpler if one could just specify the physical clock time period and all the folding factors would then be set accordingly. However, we are not going to deal with this complication other than to state that there is a significant impedance mismatch between the two world views when moving from abstracted time to physical time when you are trying to pretend that physical time does not exist or is subservient to abstracted time.

The game that the Simulink developers make you play is the requirement to use the DSP concept of "oversampling" and you have to know what is the fastest signal in your Simulink model, especially if there are rate transitions buried in the model. Below are the instructions on how to play this game.

Step 1: Determine the fastest sample rate $F_{\text{fastestSR}}$ that is in the Simulink model. You can determine this by opening the Simulink *Timing Legend* by either going to *Simulink* → *DEBUG* tab → *Information Overlays* in the *DIAGNOSTICS* section → selecting *Timing Legend* in the *SAMPLE TIME* section or simply pressing *Ctrl* + *J*. The *Timing Legend* will appear next to the Simulink model to show what the colors in the model represent in terms of sample rates. Press the $\frac{1}{P}$ button to display frequency and make note of the frequency of the fastest sample rate in the model.

Step 2: Determine the *Oversampling Factor*, which is calculated in Eq. 2.10.

$$\text{OversamplingFactor} = \frac{\text{FPGA Clock Frequency}}{F_{\text{fastestSR}}} \tag{2.10}$$

In the comb filter model, the fastest sampling frequency is 48,000 Hz and the FPGA clock frequency is 98,304,000 Hz. This means that the oversampling factor needs to be 2048. The steps to set this in Simulink are as follows:

Setting Oversampling Factor to Set Target Clock Frequency

Step 1: Open the Model Settings window (Ctrl+E).

Step 2: Select *Global Settings* under *HDL Code Generation* as seen in Fig. 2.13.

Step 3: Enter the oversampling factor.

Unfortunately, the *oversampling factor* field only accepts integers and not a variable name, so we cannot directly compute what the oversampling factor should be during initialization and set the oversampling factor as a variable entry. However, we can call the function createHdlParams during initialization after computing what the oversampling factor should be and set it using the HDL Coder function *hdlset_param* as shown in Listing 2.6.

```
hdlset_param(gcs,'Oversampling',hdlParams.↲
    ↪clockOversamplingFactor)
```

Listing 2.6: Setting the clock oversampling factor

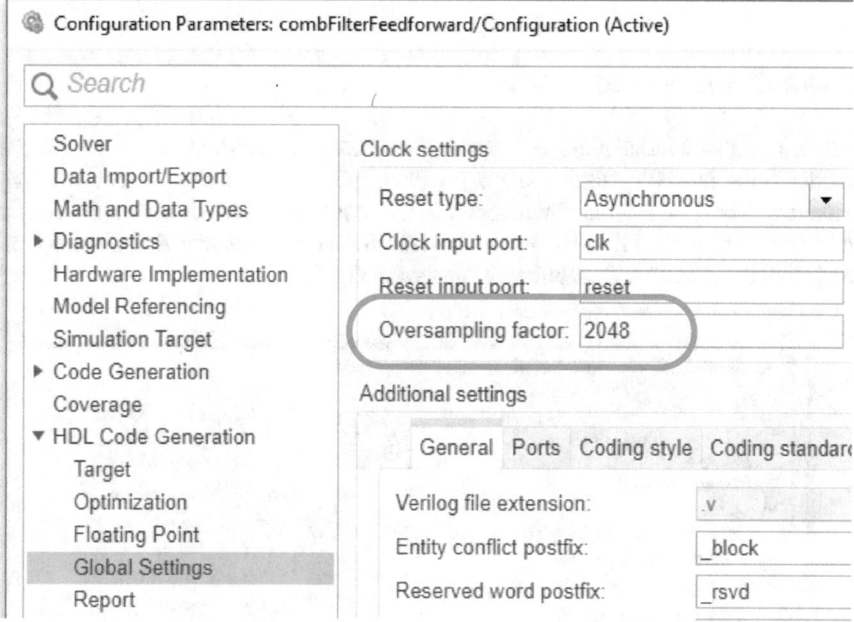

Fig. 2.13: Setting the oversampling factor to match the FPGA clock frequency

Clock Settings
Setting 9: Set *Oversampling factor* as **2048**. Note: This is actually set by the Simulink model function *createHdlParams.m*

After setting the oversampling factor, an additional setting that is good to include is to set the option to generate the resource utilization report, which gives a report on the FPGA fabric resources that are required for the generated VHDL code. With this report, you can determine if the model will fit in the targeted FPGA. To generate this report, open the Model Settings window (Ctrl+E) and select *Report* under *HDL Code Generation* and check the box next to *Generate resource utilization report*.

Reports

Setting 10: **Check** *Generate resource utilization report*.

Finally, click the *Run This Task* button in *3.1 Set HDL Options*. We are now ready to generate VHDL code.

2.4.3.2 Generating VHDL Code

Make sure that Matlab's current working directory is the Simulink folder for the model since the HDL folder *hdl_prj* will be placed under the current working directory. Then, to generate the model's VHDL code, select *3.2 Generate RTL Code and Testbench* in the HDL Workflow Advisor. Select the *Generate RTL code* option and click the *Run This Task* button as shown in Fig. 2.14.

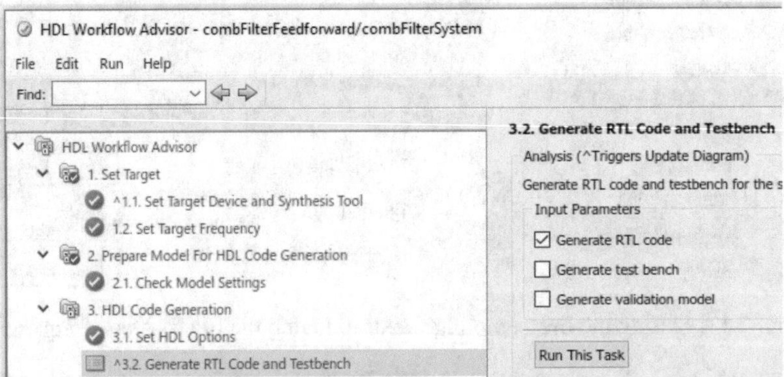

Fig. 2.14: Generating the VHDL code from the model

When the VHDL code has been generated, the Code Generation Report pops up as shown in Fig. 2.15.

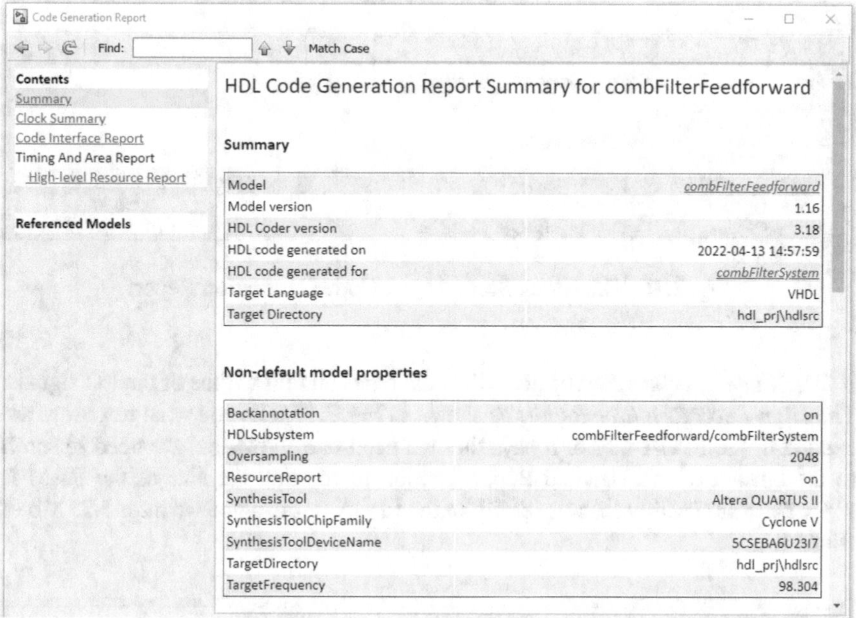

Fig. 2.15: The Code Generation Report

On the left you can view the *Clock Summary* as shown in Fig. 2.16, which gives the sample period of the audio signal and the oversampling factor. The HDL Coder developers then make you calculate what this really means, which is

$$\begin{aligned}\text{FPGA Clock Frequency} &= \frac{\text{Explicit Oversampling Request}}{\text{Model Base Rate}} \\ &= \frac{2048}{2.08333^{-05}} = 98.304\,\text{MHz}\end{aligned} \tag{2.11}$$

The HDL Coder developers do give this clock information in the Summary under the non-default model properties as the *TargetFrequency* value of 98.304. They also give this TargetFrequency value in the header of the top level VHDL file *combFilterSystem.vhd*, but they confusingly state *model base rate: 1.01725e-08*, which equals 98.304 MHz, when they use the same term in the Clock Summary page of the Code Generation Report where they state *model base rate: 2.08333e-05*, which is 48 kHz, the sample rate of the input audio signal. You just need to know from the context that these same terms mean different things.

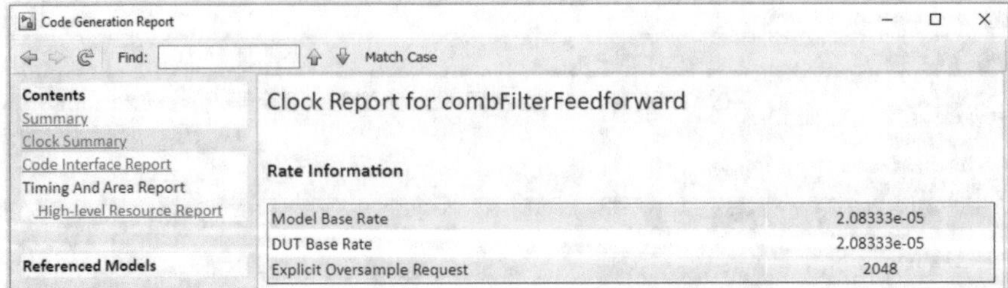

Fig. 2.16: The Clock Summary in Code Generation Report

The *Code Interface Report* gives the data types and bit widths of the I/O signals. The *High-Level Resource Report* as shown in Fig. 2.17 tells you what resources are needed in the FPGA fabric. It says that we need one RAM, but we need to scroll to the bottom of the Detailed Report section to see that the size of the RAM is $2^{16} \times 24 = 1.57$ Mb, which fits easily in the FPGA fabric since we have 5.57 Mb of BRAM.

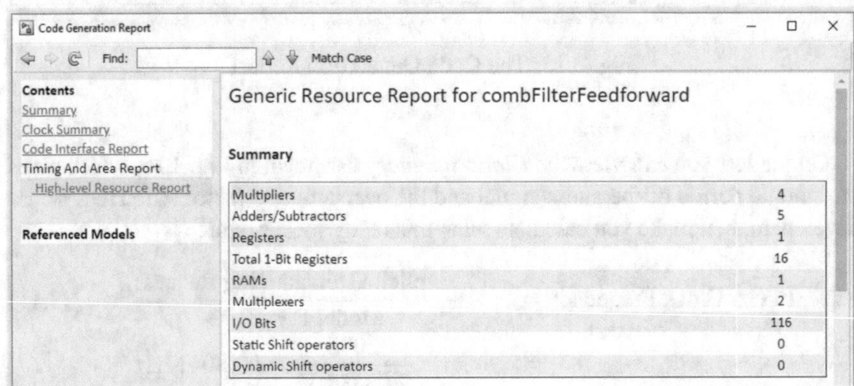

Fig. 2.17: The High-Level Resource Report tells you resources are needed in the FPGA fabric

2.4.3.3 VHDL Files

If you have accepted the default locations for the generated VHDL files in the HDL Workflow Advisor, they will be located in the Simulink model folder under \<model_folder>\hdl_prj\hdlsrc\combFilterFeedforward. The folder *combFilterFeedforward* is the name of the Simulink model and contained in this folder is the VHDL file *combFilterSystem.vhd*, which is the name of the block in the model that was converted to VHDL. Make a note where the folder *combFilterFeed-*

forward is located since we will eventually need to copy this folder to our Quartus project folder and put it in the `\ip` subfolder.

2.5 HPS Integration Using Platform Designer

We now need to create a Platform Designer component called the *combFilterProcessor* that has the three Avalon interfaces shown in Fig. 2.1. These Avalon interfaces are listed below (Table 2.2).

Table 2.2: Avalon Interfaces in the Comb Filter Processor

Avalon Interface	Purpose
Avalon Streaming Sink	Receive stereo audio from the *ad1939_subsystem*.
Avalon Streaming Source	Send stereo audio to the *ad1939_subsystem*.
Avalon Memory-Mapped	Connect registers to the ARM CPUs via the HPS lightweight bus.

The files associated with the comb filter model, which are quite similar in name and can be confused with each other, are listed in Table 2.3.

Table 2.3: Comb Filter Files

File	Purpose	Location
combFilterFeedforward.slx	Simulink Model of the comb filter audio effect.	(click here)
combFilterSystem.vhd	Top level VHDL file created by the HDL Coder. It is also the name of the top level block in the combFilterFeedforward Simulink model that was converted to VHDL. The HDL Coder placed the generated files into `\hdl_prj\hdlsrc\combFilterFeedforward` under the Simulink project folder.	(click here)
combFilterProcessor.vhd	The VHDL file for the Platform Designer component *combFilterProcessor*	(click here)

2.5.1 Hardware Integration Steps

The steps for creating the new Platform Designer component *combFilterProcessor*, which is the block called *Comb Filter Processor* in Fig. 2.1, are listed below. These steps are discussed further in the following sections:

Step 1: **Start with a known working system.** Our starting point will be the *passthrough* project where we simply copy the project into a new folder and we do not rename any files before starting out. Renaming files is tempting since this is a new project, but this opens the door for errors that would defeat the process of starting with a known working system. Renaming is the last step.

Step 2: **Create the *CombFilterProcessor* Platform Designer Component.** This will require:

1: Creating the VHDL code for the *CombFilterProcessor*.
2: Implementing the three Avalon interfaces listed in the table.
3: Converting the Avalon streaming interface to individual left/right channels that the HDL Coder expects (we purposely set the target workflow as *Generic ASIC/FPGA*).
4: Converting the left/right channels back to the Avalon streaming interface Table 2.2.
5: Instantiating the *combFilterSystem* component that the HDL Coder created from our Simulink model. We will instantiate the *combFilterSystem* component twice, once for the left channel and once for the right channel. Note: The registers will control the left and right channels in the same way.

Step 3: **Importing the *combFilterProcessor* component into Platform Designer**. The component will be imported and then added to the system.

Step 4: **Testing the *combFilterProcessor* component using System Console.** We need to verify that our hardware works before we can move on to creating the software that will use our hardware.

Step 5: **Quartus Project Cleanup**. Here we rename the files to be consistent with the project name and put files in their expected directory locations.

2.5.2 Quartus Project Setup

Our starting point will be the *Passthrough Quartus* project described in Sect. 1.3 Quartus Passthrough Project (page 267). Copy this project into a new folder. If you have not already run this project on the DE10-Nano board with the Audio Mini board, do so now before proceeding to make sure everything works.

2.5.3 Creating the Platform Designer Component

The VHDL component combFilterProcessor.vhd is essentially Avalon wrapper code that implements the three Avalon interfaces described in Table 2.2 and connects them to the VHDL component combFilterSystem.vhd that was created by the HDL Coder. The entity of *combFilterProcessor* is shown in Listing 2.7.

```
entity combfilterprocessor is
  port (
    clk                        : in    std_logic;
    reset                      : in    std_logic;
    avalon_st_sink_valid       : in    std_logic;
    avalon_st_sink_data        : in    std_logic_vector(23 downto 0);
    avalon_st_sink_channel     : in    std_logic_vector(0 downto 0);
    avalon_st_source_valid     : out   std_logic;
    avalon_st_source_data      : out   std_logic_vector(23 downto 0);
    avalon_st_source_channel   : out   std_logic_vector(0 downto 0);
    avalon_mm_address          : in    std_logic_vector(1 downto 0);
    avalon_mm_read             : in    std_logic;
    avalon_mm_readdata         : out   std_logic_vector(31 downto 0);
    avalon_mm_write            : in    std_logic;
    avalon_mm_writedata        : in    std_logic_vector(31 downto 0)
  );
end entity combfilterprocessor;
```

Listing 2.7: Entity of combFilterProcessor.vhd

Stereo audio coming from the AD1939_subsystem, which is in the data-channel-valid protocol implemented by the Avalon streaming interface, is converted by ast2lr.vhd to separate left and right audio channels. This instantiation in *combFilterProcessor.vhd* is shown in Listing 2.8.

```
  u_ast2lr : component ast2lr
    port map (
      clk                 => clk,
      avalon_sink_data    => avalon_st_sink_data,
      avalon_sink_channel => avalon_st_sink_channel(0),
      avalon_sink_valid   => avalon_st_sink_valid,
      data_left           => left_data_sink,
      data_right          => right_data_sink
    );
```

Listing 2.8: Instantiation of ast2lr.vhd

Taking the left channel as the example, the left data in signal *left_data_sink* is sent to the component combFilterSystem.vhd whose instantiation is shown in Listing 2.9. The processed audio is now in signal *left_data_source*.

```
  left_combfiltersystem : component combfiltersystem
    port map (
      clk         => clk,
      reset       => reset,
```

```
      clk_enable => '1',
      audioin    => left_data_sink,
      delaym     => delaym,
      b0         => b0,
      bm         => bm,
      wetdrymix  => wetdrymix,
      ce_out     => open,
      audioout   => left_data_source
    );
```

Listing 2.9: Instantiation of combFilterSystem.vhd

The individual left and right channels are then converted back into the Avalon streaming interface with the component lr2ast.vhd whose instantiation is shown in Listing 2.10. The Avalon streaming audio is then sent to the audio codec.

```
  u_lr2ast : component lr2ast
    port map (
      clk                    => clk,
      avalon_sink_channel    => avalon_st_sink_channel(0),
      avalon_sink_valid      => avalon_st_sink_valid,
      data_left              => left_data_source,
      data_right             => right_data_source,
      avalon_source_data     => avalon_st_source_data,
      avalon_source_channel  => avalon_st_source_channel(0),
      avalon_source_valid    => avalon_st_source_valid
    );
```

Listing 2.10: Instantiation of lr2ast.vhd

The control signals are defined in Listing 2.11 where the default values are given (0.5 s delay, setDryMix = 1) so that the echo is very apparent on power-up. This size and value of each *std_logic_vector* have to be consistent with the data type defined in the Simulink model.

```
  -- delayM - uint16 - default value = "0001001011000000" = 24000 (0.5 sec)
  signal delaym    : std_logic_vector(15 downto 0) := "0101110111000000";
  -- b0 - sfix16_En16 - default value = "0111111111111111" = ~0.5
  signal b0        : std_logic_vector(15 downto 0) := "0111111111111111";
  -- bm - sfix16_En16 - default value = "0111111111111111" = ~0.5
  signal bm        : std_logic_vector(15 downto 0) := "0111111111111111";
  -- wetDryMix - ufix16_En16 - default value = "1111111111111111" = ~1
  signal wetdrymix : std_logic_vector(15 downto 0) := "1111111111111111";
```

Listing 2.11: Control signals in combFilterProcessor.vhd

The Avalon memory-mapped interface for writing to the registers is shown in Listing 2.12. The data in bus signal *avalon_mm_write* is converted to the size of the register constrained by the data type sign consideration.

```
bus_write : process (clk, reset) is
begin

  if reset = '1' then
    delaym    <= "0101110111000000"; -- 24000
    b0        <= "0111111111111111"; -- ~0.5
    bm        <= "0111111111111111"; -- ~0.5
    wetdrymix <= "1111111111111111"; -- ~1
  elsif rising_edge(clk) and avalon_mm_write = '1' then

    case avalon_mm_address is

      when "00" =>
        delaym <= std_logic_vector(resize(unsigned(avalon_mm_writedata), 16));

      when "01" =>
        b0 <= std_logic_vector(resize(signed(avalon_mm_writedata), 16));

      when "10" =>
        bm <= std_logic_vector(resize(signed(avalon_mm_writedata), 16));

      when "11" =>
        wetdrymix <= std_logic_vector(resize(unsigned(avalon_mm_writedata), 16));

      when others =>
        null;

    end case;

  end if;

end process bus_write;
```

Listing 2.12: Writing to control registers in combFilterProcessor.vhd

The Avalon memory-mapped interface for reading from the registers is shown in Listing 2.13. The data in the control registers are converted to the size of the bus signal *avalon_mm_readdata* that is 32-bits wide, and the sign of the data type is given so that the appropriate sign extension is made when it is resized to the larger signal.

```
bus_read : process (clk) is
begin

  if rising_edge(clk) and avalon_mm_read = '1' then

    case avalon_mm_address is

      when "00" =>
        avalon_mm_readdata <= std_logic_vector(resize(unsigned(delaym), 32));

      when "01" =>
        avalon_mm_readdata <= std_logic_vector(resize(signed(b0), 32));

      when "10" =>
        avalon_mm_readdata <= std_logic_vector(resize(signed(bm), 32));

      when "11" =>
        avalon_mm_readdata <= std_logic_vector(resize(unsigned(wetdrymix), 32));

      when others =>
        avalon_mm_readdata <= (others => '0');

    end case;

  end if;

end process bus_read;
```

Listing 2.13: Reading from the control registers in combFilterProcessor.vhd

2.5.4 Importing combFilterProcessor into Platform Designer

The process of importing the combFilterProcessor.vhd into Platform Designer to create the *combFilterProcessor* component is the same process as described in Sect. 1.4.1 Creating the AD1939 Platform Designer Component (page 269). The notable differences are the name of the component and associated VHDL files, the addition of the Avalon memory-mapped interface, and no exported signals.

The process of adding the combFilterProcessor Platform Designer component to the system is similar to the process described in Sect. 1.4.3 Creating the Platform Designer Passthrough System (page 284). The resulting system that shows the interface connections is shown in Fig. 2.18. The *jtag_master* component has been added in order to test the control registers.

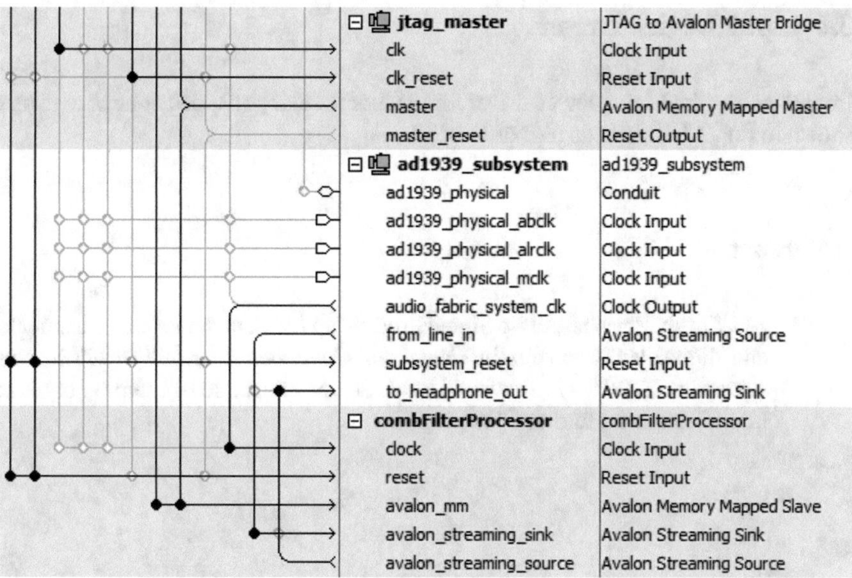

Fig. 2.18: The Platform Designer system with the combFilterProcessor component

2.5.5 Testing the combFilterProcessor Using System Console

The final hardware test is to exercise the system by placing values into the control registers using *System Console* and testing that the hardware system is working correctly. If you move on without verifying your hardware and your system does not work when you are developing software, you will waste enormous amounts of time chasing software bugs when the problem is your hardware.

Make sure that the *jtag_master* component has been added as shown in Fig. 2.18 and then following the testing procedure found in Sect. 6.3 System Console (page 69).

2.5.6 Quartus Project Cleanup

Now that we know that the system works, we can change file and variable names to be consistent with the project name. If the system breaks when doing this, we know that it is just a naming issue and not something more serious.

2.6 Linux Device Driver

Implementing the Linux device driver for the four memory-mapped control registers shown in Fig. 2.1 is done as part of Lab 13.

References

1. J.O. Smith, Physical Audio Signal Processing for virtual musical instruments and digital audio effects, in *Center for Computer Research in Music and Acoustics (CCRMA)* (Stanford University, Stanford, 2010). Online book at http://ccrma.stanford.edu/jos/pasp/

Example 3
FFT Analysis Synthesis System

3.1 Overview

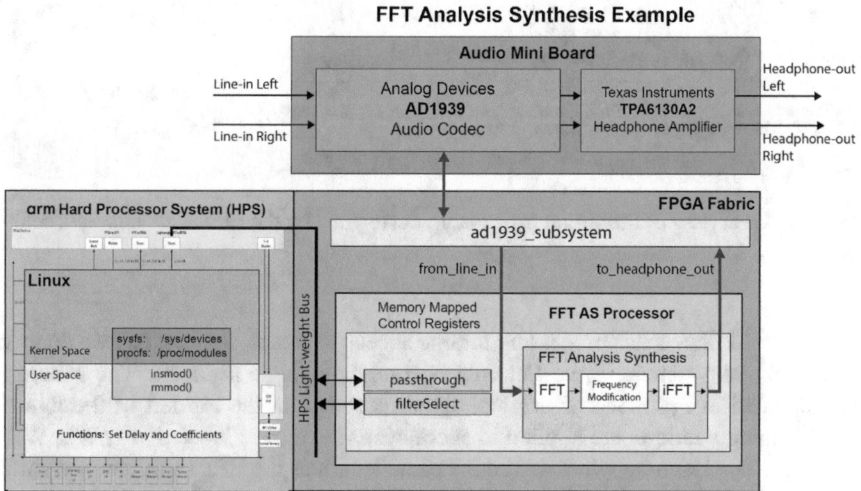

Fig. 3.1: The FFT Analysis Synthesis System. Audio signals are converted to the frequency domain by the FFT where the spectrum is modified and then converted back to the time domain by the inverse FFT (iFFT)

Converting audio signals to the frequency domain allows the spectrum to be easily modified. This is because frequency amplitudes and phases can be changed by multiplication with complex values. The FFT transforms the convolution operation in the time domain (e.g., FIR filtering) to multiplication in the frequency domain. An example of this signal processing is shown in Fig. 3.2 where a speech signal segment is converted to the frequency domain by the FFT, the high frequencies boosted like

R. K. Snider, *Advanced Digital System Design using SoC FPGAs*,
https://doi.org/10.1007/978-3-031-15416-4_15

what a hearing aid would do, and then converted back to the time domain by way of the inverse FFT. The system that can do this is shown in Fig. 3.1.

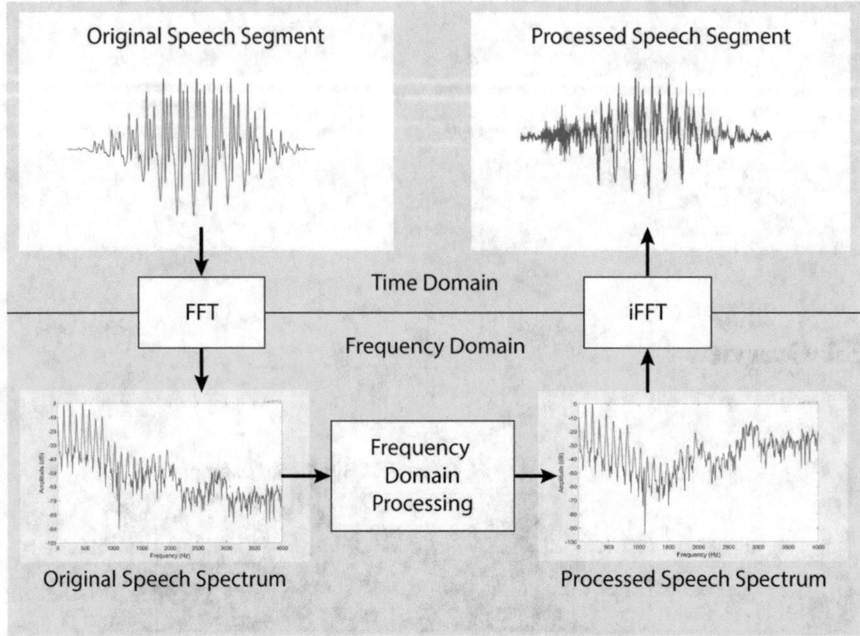

Fig. 3.2: Overview of the signal processing performed by the FFT Analysis Synthesis system

Figure 3.2 shows this processing for one speech segment, but how can we do this in a continuous manner if we can only operate on a single segment? The solution is illustrated in Fig. 3.3. The original signal is shown at the top left of the figure and Hanning windows are applied in succession (the curves labeled W1, W2, W3, . . ., W5). The Hanning windows are the same length as the FFT (Nfft) and we shift the Hanning windows by a quarter of the FFT length (Nfft/4). For example, W2 is identical to W1 except that it has been shifted in time by Nfft/4 samples.

The row in the figure that starts with W1 shows the frequency domain processing that was illustrated in Fig. 3.2, except that it has been unfolded to be on a single row. Each row starting with W1, W2, W3, etc. keeps its position in time and shows the result of multiplying the Hanning window with the original signal. Once the FFT and inverse FFT operations are done for each of the windowed segments, the processed signal appears after the delay associated with performing the FFTs. These time domain signals are kept relative to each other in terms of the Hanning window shifts (Nfft/4 sample delays), which means that they overlap with each other in time. The overlapped signals are simply added together resulting in the processed signal shown at the bottom of Fig. 3.3. This combination of output segments is known as *Overlap and Add*.

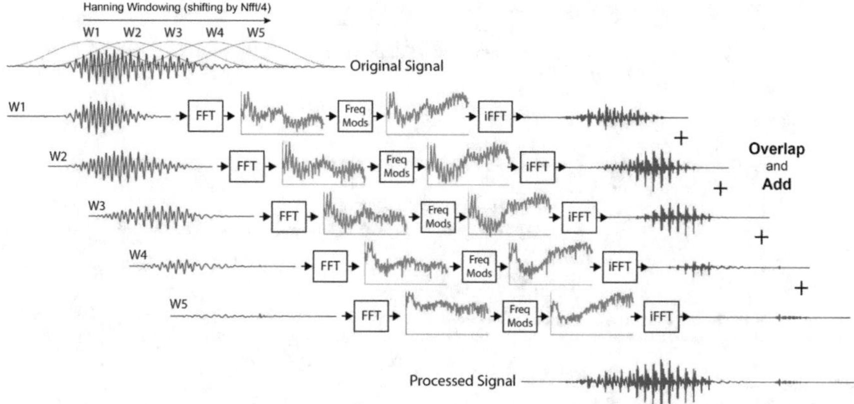

Fig. 3.3: Continuous FFT processing using the Overlap and Add method. In this example high frequencies have been given a boost similar to what a hearing aid would do, which causes the processed signal to look a bit noisier than the original signal

3.1.1 Frequency Domain Processing

Once the FFT has transformed the audio signal into the frequency domain, we can perform any frequency domain operation on the signal that we want. To illustrate this frequency domain processing, which also has distinct audio differences, we created four different filtering operations. Three of these simple filters are shown in Fig. 3.4 (low pass, bandpass, and high pass). The fourth is the all pass in which the filter mask has all ones. If we select the low pass filter option, then the low frequencies up to 4 kHz are unaffected (multiplied by a gain of one) while the frequencies above 4 kHz are all set to zero. The resulting spectrum is then sent to the inverse FFT to be converted back to the time domain. It should be noted that we only show the positive frequencies in Fig. 3.4. However, the mask, when applied to the output of the FFT, needs to have the same number of values as the length of the FFT since the mask needs to be applied to the conjugate (negative) frequencies as well. See Sect. 10.4.3 FFT (page 176) for a discussion on positive and negative frequency representation in the FFT and how frequencies are related to Matlab FFT indexing.

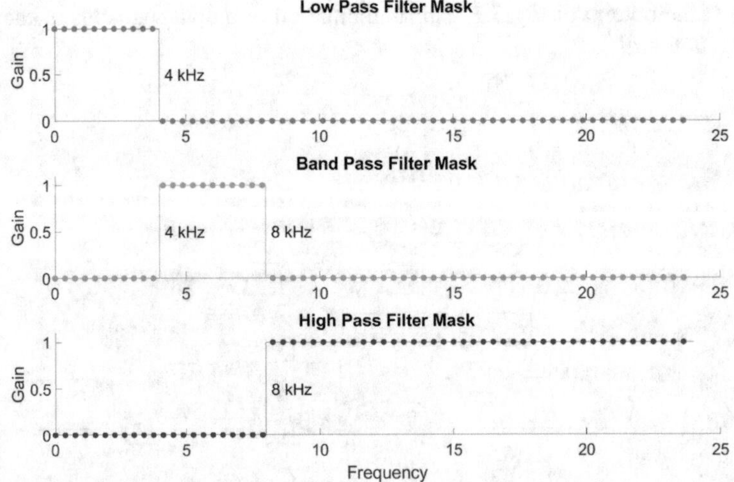

Fig. 3.4: Frequency modifications, i.e., filtering operations (low pass, bandpass, and high pass) that can be selected in the FFT Analysis Synthesis system. The all pass option is not shown

3.2 Simulink Model Files

The organization of the Simulink FFT Analysis Synthesis model conforms to the organization described in Sects. 2.2 Simulink Model (page 295)–2.2.3 Model Specific Files (page 298). This section describes the comb filter example, but the Simulink FFT Analysis Synthesis model is organized in the same way. Read these sections if you are unfamiliar with how the Simulink models are organized.

To get the files for the FFT Analysis Synthesis Simulink model, first get the *simulink-common* folder if you have not done so already, which is described in Sect. 2.2.2 Simulink Common Files (page 296). Then, go to the book's code location and clone the entire repository or click on the green Code button and download the repository .zip file, so you can just copy the files/folders you want. The Simulink model will be located at `\Code-main\examples\fftAnalysisSynthesis\simulink` (GitHub link). Copy the folder `\fftAnalysisSynthesis\simulink` to your computer. The Simulink model files in this directory are listed below:

File 1: **fftAnalysisSynthesis.slx**: This is the HDL compatible Simulink model file.

File 2: **createModelParams.m**: This file creates the model parameters that are stored in the data structure *modelParams*. These are the variable names that are used in the Simulink blocks, which allows parameter changes to be made without having to search through all the blocks in the model hierarchy. It is good programming practice to have a single place to change parameter values rather than searching everywhere in the Simulink model for hard coded values.

File 3: **createFFTFilters.m**: This file, used by *createModelParams.m*, creates the filter masks shown in Fig. 3.4.

File 4: **createSimParams.m**: This file creates the simulation specific parameters that are stored in the data structure *simParams*. These are parameters such as how long the simulation will run, which verification to run, and it defines the data types of the signals that are fed as input into the model.

File 5: **createHdlParams.m**: This file creates the HDL Coder parameters that are stored in the data structure *hdlParams*. This sets parameters such as the targeted FPGA clock speed.

File 6: **playOutput.m**: This file plays the created sound effect.

File 7: **verifySimulation.m**: This file verifies that the simulation output is correct. The type of verification is associated with the type of input signal.

3.3 Creating the Simulink Model

The FFT Analysis Synthesis Simulink model is a hierarchical model that divides the processing into three parts as shown in Fig. 3.5. These three parts are (1) the analysis part that converts the time domain signal into the frequency domain and contains the FFT engine, (2) the frequency domain processing, and (3) the synthesis part that converts the signal back to the time domain and contains the inverse FFT engine.

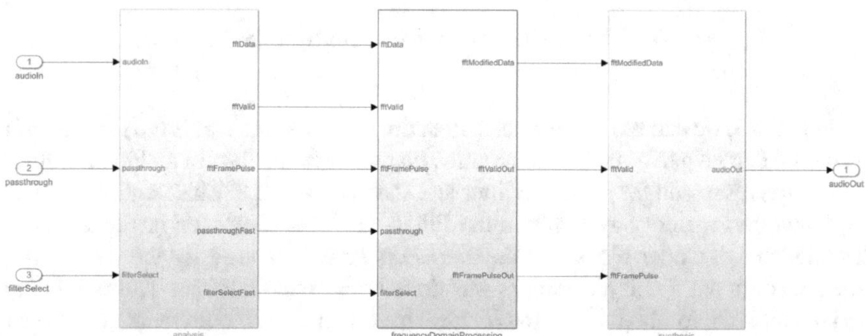

Fig. 3.5: The FFT Analysis Synthesis system is developed as a hierarchical model with three blocks for each of the subsystems, which are the analysis, frequency domain processing, and synthesis blocks

3.3.1 Analysis

The heart of the *analysis* part of the model is the FFT engine as shown in Fig. 3.6. The FFT engine is a Simulink block found in the HDL coder library that is optimized for HDL code generation. The inputs to the FFT block are the audio data and valid signals. The length of the FFT that the block performs is set by the block parameter *FFT length*, which we enter as the variable *modelParams.fft.size* so that we can easily change the FFT size in the initialization file *createModelParams.m*.

When the valid signal into the FFT block is asserted, it tells the FFT block to load data present on the data signal line. As soon as the FFT engine loads *FFT length* samples, it starts the FFT computations. When the FFT engine is finished computing the FFT, it streams the resulting data out while asserting the output valid signal. We ignore the ready (not busy) signal since by design we will not be sending data to the FFT engine when it is busy. The *fftFrameBuffering* block performs all the operations necessary to get the FFT frames to the FFT block at the correct timings (which is why we can ignore the ready signal).

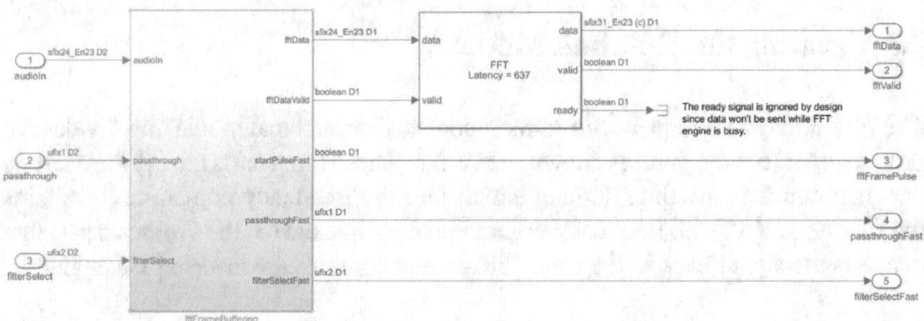

Fig. 3.6: The FFT analysis section that contains the FFT engine

In Fig. 3.3, we can see that we need an entire *frame* of data that is the same length as the FFT to be performed. This requires that we save the data in a circular buffer, which was discussed for the comb filter and shown in Fig. 2.3. This required the use of a dual port memory available in the FPGA fabric as illustrated in Fig. 2.4. The Simulink HDL Coder block *Simple Dual Port RAM* was used for the comb filter since reading out a delayed sample was done at the same rate that data was being written into the circular buffer. However, we have a different requirement for the dual port memory when it comes to performing the FFT. Audio samples need to be fed into the circular buffer at the audio sample rate, but we need to get a frame of data to the FFT engine very quickly for two reasons. The first reason is that we want to be careful with FPGA resources, which means that we want to implement only a single FFT engine. Thus it needs to run fast enough to keep up with the data being supplied to it. The second reason is that we want to run the FFT engine as fast as possible in the FPGA fabric to reduce the latency through the FFT and inverse FFT engines.

This means that we need to use the Simulink HDL Coder block *Dual Rate Dual Port RAM* that can run the two ports at different rates. This dual rate dual port memory is contained in the *fftFrameBuffering* block whose content is shown in Fig. 3.7.

Simulink color codes the sample rates of the signals, so the signals going into the *fftFrameBuffering* block in Fig. 3.6 are colored green signifying the 48 kHz audio sample rate. Signals are leaving the block in red signifying the FPGA fabric clock rate, which is controlled by the model parameter *modelParams.system.upsampleFactor*. There is a minimum speed the FFT block needs to run at in order to keep up with the audio data. Further increasing the speed (upsample factor) will decrease the latency through the FFT engines. The fast system clock is generated by an on-board FPGA PLL clock generator and the limitation of how fast the clock can run is ultimately constrained by timing closure when compiled in Quartus.

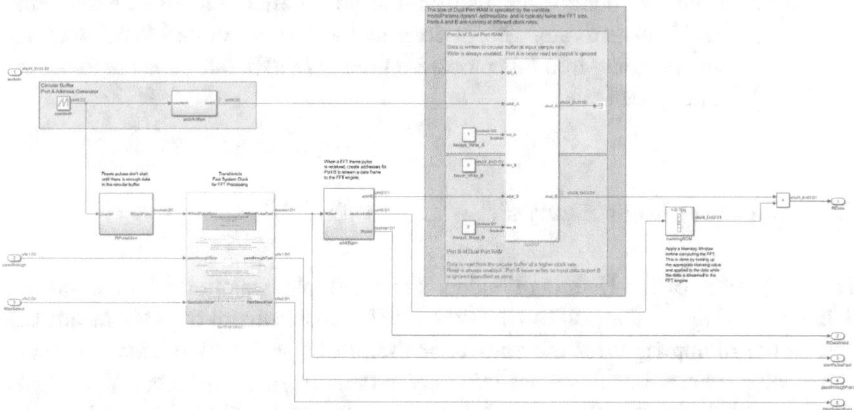

Fig. 3.7: The contents of the *fftFrameBuffering* block that implements the dual rate dual port memory for the circular buffer and transitions the Simulink processing to the FPGA fabric system rate. Green signals are running at the audio sample rate and red signals are running at the fast system rate. Note: Open the model *fftAnalysisSynthesis.slx* in Simulink for better viewing (GitHub link)

The tasks that are performed in Fig. 3.7 by the *fftFrameBuffering* block are the following:

Task 1: Collect audio data in the circular buffer in order to create the FFT data frames. Data is collected in the HDL Coder *dual rate dual port RAM* block, the size of which is set by *modelParams.dpram1.addressSize* in *createModelParams.m*.

Task 2: Apply a Hanning window to the audio data in the frame as it streams to the FFT engine. The Hanning coefficients are stored in the *hanningROM* block. The choice of window is discussed in Sect. 10.5.4 and the Hanning window compared to other windows is shown in Fig. 10.18. The Hanning window is a good trade-off between main-lobe width and side-lobe attenuation (see Fig. 10.17) if you are unsure what window to use for the FFT processing.

Task 3: Shift the data frame by FFTlength/4 after each frame has been generated, but first wait until there is enough data in the circular buffer before generating the FFT frames. This is done by the *fftPulseGen* block that creates a signaling pulse at the start of each frame. Note: The FFT length can be easily changed, but changing the shift amount will require substantial changes to the Simulink model since it will change the number of segments that need to be stored for the overlap and add operation.

Task 4: Generate the addressing required for both dual memory ports and for the Hanning ROM. This is done by the *counterA* block and the *addrBgen* block.

Task 5: Transition control signals from the slower audio sample rate to the fast FPGA system rate. This is done by the *fastTransition* block. Note that signals going into this block are green (48 kHz rate) and come out red (fast system rate).

3.3.2 Simulink Logic Analyzer

The application of the Hanning window and the implementation of the frequency domain filtering are done on the fly as the FFT frames stream by. This means that getting the timing right for the outputs of the blocks such as the *addrBgen* block can involve some debugging since it is easy to create *off-by-one* errors. When these errors occur during development, it is very useful to visualize the timing of the relevant signals using Simulink's Logic Analyzer. You can open up the Simulink Logic Analyzer by clicking its button in the *SIMULATION* panel in Simulink as shown in Fig. 3.8.

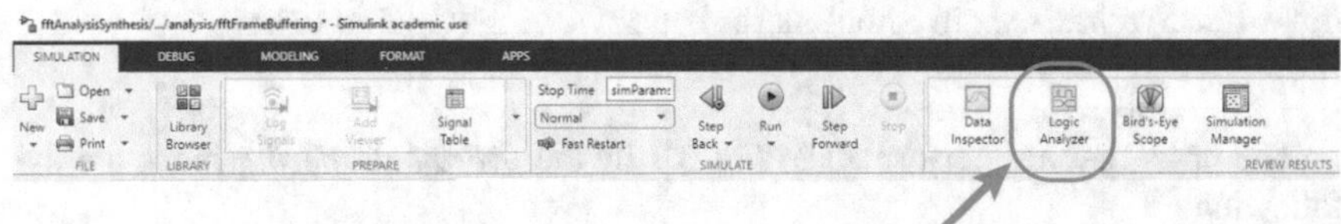

Fig. 3.8: Accessing the Logic Analyzer in Simulink

To add a signal line to the logic analyzer in order to monitor it and compare its timing to other signals, click on the line you want to visualize so that three dots appear above the line as in Fig. 3.9 (step 1). Then move the cursor to the three dots and select *Log Selected Signal*. Selecting it again will remove the signal from the logic analyzer.

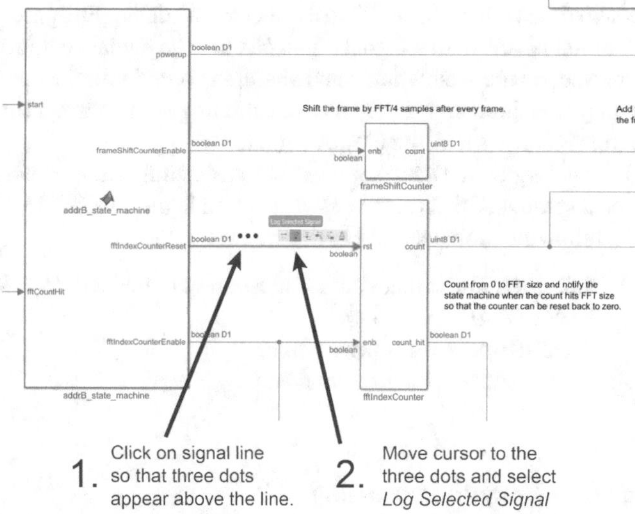

Fig. 3.9: Adding a signal to the Logic Analyzer in Simulink

An example of a logic analyzer session is shown in Fig. 3.10 that includes signals inside of the *addrBgen* block and the output of the *hanningROM* block.

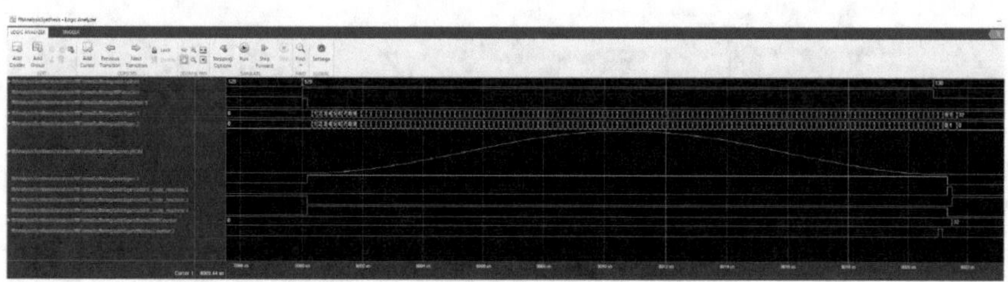

Fig. 3.10: Example Logic Analyzer session

To **view waveforms**, such as the Hanning window coefficients as shown in the figure, perform the following steps:

Step 1: Click on the signal name in Simulink's Logic Analyzer.
Step 2: Click on the *WAVE* panel.
Step 3: Set the *Format* setting to **Analog**. This renders the signal as a waveform.
Step 4: Set the *Height* setting to **100**, which was done for the *hanningROM* signal shown in Fig. 3.10, which allows you to see the waveform better.

Another common setting is to **view signals as decimal values** (such as counters) rather than hexadecimal values. To view in this manner, perform the following steps:

Step 1: Click on the signal name.
Step 2: Click on the *WAVE* panel.
Step 3: Set the *Radix* setting to **Unsigned Decimal**. Other options are *Hexadecimal*, *Octal*, *Binary*, and *Signed Decimal*.

Once the signal views have been set up as desired, the debugging process is to run a simulation to log all the added signals (you can see the signals being logged into the Logic Analyzer as the simulation runs) and then zoom in on the control signals to see when they transition and if they are transitioning at the correct times relative to other control signals such as the *fftStart* signal.

Some addition Logic Analyzer settings that are useful and convenient are the following global settings. Click on the *Settings* button in the *LOGIC ANALYZER* tab and select the following global setting options:

Option 1: Select the *Radix* that you use most often to make it default.
Option 2: Select *Fit to view at Stop*.
Option 3: Select *Display short wave names*.
Option 4: Select *Display bus element names*.

3.3.3 Frequency Domain Processing

The analysis section converts the audio signal to the frequency domain where we can modify individual frequencies in the *frequencyDomainProcessing* block whose content is shown in Fig. 3.11

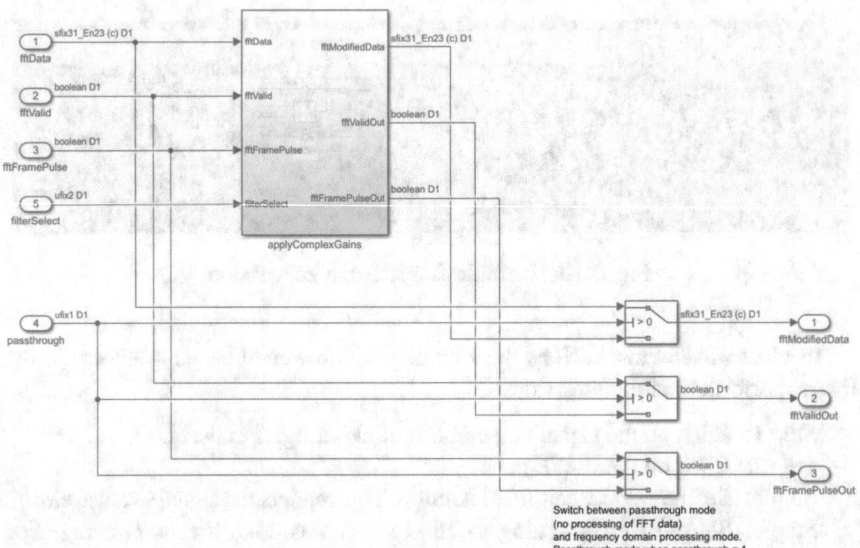

Fig. 3.11: The contents of the *frequencyDomainProcessing* block that can modify the frequency content of the audio signal. The *applyComplexGains* block can be bypassed by asserting the *passthrough* control signal

The *passthrough* control signal, when asserted, allows the audio signal to bypass any modification to its spectrum. Otherwise spectral modifications will occur in the *applyComplexGains* block, whose content is shown in Fig. 3.12.

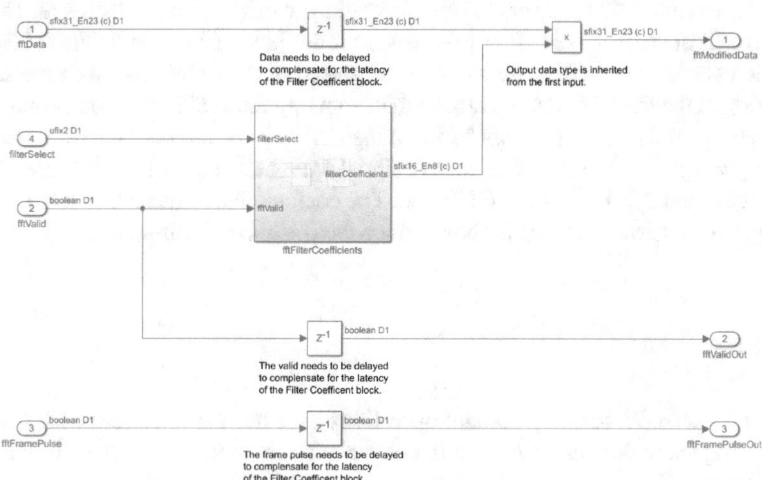

Fig. 3.12: The contents of the *applyComplexGains* block that implements the spectral filtering. The FFT frame is modified via complex multiplications as it streams through the block. The data and control signals need to be delayed to compensate for the latency of the *fftFilterCoefficients* block

The spectrum of the FFT frame is modified via complex multiplications as the FFT frame streams through the block. The filter coefficients coming out of the *fftFilterCoefficients* block are delayed relative to the data in the FFT frame, so the FFT frame and control signals need to be delayed to compensate for the latency of the *fftFilterCoefficients* block. The contents of the *fftFilterCoefficients* block are shown in Fig. 3.13.

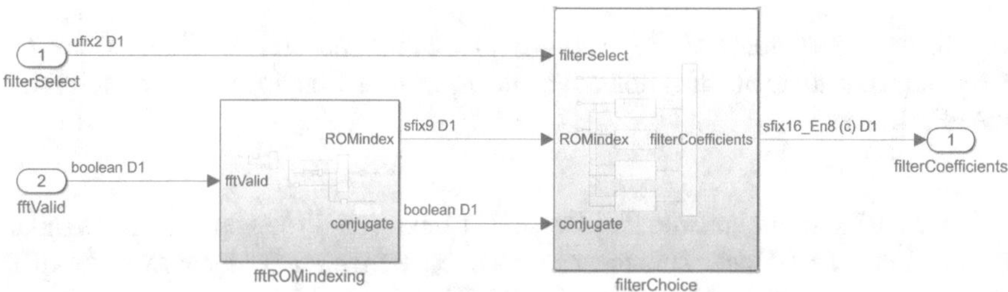

Fig. 3.13: The contents of the *fftFilterCoefficients* block

In the *fftFilterCoefficients* block, the control signal *fftValid*, when asserted, starts a counter in the *fftROMindexing* block, which provides the indexing into the ROMs that store the filter coefficients. The ROMs are implemented as *Lookup Table* blocks in the HDL Coder library. It is assumed that we are dealing with a real audio signal that has a symmetric spectrum where the conjugate negative frequencies mirror the positive frequencies. This allows us to store only half of the coefficients, i.e., the coefficients for the positive frequencies are stored in the ROMs and we reverse the indexing in the *fftROMindexing* block when applying the coefficients for the negative frequencies and change the sign of the imaginary part of the coefficients when the *conjugate* signal is asserted. The control signal *filterSelect* controls which filter ROM is selected and applied to the FFT frame. The coefficients for these filters are shown in Fig. 3.4 (only the real part is shown since the imaginary values are zero).

3.3.4 Synthesis

Once the desired frequency domain operations have been applied to the FFT frame in the frequency domain, it is time to convert the modified FFT frames back to the time domain. This is done by the *synthesis* block of Fig. 3.5, whose content is shown in Fig. 3.14.

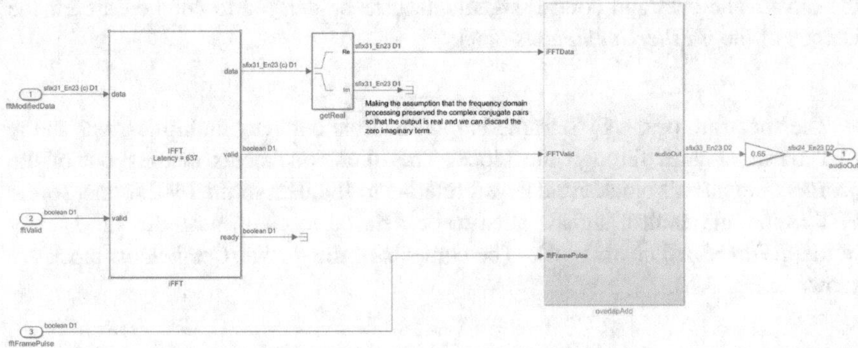

Fig. 3.14: The contents of the *synthesis* block that converts the signal from the frequency domain to the time domain by the inverse FFT engine and then assembles the FFT frames

The FFT frames stream into the inverse FFT block (iFFT) that operates in a similar manner to the FFT block. This means we can ignore the ready signal as in the FFT block since data will not be sent into the iFFT block while it is busy performing the iFFT operation. When the iFFT operation is done, that time domain data streams out of the data port while the valid signal is asserted. The data type coming out of the iFFT is a complex number data type. However, we are assuming that the following two assumptions hold in our system:

Assumption 1: The audio input signal coming into the system contains only **real numbers**. From Euler's formula (see Eq. 10.29), we can represent a real-valued signal as a sum of two complex exponential signals, i.e., $\cos(\theta) = \frac{1}{2}(\exp^{j\theta} + \exp^{-j\theta})$ since the imaginary terms cancel. This is where the two frequencies come from (one positive and the other negative).

Assumption 2: When we perform the frequency domain operations, we apply the same coefficient that we did for the positive frequency to the negative frequency, except for a conjugation operation, i.e., changing the sign of the imaginary term.

These two assumptions imply that the imaginary terms have cancelled with each other leaving only real numbers. This is why we can ignore the imaginary term of the complex numbers streaming out of the inverse FFT block as shown in Fig. 3.14.

Now that the FFT frames have been converted back to the time domain and they are real numbers, we can assemble the frames back into a continuous signal as illustrated in Fig. 3.3. This is done by the *overlapAdd* block, whose content is shown in Fig. 3.15.

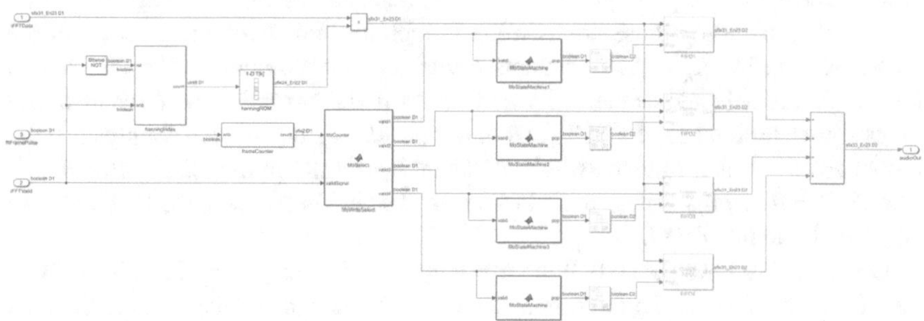

Fig. 3.15: The contents of the *overlapAdd* block that assembles the time domain FFT frames. Note: Open the model *fftAnalysisSynthesis.slx* in Simulink for better viewing (GitHub link)

We first apply a Hanning window again to make sure that the ends of the FFT frame go to zero if the frequency domain processing has changed this. We want the ends of the FFT frame to be zero since we will be splicing them together and do not want any discontinuities. As can be seen in Fig. 3.3, the FFT frames overlap in time. This means that we need to store these frames while we add them together. We need a memory that can accept a blast of data from the iFFT block that comes in very quickly at the FPGA system clock and then let the data out at the slower audio sample rate. A memory type that can do this is called a FIFO, which stands for *First*

In First Out. This memory preserves the order of the samples but allows it to be loaded and unloaded at different rates and intervals.

As can be seen on the right side of Fig. 3.15, there are four FIFOs whose outputs get added together. This architecture forms the heart of the overlap and add process. This is why if you change the frame shift value from Nfft/4, it will change the number of overlapping segments, which will necessitate a change in the number of FIFOs needed since we need a FIFO for each overlapping segment. It will also change the required control signal timings as well. Thus the architecture of this model has been built specifically for the Nfft/4 frame shifts.

You will notice that the output of the FIFO is color coded green since the data comes out at the audio sample rate. The *Pop* control signal controls when the data comes out of the FIFO, i.e., when the data is "popped" out. This means that the *Pop* control signal needs to be green, which is the audio sample rate. We will get to how we set this output control sample rate in moment. The input to the FIFO is the FFT data rate that is color coded in red signifying the fast FPGA system rate. The *Push* control signal, also in red, controls when data can be "pushed" into the FIFO. This *Push* control signal is simply the *iFFTValid* signal coming from the inverse FFT engine. Thus the data streams out of the iFFT engine, gets multiplied by a Hanning window, and then streams right into a FIFO. But which FIFO? The iFFTData line goes to all the FIFO inputs!

We need to control the FIFO *Push* signals and we do so by the *fifoWriteSelect* block. Every time a pulse on the *fftFramePulse* comes in, it enables a two-bit free-running counter whose count value is incremented by one and sent into the *fifoWriteSelect* block. This implements *round robin* scheduling for the FIFOs, i.e., the FIFOs are selected sequentially and in a circular manner. The *fifoWriteSelect* block illustrates the use of a *MATLAB Function Block* since we can simply code the output selection as a Matlab switch statement operating on the *fifoCounter* signal. The *iFFTValid* signal is then routed to one of the FIFOs allowing the FFT frame to be pushed into this FIFO.

OK, fine. We have pushed all the overlapping FFT frames into the FIFOs. How do we control when we let the frame data out? This is done by the *fifoStateMachine* block, which we implement as a *MATLAB Function Block*. The analogy regarding the timing is that we have four buckets each with their own spigots. As soon as the bucket is filled for the first time, we turn the spigot on, never to turn it off again. We just refill each bucket just as the last drop (sample) comes out of each bucket. The *fifoStateMachine* block powers up in the *ValidLow* state. As soon as it sees the *ifftValid* signal asserted, it transitions to the *ValidHigh* state and stays in this state while the *ifftValid* signal remains asserted. When the *ifftValid* signal is deasserted, we know that the FIFO has been filled with data so we open the spigot by transitioning to the *AssertPopSignal* state where it remains forever asserting the *Pop* signal of the FIFO. The *Pop* signal is run through a rate transition block where it changes rate to the audio sample rate, and thus it sets the output rate of the FIFO to the audio sample rate. Then every fourth iFFT frame refills the FIFO so that by design the FIFO never becomes empty (or overfills).

3.3.5 Verification

Once you have developed a Simulink signal processing model, how do you know it is correct since *"The problem with simulations is that they are doomed to succeed"* [1]. In our case we know that applying the forward FFT followed by the inverse FFT should result in the same signal. Thus a good signal for a quick sanity check is a simple cosine waveform, which can be selected in the initialization function *createSimParams.m* by setting *simParams.signalSelect* = *2*. This creates a cosine waveform with a frequency of 3 kHz. The system also needs to be set in *passthrough* mode so that no spectral processing occurs. One of the Simulink Logic Analyzer (see Sect. 3.3.2) debugging sessions is shown in Fig. 3.16 where the output of the Simulink model did not look at all like the cosine input (top signal), which is the signal just above the bottom trace in the figure. The bottom signal is the corrected output once the problem was identified. The debugging process was tracing the cosine waveform through the system to see where it started looking weird. The output of the iFFT engine looked fine (the second waveform line in Fig. 3.16) and the outputs of the FIFOs looked fine (waveform lines 3–6 in the figure). However, it was the output of the adder that looked strange. It turned out that the fixed-point word length was not large enough, which was causing overflow to occur where large values were suddenly turning negative when they should not. The solution was to let Simulink choose what the output data type should be and then apply a gain before setting the data type back to the default audio data type.

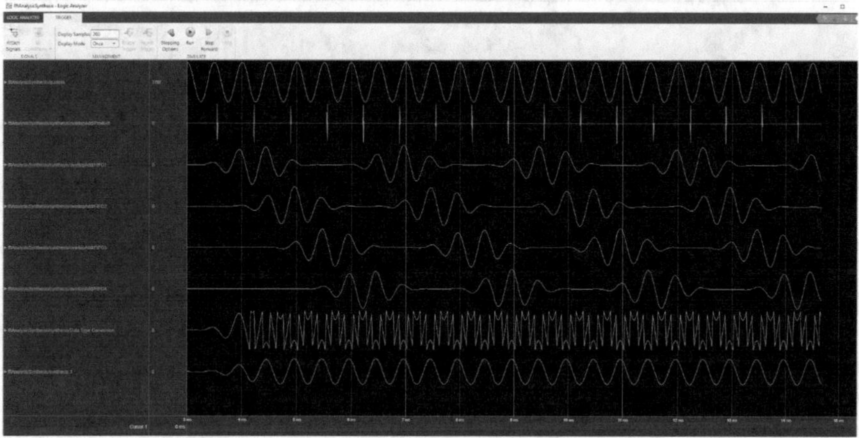

Fig. 3.16: Using the Simulink Logic Analyzer to debug the system with cosine wave. During initial development, overflow occurred in the add unit of the overlap and add section resulting in a non-sinusoidal output (the second line from bottom). The correct output is the bottom waveform

The next step was to perform a more careful comparison between the input and output than a quick check-by-eye sanity check that is shown in Fig. 3.16 where the top input waveform looks reasonably similar to the output waveform shown at the bottom. In order to make a valid comparison, we first needed to figure out what the latency is through the system since if we compare identical sinusoids that differ in phase, we will get significant errors.

A good way of determining what the system latency is to send an impulse through the system. The impulse is selected by setting *simParams.signalSelect* = *3* and the verification for the impulse will measure what the latency is through the system. This latency value is needed to align (delay) the cosine input (*simParams.signalSelect* = *2*) so that it can be compared to the output of the system, which is shown in Fig. 3.17 (upper plot). The sample errors were measured between the shifted input and the output for the samples between the two vertical red lines. A histogram of the absolute value of the sample errors is shown in the lower plot. This verification shows that the input and output waveforms closely track each other. The appropriate sample errors will ultimately be application dependent and can be reduced by increasing the precision of the internal fixed-point signals.

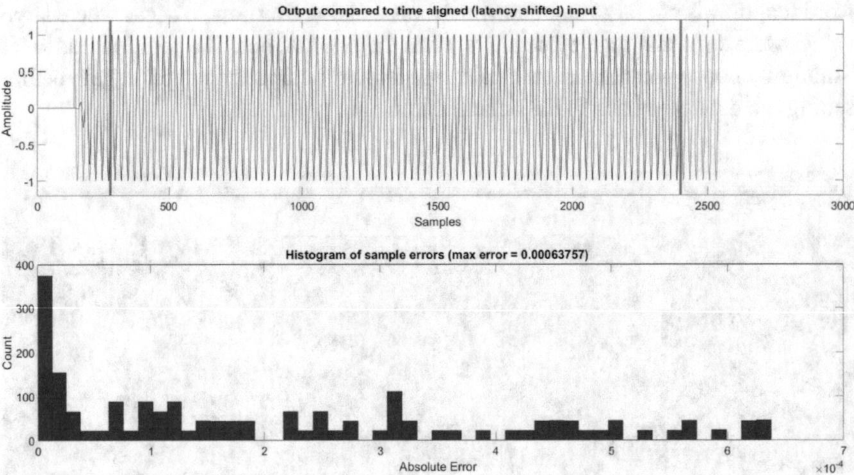

Fig. 3.17: Verifying the FFT system can reconstruct the input 2 kHz cosine signal

The final verification perform was to make sure that the frequency domain processing performed as expected. To do this, we measured the frequency response of the system to see if the filter choice (see Fig. 3.4) performed as expected. The simulation settings, found in *createSimParams.m*, to check the bandpass filter were set as follows:

Setting 1: Select the chirp signal by choosing *simParams.signalSelect* = *4*. This chirp signal will ramp from 0 to 24 kHz over the duration of the simulation.

Setting 2: *simParams.stopTime = 1.0*. If the simulation time is too short, the resolution of the frequencies in the chirp will result in a poor frequency response.

Setting 3: *passthrough = 0*. The passthrough needs to be disabled so that the FFT frames go through the *frequencyDomainProcessing* block.

Setting 4: *filterSelect = 1*. This selects the 4–8 kHz bandpass filter as shown in Fig. 3.4.

Setting 5: *simParams.verifySimulation = true*.

This verification resulted in the acceptable frequency response shown in Fig. 3.18 for the system using a 128-point FFT. Higher frequency resolutions can be obtain by setting the FFT size to greater values (e.g., 512-point FFT), but at the expense of using more FPGA resources and a longer latency through the system.

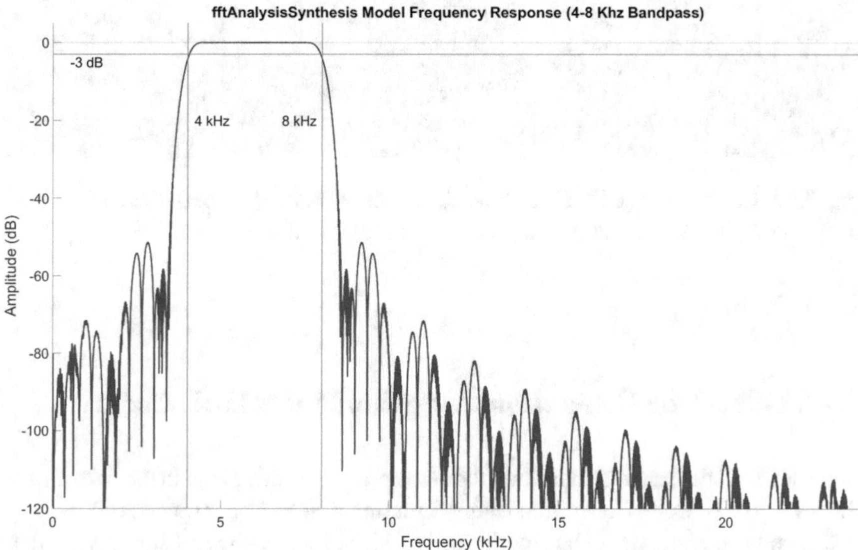

Fig. 3.18: Frequency response of the 4–8 kHz bandpass filter applied in the frequency domain processing block

3.3.6 System Latency

The latency through the system is shown in Fig. 3.19 for a model configured to operate using 256-point FFTs and running at an FPGA fabric speed of 49.152 MHz (1024 upsample factor). The circular buffer adds 5.3 milliseconds of latency, while the FFT engines take only 26.3 microseconds to compute 256-point FFTs. Since the frequency domain modifications of Fig. 3.4 are done on the fly as the FFT frame is transmitted from the forward FFT engine to the inverse FFT engine, there is negligible latency added for the frequency domain operations.

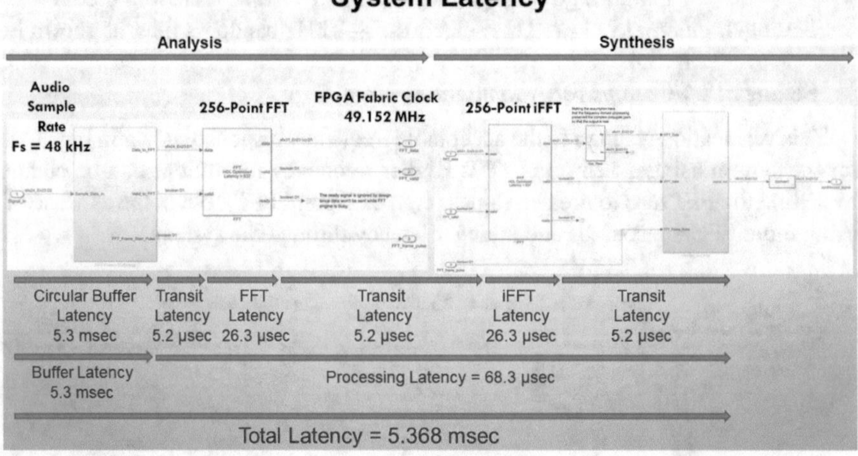

Fig. 3.19: Latency of the FFT Analysis Synthesis system. Most of the latency comes from waiting for audio samples in the input circular buffer

3.4 VHDL Code Generation Using Simulink's HDL Coder

The VHDL Code Generation follows the same process as described in the *combFilter* Sect. 2.4 VHDL Code Generation Using Simulink's HDL Coder (page 305). Follow this section to generate VHDL code for the *fftAnalysisSynthesis* Simulink model. Only the notable differences from the *combFilter* VHDL code generation will be mentioned here.

Difference 1: The **Oversampling factor** is **1** rather than 2048. This is because the *modelParams.system.upsampleFactor* is 2048 for the rate transition blocks to make the FFT engines run at 98.304 MHz. Thus $F_{\text{fastestSR}}$ is 98,304,000 Hz, which makes the *Oversampling factor* one (see Eq. 2.10).

Difference 2: The VHDL files have been placed in the `\simulink\hdl_prj\hdlsrc\fftAnalysisSynthesis` folder by the HDL Coder, and this folder needs to be copied to the *ip* folder under the *fftAnalysisSynthesis* Quartus project folder.

3.5 HPS Integration Using Platform Designer

We now need to create a Platform Designer component called the *fftAnalysisSynthesisProcessor* as seen in Fig. 3.1 (labeled *FFT AS Processor*). It requires the same three Avalon interfaces as the *combFilterProcessor*, which are listed in Table 2.2.

The files associated with the *fftAnalysisSynthesisProcessor* are listed in Table 3.1.

Table 3.1: FFT Analysis Synthesis System Files

File	Purpose	Location
fftAnalysisSynthesis.slx	Simulink Model of the FFT Analysis Synthesis system.	(click here)
fftAnalysisSynthesis.vhd	Top level VHDL file created by the HDL Coder. It is also the name of the top level block in the FT Analysis Synthesis Simulink model that was converted to VHDL. The HDL Coder placed the generated files into `\hdl_prj\hdlsrc\fftAnalysisSynthesis` under the Simulink project folder.	(Created in Lab 14)
fftAnalysisSynthesisProcessor.vhd	The VHDL file for the Platform Designer component *fftAnalysisSynthesisProcessor*. This VHDL file instantiates the *fftAnalysisSynthesis.vhd* component.	(Created in Lab 14)

3.5.1 Hardware Integration Steps

The steps for creating the new Platform Designer component *fftAnalysisSynthesisProcessor*, which is the block labeled *FFT AS Processor* in Fig. 3.1, are listed below:

Step 1: **Start with a known working system.** Our starting point will be the *combFilter* project that you created. Simply copy the project into a new folder and do not rename any files before starting out. Renaming files is tempting since this is a new project, but this opens the door for errors that would defeat the process of starting with a known working system. Renaming is the last step. See Sect. 2.5.2 Quartus Project Setup (page 318).

Step 2: **Create the *fftAnalysisSynthesisProcessor* Platform Designer Component.** Follow the steps and process that was done for the combFilter project, which is found in Sects. Step 2: Hardware Integration Steps (page 318) and 2.5.3 Creating the Platform Designer Component (page 318).

Step 3: **Importing the *fftAnalysisSynthesisProcessor* component into Platform Designer**. This process is described for the combFilter in Sect. 2.5.4

Importing combFilterProcessor into Platform Designer (page 322). Just delete the combFilter Platform Designer component and replace it with the *fftAnalysisSynthesisProcessor*.

Step 4: **Testing the *combFilterProcessor* component using System Console.** We need to verify that our hardware works before we can move on to creating the software that will use our hardware. This process is described for the combFilter in Sect. 2.5.5 Testing the combFilterProcessor Using System Console (page 323).

Step 5: **Quartus Project Cleanup**. Here we rename the files to be consistent with the project name and put files in their expected directory locations.

3.5.2 Testing the fftAnalysisSynthesisProcessor Using System Console

The final hardware test is to exercise the system by placing values into the control registers using *System Console* and testing that the hardware system is working correctly. If you move on without verifying your hardware and your system does not work when you are developing software, you will waste enormous amounts of time chasing software bugs when the problem is your hardware.

Make sure that the *jtag_master* component has been added to Platform Designer and then following the testing procedure found in Sect. 6.3 System Console (page 69).

3.6 Linux Device Driver

Implementing the Linux device driver for the two memory-mapped control registers shown in Fig. 3.1 is done as part of Lab 14.

Reference

1. P. Grim et al., How simulations fail. Synthese **190**(12), 2367–2390 (2013)

Part IV
Labs

Normally, when one thinks of an engineering course that has a laboratory associated with it, a two-hour lab that occurs once a week comes to mind. And because of the two-hour time limit, many labs will have a *Prelab* assignment to be completed before coming to lab. The labs in this book are different. *The expected time for a student to complete a lab in this book is* ***one week***. Even given a week, some students find that finishing a lab in a week's time frame can be challenging, depending on the lab, how well they have learned prerequisite material, and their prior coding experience. There are no Prelabs in this book. Rather, the labs have reading assignments that cover relevant material needed for the lab, and since students have a week to finish the lab, they have ample time to read the material. The weekly two-hour lab time slots are used as help sessions and are a good time for students to demonstration their working systems for the lab assignment given the prior week.

Another difference between typical engineering labs and labs in this book is that the labs in this book build on each other. This means that a student who gets behind will struggle all the way through because of the lab dependencies. The grading scheme that I have found that works best for this type of material is full credit for labs turned in on time and that works correctly. However, giving a zero for a late lab does not work well for this material because the lab still needs to be done and a zero ends up being a disincentive. What seems to work reasonably well is to mark late labs off a significant amount each week (e.g., ~25% *per week*).

The philosophy taken in this book is that you do not understand a system until you actually build it. Thus, the labs take a central focus in this book.

Labs 1-11 can be done using just the DE10-Nano board since it covers SoC FPGA development using the LEDs on the DE10-Nano board. Labs 12 and onward require the Audio Mini board to implement audio signal processing.

Lab 1
Setting Up the Ubuntu Virtual Machine

The DE10-Nano board contains a Cyclone V SoC FPGA that has ARM CPUs running Linux. This means that we will be developing Linux software for these ARM CPUs and we would like a Linux computer to develop on. Instead of using two computers (one for Windows 10 and another for Linux), we will create a *virtual machine* (**VM**) in Windows 10 using **VirtualBox** to run Linux. The Linux distribution that we will use will be Ubuntu 20.04 LTS . For the lab, it is assumed that you have your own computer with the capability described in Sect. 1.4.1 Laptop (page 9).

1.1 Items Needed

The items needed for Lab 1 are:

Item 1: A Windows 10 Laptop or PC.

1.2 Lab Steps

The steps for Lab 1 are:

Step 1: **Install and setup and the Ubuntu VM** by following the instructions found in Sect. 11.1.2 VirtualBox Ubuntu Virtual Machine Setup (page 197).

Step 2: **Compile and run the "Hello World" program** as described in subsubsection 11.1.2.11 Testing the Ubuntu VM by Compiling "Hello World" (page 211).

R. K. Snider, *Advanced Digital System Design using SoC FPGAs*,
https://doi.org/10.1007/978-3-031-15416-4_16

1.3 Demonstration

Print out and get Lab 1 **Instructor Verification Sheet signed off** (page 349) by demonstrating to the course TA the following:

Demo 1: Show that you can compile and run the *Hello World* program in the Ubuntu VM.

1.4 Deliverables

The **Deliverables** to be turned in or upload to the Lab 1 assignment folder are:

1: The Instructor Verification Sheet signed and turned in
2: Your "Hello World" source code, i.e., hello.c
3: A screen shot from the Ubuntu VM terminal window showing that your program ran correctly with your full name being printed out

Instructor Verification Sheet
Lab # 1
Setting Up the Ubuntu Virtual Machine

Print this page and get it signed by the Course Instructor or lab Teaching Assistant. Make sure you turn in this signed page to get credit for your lab.

Name: ____

Verified by: _____________________ **Date:** ______________

Lab Demonstration. Demonstrate to the Lab TA the following:

Demo 1: Show that you can compile and run the *Hello World* program in the Ubuntu VM.

Deliverables. Turn in or upload to your Lab 1 assignment folder the following items:

1: The Instructor Verification Sheet signed and turned in
2: Your "Hello World" source code, i.e., hello.c
3: A screen shot from the Ubuntu VM terminal window showing that your program ran correctly with your full name being printed out

Lab 2
Hardware Hello World

Lab 2 will be a simple warm-up exercise where you will use Quartus Prime Lite Edition 20.1 to implement the *hardware* version of "Hello World," which is turning on and off LEDs. This will be done by simply connecting four switches on the DE10-Nano board to four LEDs using the FPGA fabric and using simple VHDL code. The primary intent of this lab is to make sure that your DE10-Nano board and the Quartus Prime Lite toolchain work for you. During this lab you will install the free version of Quartus Prime Lite on your computer/laptop.

2.1 Items Needed

The items needed for Lab 2 are:

Item 1: A **Laptop** or PC

Item 2: **DE10-Nano Kit**. Purchase the kit from the ECE stockroom. The kit can also be purchased from the manufacturer terasic.com or from digikey.com (part# P0496-ND).

2.2 Lab Steps

The steps for Lab 2 are:

Step 1: **Read** Sect. 4 Introduction to the DE10-Nano Board (page 33) to familiarize yourself with the DE10-Nano board. Pay particular attention to the following:

R. K. Snider, *Advanced Digital System Design using SoC FPGAs*,
https://doi.org/10.1007/978-3-031-15416-4_17

- **Determine the DE10-Nano board revision that you have** by reading Sect. 4.1.1 Determining DE10-Nano Board Revision (page 34).
- **Become familiar with DE10-Nano resources** such as the DE10-Nano:
  - *User Manual*
  - *Getting Started Guide*

  See Sect. 4.1.2 DE10-Nano Information (page 34) on where these resources can be found.
- Read Sect. 4.1.3 DE10-Nano Cyclone V SoC FPGA (page 34) to understand the particular SoC FPGA device that is on the DE10-Nano board.

Step 2: **Read pages 24–26 of the DE10-Nano** *User Manual* (i.e., Section 3.6 Peripherals Connected to the FPGA in the User Manual) for information about the slide switches (SWs) and LEDs that you will be using. See Step 1 on how to get the DE10-Nano User Manual.

Step 3: **Install Quartus Prime Lite.** See Sect. 6.1.2 Download and Install Intel's Quartus Prime Lite (page 56) for installation instructions.

Step 4: **Become familiar with Quartus file types** by reading Sect. 6.1.3 Quartus File Types (page 56).

Step 5: **Know what Quartus training resources are available** by reading Sect. 6.1.6 Learning Quartus (page 60).

Step 6: In Windows, **create a `\lab2` project directory** and **copy the following three files** to this directory.

- **DE10_top_level.vhd** (click here for file)
- **DE10_nano.qsf** (click here for file)
- **DE10_nano.sdc** (click here for file)

Step 7: **Create a Quartus Project.** Open Quartus and select New Project Wizard, which goes through the following selections:

a. **Intro** (click Next)
b. **Directory, Name, Top Level Entity**
   i. Browse to your `\Lab2` project directory and set it as the Working Directory for the project.
   ii. Name of Project: Lab2
   iii. Name of Top Level Design Entity: DE10_Top_Level
   iv. Click Next
c. **Project Type**. Select *Empty Project* and click Next.
d. **Add Files**. Add the following files to the project:
   - DE10_Top_Level.vhd
   - DE10_Nano.sdc

   and Click Next
e. **Family, Device, and Board Settings**. Select the device **5CSEBA6U23I7** and click next. This is the Cyclone V SoC with 110K LEs that is on the DE10-Nano board.

Hint: To find this device, start typing 5CSEBA6U23I7 in the *Name* filter box (right side) or cut and past this device name and to narrow the device selections to this specific FPGA.
Note: To interpret the device labeling, see Sect. 4.1.3 DE10-Nano Cyclone V SoC FPGA (page 34).

f. **EDA Tool Settings**. Skip this by clicking Next.
g. **Click Finish**.

Step 8: **Load Pin Assignments**. We need to connect the signal names in the top level VHDL file with their FPGA pins. This has already been setup in the Quartus setting file that we need to load. In Quartus select, Assignments → Import Assignments → Browse to **DE10_Nano.qsf** → open → OK.

Step 9: **View Files**. In the Project Navigator bar, select the pull down menu found at the top of the left panel and select *files*. You should now see your .vhd and .sdc files listed. Double click on them to open them up in the editor window.

Step 10: **Write VHDL Code**. In the architecture section of DE10_Top_Level.vhd (starting around line 88), add VHDL code to connect the four switches (SW [3:0]) on the DE0-Nano-SoC board to the four LEDs (LED [3:0]). This can be done with one line of VHDL code. Add another VHDL statement to drive the rest of the LEDs that you are not using to zero. As a general rule, you should not leave output pins floating.

Step 11: **Compile Your Design**. Compile your design by selecting Processing → Start Compilation. There is also an icon to press on the tool bar or you can type Ctrl+L. If you get errors, you will need to fix them. You can ignore warning messages for now.

Step 12: **Check Timing**. Remember the two things that need to be correct for a correct design? (1) Your logic (VHDL) needs to be correct. (2) The timing needs to be correct. Can the bit stream that was just created actually run at the driven clock speeds? This should be checked after each compile. However, in this lab since we are just connecting Switches to LEDs and there is no clock involved, the Timing Analyzer has nothing to report. Practically, in all other designs, clocks will be used to capture values to be used the next clock cycle. To check timing, see Sect. 6.1.5 Timing (page 58).

Step 13: **Download the bit stream to the FPGA**.

a. Make sure that the USB cable is connected to the *USB Blaster* II port on the DE10-Nano board (left side below power jack).
b. Open the Programmer found in Quartus under Tools → Programmer. If the programmer says "No Hardware," click the Hardware Setup button, then in the currently selected hardware box, double click on DE-Soc [USB-1].
c. Click *Auto Detect*.
d. Select 5CSEBA6.

e. Click “Add File” and select the .sof file that was created when the project compiled as the programming file. The .sof file will be in the `\lab2\output_files` directory.
f. If there are three devices that show up on the JTAG chain, there should only be two: SOCVHPS and 5CSEBA6U23. If there are any extra blocks on the chain, right click it, select edit, and select delete. Your configuration should be like that in Figure 3–10 on Page 18 of the User Manual. See Step 1 on how to get the DE10-Nano User Manual.
g. Click “Start” to program the FPGA.

2.3 Demonstration

Print out and get Lab 2 **Instructor Verification Sheet signed off** (page 355) by demonstrating to the course TA the following:

Demo 1: Show that you can turn on and off the four LEDs [3:0] by sliding the switches.

Note 1: In Lab 2, it is known that the DE10-Nano board can reset randomly for some reason. This is likely because we are not fully configuring the FPGA since we are not configuring the HPS (ARM CPU system) and ignoring it at this time. It is possible there are HPS project settings not in place this is causing the instability. In any case, your switches should be able to drive the LEDs.
Note 2: What appears to solve the random resets in to remove the microSD card, which is not needed for the lab.

2.4 Deliverables

The **Deliverables** to be turned in or uploaded to the Lab 2 assignment folder are:

1: The Instructor Verification Sheet signed and turned in
2: Your **VHDL Code**

Instructor Verification Sheet
Lab # 2
Hardware Hello World

Print this page and get it signed by the Course Instructor or lab Teaching Assistant.
Make sure you turn in this signed page to get credit for your lab.

Name: ______________________________________

Verified by: ____________________ **Date:** ______________

Lab Demonstration. Demonstrate to the Lab TA the following:

Demo 1: Show that you can turn on and off the four LEDs [3:0] by sliding the switches.

Deliverables. Turn in or upload to your Lab 2 assignment folder the following items:

1: The Instructor Verification Sheet signed and turned in
2: Your **VHDL Code**

Lab 3
Developer's Setup

In Lab 3, you will configure your environment to allow the DE10-Nano board to boot from the Ubuntu VM that you created in Lab 1. This is called the **Developer's Setup** since it allows development without having to remove and modify the microSD card for every little modification. Instead, the DE10-Nano board will boot and use files served from the Ubuntu VM. This requires setting up two servers. The **TFTP server** and the **NFS server**. See Fig. 11.13 in Sect. 11.1.3 Developer's Boot Mode Setup (page 212) on how these servers will be used to allow the DE10-Nano board to boot over a local network.

3.1 Items Needed

The items needed for Lab 3 are:

- Item 1: A **Laptop** or PC
- Item 2: A USB Ethernet adapter that allows an **Ethernet Cable** to be plugged into it
- Item 3: The **Ubuntu VM** that was setup in Lab 1
- Item 4: **DE10-Nano Kit**
- Item 5: A **microSD Card** that is at least 8 GB. A new one or one you don't mind overwriting, but one that is different from the one that shipped with the DE10-Nano board. We will assume that the size is 16 GB for this lab.
- Item 6: A short **Ethernet Cable** to connect the DE10-Nano board to your USB Ethernet adapter

R. K. Snider, *Advanced Digital System Design using SoC FPGAs*,
https://doi.org/10.1007/978-3-031-15416-4_18

3.2 Lab Steps

The steps for Lab 3 are:

Step 1: **Understand how the Cyclone V SoC FPGA boots** by reading Sect. 3 Introduction to the SoC FPGA Boot Process (page 25).

Step 2: **Install PuTTY** by following the instructions in Sect. 11.1.1 Setting up a PuTTY Terminal Window in Windows to Communicate with Linux on the DE10-Nano Board (page 193).

Step 3: **Setup the TFTP and NFS servers** by following the instructions found in Sect. 11.1.3 Developer's Boot Mode Setup (page 212) from the start of the section through till the end of Sect. 11.1.3.5 NFS Server Setup in the Ubuntu VM (page 224).

Step 4: **Create a new microSD card image** by following the instructions found in Sect. 11.1.3.6 Reimaging the microSD Card with the Developer's Image (page 227)

Step 5: **Change DE10-Nano board switch settings** by following the instructions found in Sect. 11.1.3.7 DE10-Nano Board Switch Settings (page 231)

Step 6: **Set the appropriate U-boot Environment Variables** by following the instructions found in Sect. 11.1.3.8 Setting U-boot Variables on the DE10-Nano Board to Boot via NFS/TFTP (page 232)

Step 7: **Set Up the Audio Mini Passthrough Example** by following the instructions found in Sect. 11.1.3.9 Setting Up the Audio Mini Passthrough Example (page 234).

Step 8: **Configure and setup the Linaro GCC tools in Ubuntu VM (for cross compilation)** by following the instructions found in Sect. 11.1.4.1 Configure and Set up the Linaro GCC ARM Tools in Ubuntu VM (page 238)

Step 9: **Manually Cross Compile Hello World** by following the instructions found in Sect. 11.1.4.2 Manual Cross Compilation of Hello World (page 239).

Step 10: **Cross Compile the Hello World program using a Makefile** by following the instructions found in Sect. 11.1.4.3 Makefile Cross Compilation of Hello World (page 241). Take a screen shot and upload it to the D2L Lab 3 Assignment folder. The third line should say "Makefile Compilation."

3.3 Demonstration

Print out and get Lab 3 **Instructor Verification Sheet signed off** (page 360) by demonstrating to the course TA the following:

Demo 1: Demonstrate that the DE10-Nano board boots from your Ubuntu VM over Ethernet.

Demo 2: Demonstrate that you can create a new file in the Ubuntu VM in the directory: **/srv/nfs/de10nano/ubuntu-rootfs/root** (e.g., use the Linux command *touch <your_name>*) and show that this new file exists on the DE10-Nano board using the Putty Terminal Window.

Demo 3: Run the *Hello World* program on the **DE10-Nano board** that was compiled in the Ubuntu VM using a Makefile. It should print the following:

```
Hello ARM CPU World
My name is <first_name> <last_name>
Makefile Compilation
```

3.4 Deliverables

The **Deliverables** to be turned in or uploaded to the Lab 3 assignment folder are:

1: The Instructor Verification Sheet signed and turned in

2: A screenshot of the Hello World program output that ran on the ARM CPU (screenshot of the PuTTY terminal window) and that was compiled using a Makefile

Instructor Verification Sheet
Lab # 3
Developer's Setup

Print this page and get it signed by the Course Instructor or lab Teaching Assistant.
Make sure you turn in this signed page to get credit for your lab.

Name: ______________________________

Verified by: ____________________ **Date:** ______________

Lab Demonstration. Demonstrate to the Lab TA the following:

Demo 1: Demonstrate that the DE10-Nano board boots from your Ubuntu VM over Ethernet.
Demo 2: Demonstrate that you can create a new file in the Ubuntu VM in the directory: **`/srv/nfs/de10nano/ubuntu-rootfs/root`** (e.g., use the Linux command *touch* *<your_name>*) and show that this new file exists on the DE10-Nano board using the Putty Terminal Window.
Demo 3: Run the *Hello World* program on the **DE10-Nano board** that was compiled in the Ubuntu VM using a Makefile. It should print the following:

```
Hello ARM CPU World
My name is <first_name> <last_name>
Makefile Compilation
```

Deliverables. Turn in or upload to your Lab 3 assignment folder the following items:

1: The Instructor Verification Sheet signed and turned in
2: A screenshot of the Hello World program output that ran on the ARM CPU (screenshot of the PuTTY terminal window) and that was compiled using a Makefile

3.5 Common Problems and Solutions

Acknowledgement: Thanks to Trevor Vannoy for his help in troubleshooting student's labs.

3.5.1 Problem: Serial Connection to DE10-Nano in PuTTY (Or Similar) is Not Working

Solution 1: Check that your UART settings are correct:

- Baud rate: 115200
- 8 bits
- 1 stop bit
- No parity

Solution 2: Ensure you are plugged into the UART port, not the USB Blaster port.

Solution 3: Ensure you are using the correct COM port (using Device Manager in Windows or using dmesg in Linux).

3.5.2 Problem: DE10-Nano Cannot Connect to the Virtual Machine

Solution 1: Ensure the "cable connected" box is checked in your Virtual Box settings for the Bridged Adapter

Solution 2: Ensure your Ethernet cable is plugged in

Solution 3: Bring your network interfaces down and then up again

Solution 4: Disable the NAT connection and keep the Bridged connection

Solution 5: Ensure your Bridged Adapter is set to use your Ethernet interface, not your wireless interface

3.5.3 Problem: Virtual Machine Does Not Boot

Solution 1: If using a USB-to-Ethernet adapter, ensure the adapter is plugged into your computer. The Virtual Machine will not start if a device it expects does not exist.

Solution 2: If your USB-to-Ethernet adapter is plugged in, ensure the Virtual Box network settings are using the correct hardware device. Some adapters seem to show up as multiple devices in Virtual Box.

3.5.4 Problem: DE10-Nano Cannot Download Files from TFTP Server

Solution 1: Ensure your Ethernet cable is plugged in and your virtual machine is running.

Solution 2: Ensure your tftp server is running. Use `systemctl status tftpd-hpa` to see if it is running. If it is not running, use `systemctl start tftpd-hpa` to start the server.

Solution 3: Maker sure read permissions are set to allow everyone to read.

3.5.5 Problem: No Internet Connection in Virtual Machine

Solution 1: Ensure your bridged adapter is not using the same hardware device that your Windows machine is using for Internet.

Solution 2: Disable your bridged adapter.

3.5.6 Problem: DE10-Nano Fails to Mount the Rootfs Over NFS, Resulting in a Kernel Panic

Solution 1: Ensure the NFS server is running in your virtual machine:

- `systemctl status nfs-kernel-server`
- `systemctl start nfs-kernel-server`

Solution 2: Ensure that your ubunut-rootfs folder is exported by typing `sudo exportfs -v`. This should print out something containing "/srv/nfs/de10nano/ubuntu-rootfs." If this does not happen, there is either a syntax error in /etc/exports or you need to restart the NFS server with sudo systemctl restart nfs-kernel-server. If restarting the server does not work, then you probably have a syntax error. Check the syntax in the lab handout, and make sure you did not include a comment (#) at the beginning of the line.

Solution 3: Restart the NFS server: `systemctl restart nfs-kernel-server`. Sometimes one of the dependencies (found by systemctl list-dependencies nfs-kernel-server) fails to start. Restarting the NFS server may correct this. If restarting the NFS server does not correct the problem, then further debugging of the failed dependent service is necessary.

3.5.7 Problem: You Tried to Copy a File to a Directory, but the Directory Shows up as a File

Solution 1: You did not create the directory before copying the file, so the copy command renamed the file to what the directory should have been. Remove the copied file, create the directory, and then recopy the original file.

3.5.8 Problem: Copying a File Results in a Permission/Access Denied Error

Solution 1: You (probably) need to use sudo to copy the file. The exception to this is if you are trying to copy a file to a folder that you should have access to, like your home folder. If that is the case, you have a more serious permissions issue, probably because you used sudo when you should not have. If that is the case, you need to give yourself ownership of the directory in question by using chown.

3.5.9 Problem: Running a Makefile or Similar Results in a Permission/Access Denied Error

Solution 1: You need to give yourself ownership of the offending files/directories by using chown. You likely used sudo when you should not have, resulting in root being the owner of the offending files/directories. If you are trying to run a file that is in a system directory (e.g., pretty much anything that is not a subdirectory of your home folder), you should not give yourself ownership; you need to use sudo to run the file.

3.5.10 Problem: I Get an "altera_load: Failed with Error Code -4" When Booting Up My Board

Solution 1: Switch SW10 on the DE10-Nano board needs to be set to be consistent with the type of .rbf file being loaded. The lab is using passive parallel x16, which means all the switches need to be "on," to make the MSEL pins all zeros. The DE10-Nano board ships with it set to FPPx32, which is the wrong SW10 setting for this lab.

Lab 4
LED Patterns

For Lab 4, you will create a custom hardware component called **LED_patterns** that will create light patterns using the LEDs on the DE10-Nano board. This component will run in the FPGA fabric and you will create this component using VHDL. We will in a later lab connect this component to the ARM CPUs to control the LED patterns from software. Thus, for now, you can ignore the registers that are shown in Fig. 12.1. In this lab you will instantiate the component LED_patterns at the top level in your Quartus project and the register-related signals will be hard coded with appropriate values in the instantiation port map. Later we will create registers where these values can be changed from the ARM CPUs.

4.1 Items Needed

The items needed for Lab 4 are:

Item 1: A **Laptop** or PC
Item 2: **DE10-Nano Kit**
Item 3: **Quartus** software that was installed as described in Sect. 6.1.2
Item 4: A short **USB Cable** with a Mini-B connection to connect the DE10-Nano board to your computer. You will configure the FPGA fabric via the JTAG USB connection.

4.2 Lab Steps

The steps for Lab 4 are:

Step 1: **Read page 24 of the DE10-Nano** *Terasic User Manual* (i.e. Section 3.6.1 User push buttons in the Terasic User Manual) for information about the push buttons that you will be using. **Note:** Make sure that you

R. K. Snider, *Advanced Digital System Design using SoC FPGAs*,
https://doi.org/10.1007/978-3-031-15416-4_19

understand what happens to the push-button signal when it is pushed and released. If you do not, your state machine will not work since it will be held in reset.

Step 2: Write VHDL code for the LED_patterns component described in Sect. 12.1 LED_Patterns Component (page 246). You will need to use the entity given in Sect. 12.1.1 LED_Patterns Entity (page 246) and you will need to implement the *functional requirements* for the LED_Patterns component found in Sect. 12.1.2 Functional Requirements for LED_Patterns (page 248).

Step 3: You can use the Quartus project you created for the Hardware Hello World Lab and build on that project.

Step 4: Compile your Quartus project.

Step 5: Download the bitstream (.sof) file using the JTAG programmer to configure the FPGA fabric.

4.3 Demonstration

Print out and get Lab 4 **Instructor Verification Sheet signed off** (page 367) by demonstrating to the course TA the following:

Demo 1: Demonstrate the 4 states (i.e., 4 specified patterns) of the LED_control component on the DE10-Nano board.

Demo 2: Demonstrate your user defined LED pattern.

4.4 Deliverables

The **Deliverables** to be turned in or uploaded to the Lab 4 assignment folder are:

1: The Instructor Verification Sheet signed and turned in.

2: A ***Lab Report*** that includes:

   1: A description of the hardware architecture you created for the LED_patterns component

   2: A description of your user LED pattern that you implemented, along with the transition rate that you implemented for this pattern

3: The **VHDL code** of all your Lab 4 components that you created. Upload these VHDL files to the Lab 4 assignment folder. **Note:** Do not upload any of the Quartus project files, just the source code that you created or modified.

Instructor Verification Sheet
Lab # 4
LED Patterns

Print this page and get it signed by the Course Instructor or lab Teaching Assistant. Make sure you turn in this signed page to get credit for your lab.

Name: ____

Verified by: ______________________ **Date:** ______________

Lab Demonstration. Demonstrate to the Lab TA the following:

Demo 1: Demonstrate the 4 states (i.e., 4 specified patterns) of the LED_control component on the DE10-Nano board.
Demo 2: Demonstrate your user defined LED pattern.

Deliverables. Turn in or upload to your Lab 4 assignment folder the following items:

1: The Instructor Verification Sheet signed and turned in.
2: A *Lab Report* that includes:

    1: A description of the hardware architecture you created for the LED_patterns component
    2: A description of your user LED pattern that you implemented, along with the transition rate that you implemented for this pattern

3: The **VHDL code** of all your Lab 4 components that you created. Upload these VHDL files to the Lab 4 assignment folder. **Note:** Do not upload any of the Quartus project files, just the source code that you created or modified.

4.5 Common Problems and Solutions

4.5.1 Problem: Hardware State Machine Works When Reset Is Held Down

Solution 1: The polarity of your reset is probably inverted. The buttons on the board are active low. Ensure that you are either checking for an active low reset or preferably you inverted the active low reset to make it active high.

4.5.2 Problem: Hardware State Machine Only Displays the Switch State, Not the Patterns

Solution 1: The polarity of your push-button signal is probably inverted. The push button is active low on the DE10-Nano.

4.5.3 Problem: Nothing Works

Solution 1: Ensure your board is programmed with the correct .sof file.

Solution 2: Check the polarity of your reset signal. Your processes might be held in a reset state.

Solution 3: Check your values for sys_clk_sec and Base_rate. If either of these are 0, your state machine will not work because you have effectively eliminated the clock that drives the pattern generators.

Solution 4: Ensure that your top level entity name matches the one in the QSF file. If this is not the case, none of your pin assignments will work.

4.5.4 Problem: I Cannot Program the Board

Solution 1: Ensure you are plugged into the USB Blaster port on the DE10-Nano (left side). This port is the JTAG port, which is what gets used for programming.

Solution 2: Ensure your programmer is set up correctly. The JTAG chain in the programmer must contain the HPS and the FPGA fabric, in that order. The "program/configure" button must be checked in the FPGA device entry.

Lab 5
Signal Tap

Signal Tap is an embedded logic analyzer tool in Quartus that you can use to debug your FPGA designs while they run at speed in the FPGA fabric. If you compile your VHDL, download the bit stream into the FPGA and then find that it is not working correctly; how do you see what is going on inside the FPGA? You can see what is going on with your signals by using Signal Tap. We will use Signal Tap in this lab to probe your design from Lab 4.

5.1 Items Needed

The items needed for Lab 5 are:

Item 1: A **Laptop** or PC
Item 2: **DE10-Nano Kit**
Item 3: **Quartus** software that was installed as described in Sect. 6.1.2.
Item 4: A short **USB Cable** with a Mini-B connection to connect the DE10-Nano board to your computer. You will configure the FPGA fabric via the JTAG USB connection and Signal Tap will use this same connection.
Item 5: A completed Lab 4

As part of this lab, you will also learn how to make use of Intel's online training courses to learn about Signal Tap. Continuing education is important in order to keep up to date with fast changing technical topics such as using FPGAs in computer engineering. You should become familiar with how to learn new skills when you get a new job or start a new project. In computer engineering, this type of learning will never stop. If you do stop learning, you will quickly become obsolete. In this lab we will show you how to acquire a new debugging skill when you use Quartus, which is the use of Signal Tap. Intel provides online training for FPGA designers that we will use to learn about Signal Tap. Point your web browser to Intel's FPGA Technical Training (Fig. 5.1).

R. K. Snider, *Advanced Digital System Design using SoC FPGAs*,
https://doi.org/10.1007/978-3-031-15416-4_20

Fig. 5.1: Intel's FPGA Technical Training Website

Scroll down to the **FPGA Designers** section and click on **Level 200** under the Course Catalog heading. Select the **Course Type** to be **Online** from the drop down menu and select your appropriate language (e.g., English). Scroll down until you see the online courses for Signal Tap (you may need to go to the next page). A faster way to find these courses is to click on the **Catalog** tab and perform a search for ***signal*** since the spelling is not consistent in their titles for Signal Tap (the older online videos spell it Signal Tap).

Click on a course and select "Register Now" at the bottom of the page. Note: You will need to register and sign into the site.

5.2 Lab Steps

The steps for Lab 5 are:

Step 1: Watch at least the first two Intel training modules for Signal Tap:

- 1: **Signal Tap Logic Analyzer: Introduction and Getting Started** (search for course ODSW1164).
- 2: **Signal Tap Logic Analyzer: Basic Configuration and Trigger Conditions** (search for course ODSW1171).

Step 2: In Signal Tap, set the **sample clock** to be the fastest clock from your clock generator block from Lab 4 (e.g., the fastest *Clock for Pattern i* in Fig. 12.2). If you use the 50 MHz system clock, you will get too many samples and you will not be able to see the LED transitions.

Step 3: Use the push button that you use to change the state machine for the LED patterns as the Signal Tap **trigger**.

Step 4: In the Signal Tap Node list, add the following signals to monitor from your design:

- 1: The state machine **state signal** that is used to change between the LED pattern signals
- 2: The **LED signal** that goes out to the LEDs
- 3: The **output from all 5 of your pattern generators**:
  - 1: Output from Generator 0 (1 LED shifting right)
  - 2: Output from Generator 1 (2 LEDs shifting left)

3: Output from Generator 2 (up counter)
4: Output from Generator 3 (down counter)
5: Output from Generator 4 (your pattern)

Step 5: Run the design and transition into State 3 (down counter). This is the state you will transition from.
Step 6: Set the switches to 1 so that you will go to State 1 (2 LEDs shifting left).
Step 7: Trigger Signal Tap by pushing the push button and it will transition to State 1 where the switch values will be displayed for 1 s.
Step 8: Take a screen shot showing the waveform data of your monitored signals and where the output of Generator 4 (your pattern) is clearly visible.

5.3 Demonstration

Print out and get Lab 5 **Instructor Verification Sheet signed off** (page 372) by demonstrating to the course TA the following:

Demo 1: Demonstrate that you can use Signal Tap by showing the signals in Step 4: being monitored.

5.4 Deliverables

The Deliverables to be turned in or uploaded to Lab 5 assignment folder are:

1: The Instructor Verification Sheet signed and turned in
2: A screen shot of the waveform view of Signal Tap monitoring the signals in Step 4: and where your patterns is clearly visible
3: A screen shot of the signal configuration pane (where it shows the sample depth)
4: The answer to this question: How much FPGA on-chip memory was required to monitor your signals?

Instructor Verification Sheet
Lab # 5
Signal Tap

Print this page and get it signed by the Course Instructor or lab Teaching Assistant. Make sure you turn in this signed page to get credit for your lab.

Name: ______________________________

Verified by: ____________________ **Date:** __________

Lab Demonstration. Demonstrate to the Lab TA the following:

Demo 1: Demonstrate that you can use Signal Tap by showing the signals in Step 4: being monitored.

Deliverables. Turn in or upload to your Lab 5 assignment folder the following items:

1: The Instructor Verification Sheet signed and turned in
2: A screen shot of the waveform view of Signal Tap monitoring the signals in Step 4: and where your patterns is clearly visible
3: A screen shot of the signal configuration pane (where it shows the sample depth)
4: The answer to this question: How much FPGA on-chip memory was required to monitor your signals?

Lab 6
Creating a Custom Hardware Component in Platform Designer

In Lab 4, you created the **LED_patterns** component that created patterns on the DE10-Nano LEDs using state machines in the FPGA fabric. In this lab, you will create a new component **HPS_LED_patterns**, shown in Fig. 12.1, which will instantiate the LED_patterns component and add **registers** that will allow you to control the LEDs from software. This lab will show you how to use *Platform Designer*, a system integration tool in Quartus, to add custom hardware in the FPGA fabric that can be controlled from the ARM CPUs.

6.1 Items Needed

The items needed for Lab 6 are:

Item 1: A **Laptop** or PC.
Item 2: **DE10-Nano Kit**.
Item 3: **Quartus** software that was installed as described in Sect. 6.1.2.
Item 4: A short **USB Cable** with a Mini-B connection to connect the DE10-Nano board to your computer. You will configure the FPGA fabric via the JTAG USB connection.
Item 5: A completed Lab 4.

6.2 Lab Steps

In this lab, you will use a system that is known to work by modifying and compiling the **AudioMini_Passthrough** project. The preparation steps for using this project are:

R. K. Snider, *Advanced Digital System Design using SoC FPGAs*,
https://doi.org/10.1007/978-3-031-15416-4_21

Prep 1: **Copy the AudioMini_Passthrough Project**

1: Get the project from GitHub by clicking on the link DE10-Nano Projects.
2: Click on the green "Clone or download" button.
3: Select the "Download ZIP" option.
4: It will download *all* the projects, so just copy the folder `/AudioMini_Passthrough` to your Lab 6 location. (In the downloaded zip file *de10nano_projects-dev.zip*, the folder is located at: `/de10nano_projects-dev/AudioMini_Passthrough`.)

Prep 2: **Modify the project using Platform Designer**

1: In Quartus, open the project file **DE10Nano_System.qpf** that is contained in `/AudioMini_Passthrough`. Ignore the warning that soc_system.qip is not found because we will recreate this file using Platform Designer.
2: Within Quartus, from the Tools menu, open **Platform Designer** and open the file **soc_system.qsys**.
3: Platform Designer will show a red Error message that states *FE_Qsys_AD1939_Audio_Mini_v1 is not found*. We will fix this by deleting this component since we will not be using it in this lab.
4: In Platform Designer and in the System Contents view, scroll down till you see the FE_Qsys_AD1939_Audio_Mini_v1 component, click on the name, and delete it (or right click and select remove). The error message should go away.
5: Regenerate the system by clicking on the "Generate HDL..." button at bottom right.
    1: In the window that pops up, select VHDL for the Synthesis language.
    2: Click Generate.
    3: Click Close, and the system will be generated (it will take a few minutes) and click close again.
    4: This generates the soc_system.qip file in the project directory under `\soc_system\sythesis` that was missing earlier.

Prep 3: **Modify the top-level VHDL file**

1: In Quartus, select *Files* in the Project Navigator pull down menu.
2: Open the top-level file **DE10Nano_System.vhd** by double clicking on it in the Files menu. **Note:** This is the new top-level file that you will be using from now on, and it is different from what you have used in the earlier labs because it now contains the soc_system component that is connected to external DRAM.
3: We need to remove the signals from the **soc_system** component declaration and instantiation that no longer exist since we deleted the AD1939 component.

1: In the DE10Nano_System.vhd file, search (Ctrl-F) for the lines that contain the prefix "**ad1939_physical**." There will be four lines in the component declaration (lines 213–216 or nearby) and four lines in the instantiation port map (lines 393–396 or nearby) that you need to delete.
2: Save your changes and compile the system. Your system should compile with no errors. **Note 1:** On a computer with a 6 core 3.4 GHz i7 with 64 GB of RAM, this took 6 minutes 29 seconds. On an AMD Ryzen 7 (8 core 3.6 GHz) and 64 GB of RAM, this took 7 minutes 45 seconds (slower because of slower hard disk I/O). **Note 2:** Quartus compilations perform many disk I/O operations, so a SSD in your computer will make compilations significantly faster.

4: **Note:** This is the system that you will now add your custom component HPS_LED_patterns to.

Step 1: Create VHDL code for the HPS_LED_patterns component. Read Sect. 6.2 Platform Designer (page 62) on how to create registers for your custom component.
Step 2: Add this component in Platform Designer. Read Sect. 6.2.2 Creating a Custom Platform Designer Component (page 65) on how to do this.

6.3 Demonstration

Print out and get the Lab 6 **Instructor Verification Sheet signed off** (page 376) by demonstrating to the course TA the following:

Demo 1: Demonstrate that the behavior of the LED patterns is similar to Lab 4.

Note 1: The demonstration just requires downloading the configuration bit stream through JTAG and showing that the hardware component runs.
Note 2: We will test the registers in the next lab.

6.4 Deliverables

The **Deliverables** to be turned in or uploaded to the Lab 6 assignment folder are:

1: The Instructor Verification Sheet is signed and turned in.
2: The **VHDL code** of the HPS_LED_patterns component that you created.
3: The **.tcl** file that Platform Designer created when you imported HPS_LED_patterns.
4: The answer to this question: What is the base address of your HPS_LED_patterns custom component in your Platform Designer system?

Instructor Verification Sheet
Lab # 6
Creating a Custom Hardware Component in Platform Designer

Print this page and get it signed by the Course Instructor or lab Teaching Assistant. Make sure you turn in this signed page to get credit for your lab.

Name: ______________________________

Verified by: ____________________ **Date:** ______________

Lab Demonstration. Demonstrate to the Lab TA the following:

Demo 1: Demonstrate that the behavior of the LED patterns is similar to Lab 4.

Deliverables. Turn in or upload to your Lab 6 assignment folder the following items:

1: The Instructor Verification Sheet is signed and turned in.
2: The **VHDL code** of the HPS_LED_patterns component that you created.
3: The **.tcl** file that Platform Designer created when you imported HPS_LED_patterns.
4: The answer to this question: What is the base address of your HPS_LED_patterns custom component in your Platform Designer system?

6.5 Common Problems and Solutions

6.5.1 Problem: Error Generating System with Platform Designer or Compiling the System in Quartus

This includes errors when compiling in Quartus where Quartus does not seem to be able to compile its own Platform Designer components.

Solution 1: Newer versions of Quartus require Windows Subsystem for Linux (WSL) to be installed. You need to use WSL 1. See the Quartus installation guide and the official WSL installation instructions.

Solution 2: Rerun the WSL commands as outlined in Intel's FPGA Knowledge Base Article 000080350.

Solution 3: After the WSL updates, rerun the system generation in Platform Designer, i.e., click the "Generate HDL. . . " button.

Solution 4: If you keep getting errors, use Quartus 18.1 that does not require WSL.

Solution 5: Make sure the soc_system.qip file has been added to your project (located in the project folder at `./soc_system/synthesis/soc_system.quip`).

Lab 7
Verifying Your Custom Component Using System Console and /dev/mem

In Lab 6, you created the custom Platform Designer component HPS_LED_patterns by developing an Avalon bus wrapper for the LED_patterns component you created from Lab 4. Registers were added via a simple Avalon memory-mapped interface in order for the component to be controlled by the HPS lightweight bus. This will allow the ARM CPUs to control the component via software.

We want to make sure that our hardware is alive and well before we start developing software to interact with it. If we push ahead and start writing software before we know our hardware is operating correctly, and it does not work, where is the problem? Is it our software or hardware? We will use **System Console**, which is a tool in Quartus, to make sure that the registers we created in Lab 6 function as expected before we start layering on software.

In order to use System Console, a new Platform Designer component needs to be added to your design, which is the **JTAG to Avalon Master Bridge** component. This is a component that Platform Designer connects to JTAG on one end so that Quartus can access it through the JTAG interface via System Console. This JTAG interface is the one you use on the DE10-Nano board when you download your bitstream to the FPGA fabric. The other end of the *JTAG to Avalon Master Bridge* component is a bus master that can initiate bus transactions. By connecting the bus master interface to the memory-mapped interface of your custom component in Platform Designer, you can access and test the registers just like the ARM CPUs would do, but without needing the HPS to do so.

Once the HPS_LED_patterns component has been verified using System Console, which shows that you can read and write the component's registers, we will take off our hardware development hat, i.e., our "hard hat," and put on our software development hat, i.e., our "soft hat." We will do this from our C code running on Linux that uses the character device file `/dev/mem`, which allows access to physical memory. This will require root access, which is fine for our testing purposes, but cannot be used to provide user space access to physical memory. Getting access to our registers from user space will require developing a Linux device driver, which will be done in Lab 11.

R. K. Snider, *Advanced Digital System Design using SoC FPGAs*,
https://doi.org/10.1007/978-3-031-15416-4_22

We will use `/dev/mem` to test the *software control mode* of your HPS_LED_Patterns hardware component. In Linux, `/dev/mem` is a character device file whose byte addresses are interpreted as **physical memory addresses**. You already know the physical address of your registers because you created them and placed them in memory with Platform Designer. Thus you can directly write to the **LED_reg** register to create patterns on the LEDs from software.

Note: In order to use `/dev/mem`, you need to be *root* when you run the software. This approach is for testing purposes since you cannot access `/dev/mem` from user space. We will later write a device driver that will allow programs running in user space to access our hardware.

Finally, be aware that there is a difference in memory addressing between the *JTAG to Avalon Master Bridge* (see Sect. 7.2 The View of Memory from System Console (page 89)) and the ARM CPUs (see Sect. 7.1.1 Memory Addressing for Registers on the HPS Lightweight Bus (page 88)), but we will get into this shortly.

7.1 Items Needed

The items needed for Lab 7 are:

Item 1: A **Laptop** or PC.
Item 2: **DE10-Nano Kit**.
Item 3: **Quartus** software that was installed as described in Sect. 6.1.2.
Item 4: A short **USB Cable** with a Mini-B connection to connect the DE10-Nano board to your computer. You will configure the FPGA fabric via the JTAG USB connection and then connect through it using System Console.
Item 5: A completed **Lab 6**.

7.2 Lab Steps

The steps for Lab 7 are:

Step 1: Using **Platform Designer**, modify your design from Lab 6 by adding a **JTAG to Avalon Master Bridge** component. Read Sect. 6.3 System Console (page 69) through Sect. 6.3.2 Modifying the Design in Platform Designer (page 71) on how to add this component.
Step 2: Understand how to calculate **register memory addresses** in System Console by reading Sect. 7.2 The View of Memory from System Console (page 89).
Step 3: Using **System Console**, read and write to registers in your **HPS_LED_patterns** custom component. Read how to do this in Sect. 6.3.3 Using System Console (page 72).

Step 4: Perform **System Console verification** of your custom component by doing the following:

1: **Write a 1 to Register 0** (HPS_LED_control) to put the component in software control mode.
2: **Write a value to Register 2** (LED_reg). The least significant bits of the value should show up on the LEDs.
3: **Write another bit pattern to Register 2**. The LEDs should reflect this new pattern.

Step 5: **Know how to calculate the physical addresses of your custom component's registers** that are connected to the HPS lightweight bus by reading Sect. 7.1.1 Memory Addressing for Registers on the HPS Lightweight Bus (page 88).

Step 6: Copy the file **devmem2.c** (click here for the source file) to your Lab 7 directory and rename it **mydevmem.c**. A couple of notes regarding devmem2.c:

Note 1: On line 74 of *devmem2*, **mmap()** is used to create the physical address mapping. The last argument (target & ~MAP_MASK) is done so that mmap() maps a full page on a page boundary that contains the physical memory mapping request.

Note 2: On line 79 of *devmem2*, it adds the appropriate offset back in when it accesses the address using the page boundary of the mapped page.

Step 7: **Cross Compile** *mydevmem.c* as you did in Sect. 11.1.4 Cross Compiling "Hello World" (page 238) and put it in the root file system served by the NFS server.

Step 8: Use *mydevmem.c* to write the same values to the same registers as you did using System Console in Step 4:.

Step 9: *devmem2* is not strictly portable because it hardcodes the page size to 4096 rather than using the system call sysconf(_SC_PAGE_SIZE) to determine what the page size is. Modify *mydevmem.c* to use sysconf(_SC_PAGE_SIZE) to determine what the page size is to make the code portable. (click here for the getpagesize and sysconf reference).

Step 10: **Cross Compile** and **test** your modified *mydevmem.c* program by doing the following:

1: **Write a new value to Register 1** (SYS_CLKS_sec) that doubles the LED pattern speeds.
2: **Write a new value to Register 3** (Base_rate) that changes the doubled LED pattern's speed back to the original speed (keeping Register 1 with the value that doubles the speed).

7.3 Demonstration

Print out and get the Lab 7 **Instructor Verification Sheet signed off** (page 383) by demonstrating to the course TA the following:

Demo 1: In **System Console**, change the value in Register 0 by first reading it, then writing a new value, and then reading this new value back.
Demo 2: In **System Console**, write values to Register 2 and show that these values (least significant bits) are shown up on the LEDs.
Demo 3: **Using mydevmem.c**, your modified devmem2.c program, show that you can double a pattern's speed by writing to Register 1 (SYS_CLKS_sec).
Demo 4: **Using mydevmem.c**, your modified devmem2.c program, show that you can put the pattern's speed back to its original rate by writing to Register 3 (Base_rate) (and keeping Register 1 with the value that doubles the speed).

7.4 Deliverables

The Deliverables to be turned in or uploaded to the Lab 7 assignment folder are:

1: The Instructor Verification Sheet is signed and turned in.
2: The Platform Designer system (.qsys) file.
3: Your modified *mydevmem.c* code.
4: The answer to these questions:
    1: What value did you write to Register 1 (SYS_CLKS_sec) that doubled the speed of the patterns?
    2: What value did you write to Register 3 (Base_rate) that changed the doubled pattern's speed back to the original speed (keeping Register 1 with the value that doubles the speed). Give this value that you wrote to Register 3 in hexadecimal.

Instructor Verification Sheet
Lab # 7

Verifying Your Custom Component Using System Console and /dev/mem

Print this page and get it signed by the Course Instructor or lab Teaching Assistant. Make sure you turn in this signed page to get credit for your lab.

Name: ____

Verified by: ____________________ **Date:** ______________

Lab Demonstration. Demonstrate to the Lab TA the following:

Demo 1: In **System Console**, change the value in Register 0 by first reading it, then writing a new value, and then reading this new value back.

Demo 2: In **System Console**, write values to Register 2 and show that these values (least significant bits) are shown up on the LEDs.

Demo 3: **Using mydevmem.c**, your modified devmem2.c program, show that you can double a pattern's speed by writing to Register 1 (SYS_CLKS_sec).

Demo 4: **Using mydevmem.c**, your modified devmem2.c program, show that you can put the pattern's speed back to its original rate by writing to Register 3 (Base_rate) (and keeping Register 1 with the value that doubles the speed).

Deliverables. Turn in or upload to your Lab 7 assignment folder the following items:

1: The Instructor Verification Sheet is signed and turned in.
2: The Platform Designer system (.qsys) file.
3: Your modified *mydevmem.c* code.
4: The answer to these questions:
    1: What value did you write to Register 1 (SYS_CLKS_sec) that doubled the speed of the patterns?
    2: What value did you write to Register 3 (Base_rate) that changed the doubled pattern's speed back to the original speed (keeping Register 1 with the value that doubles the speed). Give this value that you wrote to Register 3 in hexadecimal.

7.5 Common Problems and Solutions

7.5.1 Problem: Hardware State Machine Keeps Running the LEDs After Setting HPS_LED_control to a 1 (Software Control Mode)

Solution 1: Ensure there is an **if** statement in your VHDL code to switch between hardware and software control of the LEDs.

7.5.2 Problem: The Values Written to Registers Do Not Show Up

Solution 1: Ensure that the base address and offset you are writing to are correct.
Solution 2: Ensure that you created a read process in your VHDL code.

7.5.3 No JTAG Masters Show Up in System Console

Solution 1: Ensure you are connected to the USB Blaster port, not to the UART port.
Solution 2: Click "Load Design" to tell System Console which project you are using, and then click "Refresh Connections" to make System Console scan for JTAG connections.
Solution 3: Ensure that you have programmed your board with the latest .sof file.
Solution 4: Ensure that your Avalon memory-mapped interface is connected to an Avalon-to-JTAG-master component in Platform Designer.
Solution 5: Make sure that you download the bitstream to configure the FPGA fabric **after** the Linux kernel boots on power up or after a reset. Otherwise, the .rbf file specified in the bootscript will be loaded from the VM, overwriting the bitstream you downloaded, which will not have the hardware you expect.

7.5.4 Downloading the Bitstream When the Programmer Fails

Solution 1: Make sure that you are not holding the system in reset (a push button polarity issue).

Lab 8
Creating LED Patterns with a C Program Using /dev/mem in Linux

In Lab 7, we tested the HPS_LED_patterns component by using System Console to read and write the component's registers. Using System Console is part of the SoC FPGA developer's toolbox to make sure your hardware system is functioning without having to worry about software. We then performed a simple register test using `/dev/mem` and devmem2 to read and write the LED_reg register to verify that we can access the register from the ARM CPUs over the lightweight bus.

In this lab we will take off our hardware development hat, i.e., our "hard hat," and put on our software development hat, i.e., our "soft hat." We will still use the `/dev/mem` file to access physical memory, which requires having root access, but we write a C program that will generate LED patterns.

8.1 C Program Requirements

The C program you need to write for Lab 8 that generates LED patterns has the following requirements:

Requirement 1: Name the program **myLEDpatterns.c**.
Requirement 2: It needs to be able to access physical memory using `/dev/mem` to write to your HPS_LED_patterns component, so you can use your program mydevmem.c from Lab 7 as a starting point on how to use `/dev/mem`.
Requirement 3: It needs to process the following **Command Line Arguments**:

Arg 1: **-h** for **help**. Typing **-h** will list what all the command line arguments can be and how to use them.
Arg 2: **-v** for **verbose**. Typing **-v** will print what the LED pattern is as a binary string and how long it is being displayed for as the patterns are being written to the LEDs. For example:

R. K. Snider, *Advanced Digital System Design using SoC FPGAs*,
https://doi.org/10.1007/978-3-031-15416-4_23

LED pattern = 01010101 Display time = 500 msec
LED pattern = 00001111 Display time = 1500 msec

Arg 3: **-p** for **pattern**. Typing **-p** and the associated arguments **pattern1 time1 pattern2 time2 pattern3 time3**, etc. will display these patterns for the time specified and will loop indefinitely until **Ctrl-C** is entered.

Note 1: The pattern values should be entered as hexadecimal values, e.g., 0x0f.

Note 2: The interval values should be entered as milliseconds, e.g., 1500.

Note 3: This is a variable length argument case where any number of pattern/time pairs could be entered.
Example 1: myLEDpatterns -p 0x55 500 0x0f 1500
Example 2: myLEDpatterns -p 0x55 500 0x0f 1500 0xf0 2000

Note 4: You can assume that **-p** is the last command line argument flag given, so you can then process the variable number of pattern/time pairs last and not have to worry about any subsequent flags. This will simplify your command line argument processing. In order to handle the variable number of arguments for myLEDpatterns.c, you can use the **getopt()** function, which can be found here. An example of using getopt to get multiple arguments can be found here. For those more adventuresome when it comes to argument parsing, you might want to consider (Approach 1) or (Approach 2) or (Approach 3). The approach you take for argument parsing is up to you.

Note 5: You can implement breaking out of your C program using Ctrl-C by looking at this example. Other signals are given here.

Note 6: In order to handle the display time, you can use sleep() or usleep() .

Arg 4: **-f** for **file**. Typing **-f** followed by a filename (e.g., mypatterns.txt) will open a text file and will read the patterns and times from this text file. The patterns text file will have a pattern and time on each row:
<pattern1> <time1>
<pattern2> <time2>
<pattern3> <time3>
where <pattern> is a hexadecimal value (e.g., 0xff) and <time> is an unsigned integer with units of milliseconds (e.g., 1500). The program will read to the end of the file,

displaying the LED patterns for the time listed on each row, and when it reaches the end of the file, it will terminate.
Example: myLEDpatterns -f mypatterns.txt
This will read the file mypatterns.txt, which for example could have the following contents:
0x55 500
0x0f 1500
0xf0 2000

Note 1: You can assume that **-f** is the last command line argument flag given, so you can then process the filename and not have to worry about any subsequent flags.

Requirement 4: The program needs to **return an error** for the following conditions:

Error 1: Both **-p** and **-f** are given as arguments. These are mutually exclusive, so if both are given, the program terminates saying it can be one or the other, but not both.
Error 2: If the number of arguments following the **-p** flag is odd, the program terminates saying that each pattern value should be followed by a time value.

8.2 Items Needed

The items needed for Lab 8 are:

Item 1: A **Laptop** or PC
Item 2: **DE10-Nano Kit**
Item 3: **Quartus** software that was installed as described in Sect. 6.1.2
Item 4: A completed **Lab 7**

8.3 Lab Steps

The steps for Lab 8 are:

Step 1: In Lab 7, you programmed the FPGA fabric using the Quartus programmer where it used the .sof file. We do not want to keep programming the FPGA fabric every time the DE10_Nano powers up, so let us have U-boot do this for us now since it is loading a .rbf file anyway. We just need to have U-boot load the .rbf file we want since we are done with creating our custom hardware in the FPGA fabric. To do this, you will need to:

1: **Convert the .sof file to a .rbf file**. See Sect. 6.1.4 Converting Programming Files (page 58) on how to perform this conversion.
2: Rename the .rbf file to **soc_system.rbf** since this is the name of the .rbf that U-boot is loading after it reads the bootscript in the Ubuntu VM. Later, we will learn how to customize bootscripts and U-boot to reflect new projects. For now, we will just overwrite the .rbf file that U-boot is currently using.
3: Copy soc_system.rbf to `/srv/tftp/de10nano/audiomini_passthrough` in the Ubuntu VM so that U-boot can load it. You will be overwriting the .rbf of the same name in this directory.

Step 2: As in Lab 7, **know how to calculate the physical addresses of your custom component's registers** that are connected to the HPS lightweight bus by reading Sect. 7.1.1 Memory Addressing for Registers on the HPS Lightweight Bus (page 88).
Step 3: **Write code for myLEDpatterns.c** using the requirements given in Sect. 8.1.
Step 4: **Cross Compile** your myLEDpatterns C program as you did in Sect. 11.1.4 Cross Compiling "Hello World" (page 238) and put it in the root file system served by the NFS server.
Step 5: **Run** the myLEDpatterns program on your DE10-Nano board.

8.4 Demonstration

Print out and get Lab 8 **Instructor Verification Sheet signed off** (page 389) by demonstrating to the course TA the following:

Demo 1: Show the contents of *mypatterns.txt* (e.g., cat mypatterns.txt), and then run *myLEDpatterns -f mypatterns.txt*
Demo 2: Run it again, but with the added **-v** flag, i.e., *myLEDpatterns -v -f mypatterns.txt* that will print out what is happening.
Demo 3: Run *myLEDpatterns* with the **-p** flag and with *three* pattern/time pairs. Show that it runs in a loop until you enter Ctrl-C.

8.5 Deliverables

The **Deliverables** to be turned in or uploaded to the Lab 8 assignment folder are:

1: The Instructor Verification Sheet signed and turned in
2: Your **well commented** C program **myLEDpatterns.c**
3: Your LED patterns file **mypatterns.txt**

Instructor Verification Sheet
Lab # 8
Creating LED Patterns with a C Program Using /dev/mem in Linux

Print this page and get it signed by the Course Instructor or lab Teaching Assistant. Make sure you turn in this signed page to get credit for your lab.

Name: ____

Verified by: ____________________ **Date:** ______________

Lab Demonstration. Demonstrate to the Lab TA the following:

Demo 1: Show the contents of *mypatterns.txt* (e.g., cat mypatterns.txt), and then run *myLEDpatterns -f mypatterns.txt*

Demo 2: Run it again, but with the added **-v** flag, i.e., *myLEDpatterns -v -f mypatterns.txt* that will print out what is happening.

Demo 3: Run *myLEDpatterns* with the **-p** flag and with *three* pattern/time pairs. Show that it runs in a loop until you enter Ctrl-C.

Deliverables. Turn in or upload to your Lab 8 assignment folder the following items:

1: The Instructor Verification Sheet signed and turned in

2: Your **well commented** C program **myLEDpatterns.c**

3: Your LED patterns file **mypatterns.txt**

Lab 9
Linux Kernel Module Hello World

In Lab 8, you interacted with the HPS_LED_patterns component by using `/dev/mem` to access physical memory in Linux in order to test the software control mode of your HPS_LED_Patterns hardware component. However, this required you to have *root* access to run the software. We will now start the process of creating a device driver that will be the portal between user space and the kernel, which can directly access physical memory. This will allow you to access your custom component registers from user space where root access is not required.

This lab will introduce you to the "Hello World" kernel module, the process of cross compiling the module and inserting the kernel module into Linux running on the DE10-Nano board. However, it will not do anything useful yet like controlling your registers since the actual device driver for the HPS_LED_patterns component will be done in Lab 11.

9.1 Items Needed

The items needed for Lab 9 are:

Item 1: A **Laptop** or PC
Item 2: **DE10-Nano Kit**
Item 3: A completed **Lab 8**

9.2 Lab Steps

The steps for Lab 9 are:

Step 1: Cross Compile the Linux Kernel. Do this by following the instructions in Sect. 9.2 Cross Compiling the Linux Kernel (page 129), which includes setting up to run the new zImage on the DE10-Nano board.

R. K. Snider, *Advanced Digital System Design using SoC FPGAs*,
https://doi.org/10.1007/978-3-031-15416-4_24

Step 2: **Read the sections on Device Drivers and Loadable Kernel Modules**. (Read Sect. 9.3 Kernel Modules (page 135) through Sect. 9.3.1 Loadable Kernel Modules (page 136).)
Step 3: **Create a `~/labs/lab9` folder**.
Step 4: **Download the example "Hello World" kernel model source file** (found here). Modify the *MODULE_AUTHOR()* macro and put in your first and last names.
Step 5: **Download the associated Makefile** (found here). Modify the Makefile so that it points to where the kernel source is located (`/<path>/linux-socfpga`).
Step 6: **Cross Compile the *Hello World* Kernel Module**. Do this by following the instructions in Sect. 9.3.2 Cross Compiling the Kernel Module (page 139).
Step 7: **Insert and Remove the Kernel Module**. See Sect. 9.3.3 Inserting the Kernel Module into the Linux Kernel (page 141).

9.3 Demonstration

Print out and get Lab 9 **Instructor Verification Sheet signed off** (page 393) by demonstrating to the course TA the following:

Demo 1: Show the information related to your LKM using the **modinfo** command. Your name should be present.
Demo 2: When the LKM is inserted, show the output of the LKM initialization printk statement using the **dmesg** and **grep** commands.
Demo 3: When the LKM is removed, show the output of the LKM exit printk statement using the **dmesg** and **grep** commands.

9.4 Deliverables

The **Deliverables** to be turned in or uploaded to the Lab 9 assignment folder are:

1: The Instructor Verification Sheet signed and turned in
2: Your LKM source **hello_kernel_module.c**
3: The associated **Makefile**
4: The text file **mymodinfo.txt** that was created by piping modinfo into `mymodinfo.txt`

Instructor Verification Sheet
Lab # 9
Linux Kernel Module Hello World

Print this page and get it signed by the Course Instructor or lab Teaching Assistant. Make sure you turn in this signed page to get credit for your lab.

Name: ____

Verified by: ____________________ **Date:** _____________

Lab Demonstration. Demonstrate to the Lab TA the following:

Demo 1: Show the information related to your LKM using the **modinfo** command. Your name should be present.
Demo 2: When the LKM is inserted, show the output of the LKM initialization printk statement using the **dmesg** and **grep** commands.
Demo 3: When the LKM is removed, show the output of the LKM exit printk statement using the **dmesg** and **grep** commands.

Deliverables. Turn in or upload to your Lab 9 assignment folder the following items:

1: The Instructor Verification Sheet signed and turned in
2: Your LKM source **hello_kernel_module.c**
3: The associated **Makefile**
4: The text file **mymodinfo.txt** that was created by piping modinfo into `mymodinfo.txt`

Lab 10
Modifying the Linux Device Tree

In Lab 9, you created and inserted a simple loadable kernel module (LKM). This was possible because the Hello World LKM did not interact with hardware. In order for the Linux kernel to know what hardware it is running on, it must be informed of the hardware by the compiled device tree **.dtb** file, which is loaded as part of the boot process (see Sect. 3.1.5 Boot Step 5: Linux (page 30)). This lab will introduce you to device trees and you will create a device tree for your custom DE10-Nano system.

10.1 Items Needed

The items needed for Lab 10 are:

Item 1: A **Laptop** or PC
Item 2: **DE10-Nano Kit**
Item 3: A completed **Lab 9**

10.2 Lab Steps

The steps for Lab 10 are:

Step 1: **Read about Device Trees**. (Read Sect. 9.4 Device Trees (page 141) through Sect. 9.4.2 Device Tree Hierarchy (page 145).)
Step 2: **Create your Device Tree** by following the instructions in Sect. 9.4.3 Creating a Device Tree for Our DE10-Nano System (page 146).
In Step 2: of the section and in the include file that you created (`socfpga_cyclone5_de0.dtsi`), modify the *leds* node from:

R. K. Snider, *Advanced Digital System Design using SoC FPGAs*,
https://doi.org/10.1007/978-3-031-15416-4_25

```
    leds {
       compatible = "gpio-leds";
       hps0 {
          label = "hps_led0";
          gpios = <&portb 24 0>;
          linux,default-trigger = "heartbeat";
       };
    };
```

to:

```
    leds {
      compatible = "gpio-leds";
      hps0 {
          color = <LED_COLOR_ID_GREEN>;
          function = LED_FUNCTION_HEARTBEAT;
          gpios = <&portb 24 0>;
          linux,default-trigger = "timer";
          led-pattern = <500 250>;
      };
```

You will also need to add the include file as shown in Line 7:

```
6 #include "socfpga_cyclone5.dtsi"
7 #include <dt-bindings/leds/common.h>
```

In Step 3: of the section, when you download the example device tree source file, **modify the example .dts file by adding a node for your HPS_LED_patterns component** that you created in Platform Designer.

Step 3: Compile, move the .dtb file to the TFTP server, and boot your DE10-Nano board.

10.3 Demonstration

Print out and get Lab 10 **Instructor Verification Sheet signed off** (page 398) by demonstrating to the course TA the following:

Demo 1: Show that you can change the blinking LED times on the DE10-Nano board by changing to the directory `/sys/class/leds/hps0` and writing values to the `delay_on` and `delay_off` files (times are in milliseconds). For example:
`echo 1000 > delay_on`
`echo 100 > delay_off`

Demo 2: Show that your *HPS_LED_patterns* component shows up on the DE10-Nano board and in the directory `/proc/device-tree`. Go into the component directory and cat the compatible file that will display the compatible string associated with this hardware node.

10.4 Deliverables

The **Deliverables** to be turned in or uploaded to the Lab 10 assignment folder are:

1: The Instructor Verification Sheet signed and turned in
2: Your include file **socfpga_cyclone5_de0.dtsi**
3: Your device tree source file **socfpga_cyclone5_de10nano_mysystem.dts**
4: A screen shot that shows that your HPS_LED_patterns hardware shows up in `/proc/device-tree` along with the output of cat'ing the compatible file that shows the compatible string that contains your initials.

Instructor Verification Sheet
Lab # 10
Modifying the Linux Device Tree

Print this page and get it signed by the Course Instructor or lab Teaching Assistant. Make sure you turn in this signed page to get credit for your lab.

Name: ______________________________

Verified by: ____________________ **Date:** ______________

Lab Demonstration. Demonstrate to the Lab TA the following:

Demo 1: Show that you can change the blinking LED times on the DE10-Nano board by changing to the directory `/sys/class/leds/hps0` and writing values to the `delay_on` and `delay_off` files (times are in milliseconds). For example:
`echo 1000 > delay_on`
`echo 100 > delay_off`

Demo 2: Show that your *HPS_LED_patterns* component shows up on the DE10-Nano board and in the directory `/proc/device-tree`. Go into the component directory and cat the compatible file that will display the compatible string associated with this hardware node.

Deliverables. Turn in or upload to your Lab 10 assignment folder the following items:

1: The Instructor Verification Sheet signed and turned in
2: Your include file **socfpga_cyclone5_de0.dtsi**
3: Your device tree source file **socfpga_cyclone5_de10nano_mysystem.dts**
4: A screen shot that shows that your HPS_LED_patterns hardware shows up in `/proc/device-tree` along with the output of cat'ing the compatible file that shows the compatible string that contains your initials.

Lab 11
Creating a Platform Device Driver for the HPS_LED_Patterns Component

In Lab 10, you created a node in the Linux device tree for the HPS_LEDS_patterns component. We needed to do this, otherwise Linux would not know that this hardware existed. In this lab you will create a Linux platform device driver for HPS_LEDS_patterns that will allow you to read and write the component registers from user space. These registers will show up as files in Linux, which provides the user space access.

11.1 Items Needed

The items needed for Lab 11 are:

Item 1: A **Laptop** or PC
Item 2: **DE10-Nano Kit**
Item 3: A completed **Lab 10**

11.2 Lab Steps

The steps for Lab 11 are:

Step 1: **Read about Platform Device Drivers**. (Read Sect. 9.5 Platform Device Driver (page 148).)
Step 2: **Create your Platform Driver** by following the instructions in Sect. 9.6 Steps for Creating a Platform Device Driver for Your Custom Component in Platform Designer (page 150). The code provided only implements two of the four registers. Complete the driver so that all registers are exported to sysfs. **Note:** In the instructions, it says to perform a string

R. K. Snider, *Advanced Digital System Design using SoC FPGAs*,
https://doi.org/10.1007/978-3-031-15416-4_26

replacement to change instances of *hps_led_patterns* to *my_component*, which would be the process for using the Platform Driver code as a starting point for a new component. However, in this lab, the component name is still *hps_led_patterns*, so skip the string replacement step in this case.

Step 3: **Test your Platform Driver using a bash script:** The example below only uses one register. Modify the script so that all registers are used.

```
#!/bin/bash
LEDS=/sys/devices/platform/<base-address>.hps_led_patterns/led_reg

while true
do
    echo 0x55 > $LEDS
    sleep 0.25
    echo 0xaa > $LEDS
    sleep 0.25
done
```

Step 4: **Test your Platform Driver using a C program**. Using the provided C program outline (click here for the file), add tests for all the registers. Then cross compile and run the test program.

11.3 Demonstration

Print out and get Lab 11 **Instructor Verification Sheet signed off** (page 401) by demonstrating to the course TA the following:

Demo 1: Show that you can create LED patterns from the bash script.
Demo 2: Show the output of the test program.

11.4 Deliverables

The **Deliverables** to be turned in or uploaded to the Lab 11 assignment folder are:

1: The Instructor Verification Sheet signed and turned in
2: Your platform driver source code hps_led_patterns.c
3: Your test bash script
4: Your test C program

Instructor Verification Sheet
Lab # 11

Creating a Platform Device Driver for the HPS_LED_Patterns Component

Print this page and get it signed by the Course Instructor or lab Teaching Assistant. Make sure you turn in this signed page to get credit for your lab.

Name: ____

Verified by: _____________________ **Date:** _____________

Lab Demonstration. Demonstrate to the Lab TA the following:

Demo 1: Show that you can create LED patterns from the bash script.
Demo 2: Show the output of the test program.

Deliverables. Turn in or upload to your Lab 11 assignment folder the following items:

1: The Instructor Verification Sheet signed and turned in
2: Your platform driver source code hps_led_patterns.c
3: Your test bash script
4: Your test C program

Lab 12
Implementing the Passthrough Project

In this lab we make the transition to audio signal processing. The previous labs were solely based on the DE10-Nano board. In this lab we will add the Audio Mini board that contains the AD1939 audio codec and implement the *Audio FPGA Passthrough* example. We will not perform any audio signal processing, but rather we will just get the audio signal into the FPGA fabric and back out again. This will provide a working system that we will use in the next lab to implement audio signal processing.

12.1 Items Needed

The items needed for Lab 12 are:

Item 1: A **Laptop** or PC
Item 2: **DE10-Nano Kit**
Item 3: An **Audio Mini board** that plugs into the DE10-Nano board. The Audio Mini board can be acquired here.
Item 4: A **Sound Source** such as your smartphone or laptop that can connect to an audio cable
Item 5: An **Audio Cable** that connects the sound source to the *Line In* audio input of the Audio Mini board (example cable).
Item 6: **Wired Earbuds** (or a speaker) that can connect to the *Audio Out* or headphone/line-out connection of the Audio Mini board
Item 7: A completed **Lab 11**

R. K. Snider, *Advanced Digital System Design using SoC FPGAs*,
https://doi.org/10.1007/978-3-031-15416-4_27

12.2 Lab Steps

The steps for Lab 12 are:

Step 1: **Read about the Audio Mini board**. (Read Sect. 5 Introduction to the Audio Mini Board (page 41).)

Step 2: **Read about the Audio Passthrough System example**. (Read Sect. 1.1 Audio Passthrough System Overview (page 255).)

Step 3: **Implement the Quartus Passthrough Project**. (Read Sect. 1.3 Quartus Passthrough Project (page 267).)

Step 4: **Compile the Linux Device Drivers**. (Read Sect. 1.5 Linux Device Drivers (page 289)). This involves compiling the ad1939 driver (see Sect. 1.5.1), the tpa613a2 driver (see Sect. 1.5.2), and the device tree (see Sect. 1.5.3) and implementing the systemd service (see Sect. 1.5.4).

12.3 Demonstration

Print out and get the Lab 12 **Instructor Verification Sheet signed off** (page 405) by demonstrating to the course TA the following:

Demo 1: Show that you have implemented the Audio Passthrough System example where the audio being sent into the Audio Mini input can be heard coming out of the Audio Mini output.

12.4 Deliverables

The **Deliverables** to be turned in or uploaded to Lab 12 assignment folder are:

1: The Instructor Verification Sheet signed and turned in

2: A write-up of all the steps you had to implement to get the system working. Comment on what part of the passthrough project you liked the best and what you thought was the hardest part of the project.

Instructor Verification Sheet
Lab # 12
Implementing the Passthrough Project

Print this page and get it signed by the Course Instructor or lab Teaching Assistant. Make sure you turn in this signed page to get credit for your lab.

Name: ____

Verified by: ____________________ **Date:** ____________

Lab Demonstration. Demonstrate to the Lab TA the following:

Demo 1: Show that you have implemented the Audio Passthrough System example where the audio being sent into the Audio Mini input can be heard coming out of the Audio Mini output.

Deliverables. Turn in or upload to your Lab 12 assignment folder the following items:

1: The Instructor Verification Sheet signed and turned in
2: A write-up of all the steps you had to implement to get the system working. Comment on what part of the passthrough project you liked the best and what you thought was the hardest part of the project.

Lab 13
Implementing the Comb Filter Project

In this lab we implement the comb filter example that produces an echo and can be used as the foundation for other sound effects such as *reverberation*, *flanging*, and *chorus*. This lab builds on the audio passthrough example where we add the comb filter processor to the audio path in the FPGA fabric.

13.1 Items Needed

The items needed for Lab 13 are the same as what was needed for the audio passthrough example that was implemented in Lab 12.

13.2 Lab Steps

The steps for Lab 13 are:

Step 1: **Read about the Feedforward Comb Filter System example** in Sect. 2.1 Overview (page 293).

Step 2: **Run a simulation of the Feedforward Comb Filter model**. Download the model (see Sect. 2.2 Simulink Model (page 295) and Sect. 2.3 Creating the Simulink Model (page 299)) and run the simulation in Simulink. Become familiar with the different control signals in the model by experimenting with different values. Determine good values for the control signals *delayM* so that the echo is very audible. Make note of these values since you will need to use these values for the lab verification.

Step 3: **Generate VHDL code for the model** by following the instructions in Sect. 2.4 VHDL Code Generation Using Simulink's HDL Coder (page 305).

R. K. Snider, *Advanced Digital System Design using SoC FPGAs*,
https://doi.org/10.1007/978-3-031-15416-4_28

Step 4: **Understand how the *combFilterProcessor* integrates the model VHDL code with Platform Designer** by reading Sect. 2.5 HPS Integration Using Platform Designer (page 317).

Step 5: **Implement the Quartus Combfilter Project** by reading Sect. 2.5 HPS Integration Using Platform Designer (page 317) (again).

Step 6: **Test your hardware using System Console** (see Sect. 2.5.5 Testing the combFilterProcessor Using System Console (page 323)).

Step 7: **Update the Linux Device Tree** to include the *combFilterProcessor* component that was added to Platform Designer. Use your knowledge gained from Lab 10 to do this.

Step 8: **Create the Linux Device Driver** for the four control registers. Use your knowledge gained from Lab 11 to do this.

13.3 Demonstration

Print out and get Lab 13 **Instructor Verification Sheet signed off** (page 409) by demonstrating to the course TA the following:

Demo 1: Show that you can change the echo parameters from Linux user space. Demonstrate several echo delay values that are audibly very different.

13.4 Deliverables

The **Deliverables** to be turned in or uploaded to Lab 13 assignment folder are:

1: The Instructor Verification Sheet signed and turned in
2: Your Linux Device Tree source file (.dts)
3: Your platform driver source code

Instructor Verification Sheet
Lab # 13
Implementing the Comb Filter Project

Print this page and get it signed by the Course Instructor or lab Teaching Assistant.
Make sure you turn in this signed page to get credit for your lab.

Name: ____

Verified by: ____________________ **Date:** ____________

Lab Demonstration. Demonstrate to the Lab TA the following:

Demo 1: Show that you can change the echo parameters from Linux user space. Demonstrate several echo delay values that are audibly very different.

Deliverables. Turn in or upload to your Lab 13 assignment folder the following items:

1: The Instructor Verification Sheet signed and turned in
2: Your Linux Device Tree source file (.dts)
3: Your platform driver source code

Lab 14
Implementing the FFT Analysis Synthesis Project

In this lab we implement the FFT Analysis Synthesis example that transforms the audio signal into the frequency domain, performs frequency domain processing, and then transforms the signal back to the time domain.

14.1 Items Needed

The items needed for Lab 14 are the same as what was needed for the audio passthrough example that was implemented in Lab 12.

14.2 Lab Steps

The steps for Lab 14 are:

Step 1: **Read about the FFT Analysis Synthesis System example** in Sect. 3.1 Overview (page 325).

Step 2: **Run a simulation of the FFT Analysis Synthesis model**. Download the model (see Sect. 3.2 Simulink Model Files (page 328) and Sect. 3.3 Creating the Simulink Model (page 329)) and run the simulation in Simulink.

Step 3: **Change the all pass FFT filter to a band pass filter**. The all pass filter is not shown in Fig. 3.4, but it is created in the Simulink file *createFFTFilters.m* in lines 54–57. Note: It is filter 4, but control word *filterSelect=3*. The low and high cutoff frequencies for this new bandpass filter will be individually assigned to you. If not, use 3 kHz as the lower cutoff frequency and 5 kHz as the upper cutoff frequency.

R. K. Snider, *Advanced Digital System Design using SoC FPGAs*,
https://doi.org/10.1007/978-3-031-15416-4_29

Step 4: **Run a simulation using your new bandpass filter** by following the instructions in Sect. 3.3.5 Verification (page 339). Recreate Fig. 3.18 using your assigned bandpass cutoff values to show the simulated frequency response of the system with the new bandpass filter.

Step 5: **Generate VHDL code for the model** by following the instructions in Sect. 3.4 VHDL Code Generation Using Simulink's HDL Coder (page 342).

Step 6: **Implement the Quartus FFT Project** by reading Sect. 3.5 HPS Integration Using Platform Designer (page 343).

Step 7: **Understand how the *fftAnalysisSynthesisProcessor* integrates the model VHDL code with Platform Designer** by reading Sect. 3.5 HPS Integration Using Platform Designer (page 343). You will need to create this VHDL code (similar to the *combFilterProcessor.vhd* code found in Table 2.3) and create a Platform Designer component by importing it into Platform Designer (see Sect. 1.4.1 Creating the AD1939 Platform Designer Component (page 269)).

Step 8: **Test your hardware using System Console** (see Sect. 3.5.2 Testing the fftAnalysisSynthesisProcessor Using System Console (page 344)).

Step 9: **Update the Linux Device Tree** to include the *fftAnalysisSynthesisProcessor* component that was added to Platform Designer. Use your knowledge gained from Labs 10 and 13 to do this.

Step 10: **Create the Linux Device Driver** for the two control registers. Use your knowledge gained from Labs 11 and 13 to do this.

14.3 Demonstration

Print out and get Lab 14 **Instructor Verification Sheet signed off** (page 414) by demonstrating to the course TA the following:

Demo 1: Show the figure of the simulated frequency response of the system using the bandpass filter with your assigned cutoff values (see Step 4:).

Demo 2: Demonstrate the frequency response of the system by playing audio through the system and selecting the various FFT filters from Linux user space, including your new bandpass filter.

14.4 Deliverables

The **Deliverables** to be turned in or uploaded to the Lab 14 assignment folder are:

1: The Instructor Verification Sheet signed and turned in

2: The modified *createFFTFilters.m* file

3: The figure similar to Fig. 3.18 that you created using your assigned bandpass cutoff values to show the frequency response of the system with the new bandpass filter
4: Your Linux Device Tree source file (.dts)
5: Your platform driver source code

Instructor Verification Sheet
Lab # 14
Implementing the FFT Analysis Synthesis Project

Print this page and get it signed by the Course Instructor or lab Teaching Assistant. Make sure you turn in this signed page to get credit for your lab.

Name: ______________________________________

Verified by: ______________________ **Date:** ______________

Lab Demonstration. Demonstrate to the Lab TA the following:

Demo 1: Show the figure of the simulated frequency response of the system using the bandpass filter with your assigned cutoff values (see Step 4:).

Demo 2: Demonstrate the frequency response of the system by playing audio through the system and selecting the various FFT filters from Linux user space, including your new bandpass filter.

Deliverables. Turn in or upload to your Lab 14 assignment folder the following items:

1: The Instructor Verification Sheet signed and turned in

2: The modified *createFFTFilters.m* file

3: The figure similar to Fig. 3.18 that you created using your assigned bandpass cutoff values to show the frequency response of the system with the new bandpass filter

4: Your Linux Device Tree source file (.dts)

5: Your platform driver source code

Lab 15
Creating Your Sound Effect in Simulink

In this lab we start the process of creating your sound effect final project. The first step is to create a Simulink model that implements the sound effect of your choice.

15.1 Items Needed

The items needed for Lab 15 are:

Item 1: A **Laptop** or PC
Item 2: **Matlab and Simulink**

15.2 Lab Steps

The steps for Lab 15 are:

Step 1: **Propose a Sound Effect** to the instructor. Your sound effect needs prior approval so that the scope of the project is suitable for the course and it aligns with the requirements of the final project.

- Choose either a **Time Domain** sound effect or a **Frequency Domain** sound effect.
- If your choice is a **Time Domain** sound effect, then your example that you will follow is the Comb Filter example in Sect. 2.1 Overview (page 293) where you will replace the *combFilterProcessor* with your sound effect processor.
- If your choice is a **Frequency Domain** sound effect, then your example that you will follow is the FFT Analysis Synthesis example in Sect. 3.1 Overview (page 325) where you will replace or modify the Simulink block *applyComplexGains* found in the model hier-

R. K. Snider, *Advanced Digital System Design using SoC FPGAs*,
https://doi.org/10.1007/978-3-031-15416-4_30

archy at `fftAnalysisSynthesis/frequencyDomainProcessing/applyComplexGains`.

Step 2: Design the **Control Signals** of your Simulink sound effect model, which at a minimum must contain the following signals list below. You will need to create the appropriate data type for each control signal. **Note:** The number of control signals is actually double this number because you will need both left and right channel control signals.

Signal 1: The **Enable** signal that controls turning on and off the sound effect.

- *Enable* = *1* means the sound effect is turned on.
- *Enable* = *0* means the sound effect is turned off. This is considered a bypass where the input signal becomes the output signal.

Signal 2: The **Volume** signal that controls the volume (attenuation) coming out of the sound effect block. The volume control signal should be an unsigned fixed-point signal that has a maximum value of 1. *The range should be [0 1]*. **Note:** Volume should control the attenuation of the output signal, regardless of whether the sound effect is enabled (present) or not.

- *Volume* = *1.0* means no attenuation.
- *Volume* = *0.0* means no signal is present.

Signal 3: **Sound Effect Control Parameter 1** that controls the sound effect. Changing the sound effect control word should cause an easily identifiable (audible) change to the audio signal. The control signal names for your sound effect should be descriptive as to what the control signal does. **Note:** Extra control parameters that are implemented and that can be changed in real time will be given bonus credit. These extra signals need to be present in the final implementation and connected to a Linux device driver to get the bonus credit.

Step 3: **Create Your Simulink Sound Effect Model**

- Make sure that you use only **HDL compatible** blocks for the sound effect model that will be converted to VHDL (see Sect. 2.4).
- The sound effect needs to operate on the stereo audio signal, i.e., the same effect on both the left and right channels (unless the sound effect requires some type of mixture). **Hint:** Create the sound effect for one channel to get it working first. Then, expand to stereo by replicating the sound effect block in a new stereo block that has both left and right control signals coming into it.

Step 4: **Test Your Simulink Sound Effect Model**. The sound effect should be easy to discern. Using Matlab's *audiowrite*, create a WAVE file called *before.wav* that creates a copy of the input audio. Then create a WAVE file called *after.wav* that saves the output of your sound effect simulation. You will need both of these files for your lab verification. The *after.wav* file is created so that verification can be streamlined without having to wait for the simulation to run, which can take a long time in certain cases.

15.3 Demonstration

Print out and get the Lab 15 **Instructor Verification Sheet signed off** (page 418) by demonstrating to the course TA the following:

Demo 1: Have your Simulink model ready and open on your laptop. Have the associated Matlab files all open in an editor. Explain how your Simulink model works as you go through the model.
Demo 2: Play the *before.wav* audio file.
Demo 3: Play the *after.wav* audio file.

15.4 Deliverables

The **Deliverables** to be turned in or uploaded to the Lab 15 assignment folder are:

1: The Instructor Verification Sheet signed and turned in
2: The Simulink sound effect model (.slx) file
3: All associated Matlab model files:

- createModelParams.m
- createSimParams.m
- createHdlParams.m
- verifySimulation.m
- Any other Matlab files

4: The *before.wav* audio file
5: The *after.wav* audio file

Instructor Verification Sheet
Lab # 15
Creating Your Sound Effect in Simulink

Print this page and get it signed by the Course Instructor or lab Teaching Assistant. Make sure you turn in this signed page to get credit for your lab.

Name: ______________________________

Verified by: ____________________ **Date:** ____________

Lab Demonstration. Demonstrate to the Lab TA the following:

Demo 1: Have your Simulink model ready and open on your laptop. Have the associated Matlab files all open in an editor. Explain how your Simulink model works as you go through the model.
Demo 2: Play the *before.wav* audio file.
Demo 3: Play the *after.wav* audio file.

Deliverables. Turn in or upload to your Lab 15 assignment folder the following items:

1: The Instructor Verification Sheet signed and turned in
2: The Simulink sound effect model (.slx) file
3: All associated Matlab model files:

- createModelParams.m
- createSimParams.m
- createHdlParams.m
- verifySimulation.m
- Any other Matlab files

4: The *before.wav* audio file
5: The *after.wav* audio file

Lab 16
Implementing Your Sound Effect in the FPGA Fabric

In this lab we continue with your sound effect final project. The next step is to convert the Simulink model to VHDL and get it running in the FPGA fabric.

16.1 Items Needed

The items needed for Lab 16 are the same as what was needed for the audio passthrough example that was implemented in Lab 12.

16.2 Lab Steps

The steps for Lab 16 are:

Step 1: **Convert your Sound Effect Simulink Model to VHDL** by following the process outlined in Sect. 2.4 VHDL Code Generation Using Simulink's HDL Coder (page 305) and Sect. 3.4 VHDL Code Generation Using Simulink's HDL Coder (page 342).

Step 2: **Create the VHDL code that implements your control registers**. We will refer to this as the code *soundEffectProcessor.vhd* but rename it so that "soundEffect" is the actual name of your sound effect. An example of this code is the file *combFilterProcessor.vhd* found in Table 2.3 and is what you did in Step 7: of Lab 14.

Step 3: **Create a new Quartus Project** by copying either the combFilter or FFT Analysis Synthesis project.

Step 4: **Create a Platform Designer component for your sound effect**. Do this by importing *soundEffectProcessor.vhd* into Platform Designer as outline in Sect. 1.4.1 Creating the AD1939 Platform Designer Component (page 269).

R. K. Snider, *Advanced Digital System Design using SoC FPGAs*,
https://doi.org/10.1007/978-3-031-15416-4_31

Step 5: **Compile Quartus** and load the FPGA bitstream into the FPGA.

Step 6: **Test your sound effect hardware**. Test your sound effect in the FPGA fabric by using System Console (see Sect. 2.5.5 Testing the combFilterProcessor Using System Console (page 323)).

16.3 Demonstration

Print out and get Lab 16 **Instructor Verification Sheet signed off** (page 421) by demonstrating to the course TA the following:

Demo 1: Demonstrate that your sound effect is working in the FPGA fabric by changing values in the control registers using System Console.

16.4 Deliverables

The **Deliverables** to be turned in or uploaded to the Lab 16 assignment folder are:

1: The Instructor Verification Sheet signed and turned in
2: The *soundEffectProcessor.vhd* file (renamed after your sound effect, i.e., <soundEffect>Processor.vhd)
3: The associated TCL file (.tcl) that Platform Designer created when you imported *soundEffectProcessor.vhd* into Platform Designer
4: The Platform Designer system file *soc_system.qsys*

Instructor Verification Sheet
Lab # 16
Implementing Your Sound Effect in the FPGA Fabric

Print this page and get it signed by the Course Instructor or lab Teaching Assistant. Make sure you turn in this signed page to get credit for your lab.

Name: ______________________________________

Verified by: ____________________ **Date:** ____________

Lab Demonstration. Demonstrate to the Lab TA the following:

Demo 1: Demonstrate that your sound effect is working in the FPGA fabric by changing values in the control registers using System Console.

Deliverables. Turn in or upload to your Lab 16 assignment folder the following items:

1: The Instructor Verification Sheet signed and turned in
2: The *soundEffectProcessor.vhd* file (renamed after your sound effect, i.e., <soundEffect>Processor.vhd)
3: The associated TCL file (.tcl) that Platform Designer created when you imported *soundEffectProcessor.vhd* into Platform Designer
4: The Platform Designer system file *soc_system.qsys*

Lab 17
Writing a Linux Device Driver to Control Your Sound Effect Processor

In this lab we complete the sound effect final project. The final step is to create the Linux device driver for the sound effect processor that can read and write to the memory-mapped control registers.

17.1 Items Needed

The items needed for Lab 17 are the same as what was needed for the audio passthrough example that was implemented in Lab 12.

17.2 Lab Steps

The steps for Lab 17 are:

Step 1: **Update the Linux Device Tree** to include the *soundEffectProcessor* component that was added to Platform Designer. Use your knowledge gained from Lab 10 to do this.

Step 2: **Create the Linux Device Driver** for the control registers. Use your knowledge gained from Lab 11 to do this. When done, you should be able to list the contents of `/sys/class/misc/soundeffectname` and see files associated with each of your control signals. The device driver should have your **lastname** in the compatible string.

Step 3: **Create two shell scripts** that are listed below:

1: **effectDisable.sh** When run, the script disables the sound effect and places your system in bypass mode where the sound effect is disabled and volume=1.0, i.e., output = input.

2: **effectEnable.sh** When run, the script enables the sound effect with parameter settings that make the sound effect easy to hear.

R. K. Snider, *Advanced Digital System Design using SoC FPGAs*,
https://doi.org/10.1007/978-3-031-15416-4_32

Step 4: **Create a Linux user space program called effectShow.c** When run, this program will illustrate and show off your sound effect. The parameters chosen, which need to change in time, should allow a listener to easily hear and experience the sound effect.

17.3 Demonstration

Print out and get Lab 17 **Instructor Verification Sheet signed off** (page 425) by demonstrating to the course TA the following:

Demo 1: Demonstrate that the script *effectEnable.sh* turns on your sound effect.
Demo 2: Demonstrate that the script *effectDisable.sh* turns off your sound effect.
Demo 3: Demonstrate that the program *effectShow* cycles through control parameters that shows off your sound effect.

17.4 Deliverables

The **Deliverables** to be turned in or uploaded to the Lab 17 assignment folder are:

1: The Instructor Verification Sheet signed and turned in
2: Your Linux Device Tree source file (.dts)
3: Your device driver source code that has your **lastname** in the compatible string
4: The source file *effectShow.c*
5: The source file *effectEnable.sh*
6: The source file *effectDisable.sh*

Instructor Verification Sheet
Lab # 17

Writing a Linux Device Driver to Control Your Sound Effect Processor

Print this page and get it signed by the Course Instructor or lab Teaching Assistant. Make sure you turn in this signed page to get credit for your lab.

Name: ____

Verified by: ____________________ **Date:** _____________

Lab Demonstration. Demonstrate to the Lab TA the following:

Demo 1: Demonstrate that the script *effectEnable.sh* turns on your sound effect.
Demo 2: Demonstrate that the script *effectDisable.sh* turns off your sound effect.
Demo 3: Demonstrate that the program *effectShow* cycles through control parameters that shows off your sound effect.

Deliverables. Turn in or upload to your Lab 17 assignment folder the following items:

1: The Instructor Verification Sheet signed and turned in
2: Your Linux Device Tree source file (.dts)
3: Your device driver source code that has your **lastname** in the compatible string
4: The source file *effectShow.c*
5: The source file *effectEnable.sh*
6: The source file *effectDisable.sh*

Index

R. K. Snider, *Advanced Digital System Design using SoC FPGAs*,
https://doi.org/10.1007/978-3-031-15416-4